FRENCH SECRET PROJECTS

FRENCH SECRET PROJECTS

French and European Spaceplane Designs 1964-1994

J-C CARBONEL

Crécy Publishing Ltd

French Secret Projects
French and European Spaceplane Designs 1964-1994
J-C Carbonel

First published in 2021 by Crécy Publishing

All rights reserved. No part of this book may be reproduced or transmitted in any form or by any means electronic or mechanical, including photocopying, recording or by any information storage without permission from the Publisher in writing. All enquiries should be directed to the Publisher.

© J-C Carbonel 2021

A CIP record for this book is available from the British Library

Printed in Bulgaria by Multiprint

ISBN 9781910809914

Crécy Publishing Ltd
1a Ringway Trading Estate, Shadowmoss Rd
Manchester M22 5LH
Tel (0044) 161 499 0024
www.crecy.co.uk

FRONT COVER
Nord-Aviation EUROSPACE Aerospace Transporter at separation point. The first stage has cut its engines and is about to glide down to its base while the second stage has just lighted its rockets and will soon begin its climb to orbit. Altitude is 21.7 miles [35 km]. *Art by David Uhr*

REAR COVER
TOP: Artist's impression of interim HOTOL being launched from an Antonov An-225 Mriya. *BAE Systems via Ron Miller*
MIDDLE This drawing illustrates a configuration very close to the final Hermes version. *ESA*
BOTTOM SSTO STS-2000 in orbit as viewed by an Aerospatiale artist. *Aerospatiale*

REAR FLAP Line drawing of HYTEX, the intended demonstrator for the German Sänger II spaceplane. For once, a human figure gives an impression of size to a project. *MBB*

Contents

Introduction 6
Acknowledgements 9

Chapter One   In the Beginning 10
Chapter Two   Eurospace Ideas 29
Chapter Three   The Way of the Ramjet 46
Chapter Four   Going into Space on a Set of Wings 63
Chapter Five   French Hypersonics 89
Chapter Six   Getting into Hardware: Nord Véras 98
Chapter Seven   Competing for Hermes 104
Chapter Eight   Designing Hermes 132
Chapter Nine   Competing with Hermes 159
Chapter Ten   Co-operation Around the World 179
Chapter Eleven   End of Hermes 189
Chapter Twelve   Around and Beyond Hermes 196

Sources 227
Index 229

Introduction

France's interest in manned spaceflight began with Jules Verne's novel *De la Terre à la Lune* (1865) followed by *Autour de la Lune* (1869) in which Verne tried to go beyond the adventure story by having the calculus checked by his cousin Henri Garcet, a collaborator to the then-famous mathematician Professor Joseph Bertrand. Criticisms of the novel were soon to be voiced by astronomers – notably Camille Flammarion – and the gun-launch idea for spacecraft was abandoned.

By the turn of the century, fiction authors were concentrating on another unusual launch system: the mechanical slingshot. This was promoted by various people, some true proponents of the sciences, and even experimenters like Henri de Graffigny (real name Raoul Marquis). But mechanically accelerating a vehicle to escape velocity while on Earth was barely practical.

Unfortunately, the rocket itself (not including the spaceship) had few sponsors in France and practical rocketry was in its infancy before the Second World War, although there were a few renowned engineers leading the field, like Robert Esnault-Pelterie (who presented a 1927 conference on *'the exploration of high atmosphere through the use of rocketships and the possibility of interplanetary travel'*); René Damblanc (who patented the staged rocket in 1936, an idea quite important for the rest of the present story); and Colonel Jean-Jacques Barré (who flew with mixed success the first French liquid-fuelled rocket in 1945). However, it was left to a science-fiction author, 'J-H Rosny aîné' (Joseph Henri Honoré Boex), to invent the word *'astronautique'* (astronautics) to name the science of space travel.

After the Second World War, French interest in space was revived by Alexandre Ananoff, a self-made promoter of interplanetary travel since the 'thirties who in 1950 wrote *l'Astronautique* in which he postulated a future rocketship. His design work for this book became known the World over when Hergé took him as advisor for the Tintin stories *Objectif Lune* (Destination Moon) and *On a Marché sur la Lune* (Explorers on the Moon).

But space travel became science fact during the early 'sixties with the first two French 'astronauts': Hector, a rat (1961) and Félicette, a cat (1963), who flew aboard Véronique exploratory ('sounding') rockets.

So that was the situation in France in 1963.

However, in parallel, something had happened in Germany. Between 1933 and 1944, Austrian-born Eugen Sänger had studied the feasibility of building a large rocketplane. In 1944 his work culminated in a report on an 'Antipodal Bomber' that influenced research on early spacecraft…which were, indeed spaceplanes (Chapter 1). In the early 'sixties, Sänger became a leading exponent of the 'Aerospace Transporter' that is to say, a winged vehicle able to put a man into orbit. This concept gained industrial traction with the setting-up of Eurospace, the trade association of the European Space Industry which asked its participants to research the subject. Many companies responded and came up with spaceplane concepts, among them Junkers, ERNO in Germany, Bristol Siddeley, Hawker Siddeley in UK…and Nord-Aviation in France (Chapter 2).

FAR LEFT Jules Verne (depicted here on a *chromo* [collector's card] from the Belgian Jacques Super-chocolate) was the first Frenchman to popularise the idea of manned space travel in a believable – if not completely realistic – way. *Author's collection*

LEFT Robert Esnault-Pelterie (from another *chromo*) was a true scientist who gained fame in aviation (he patented the 'control column') before venturing into spaceflight research and promotion. *Author's collection*

INTRODUCTION

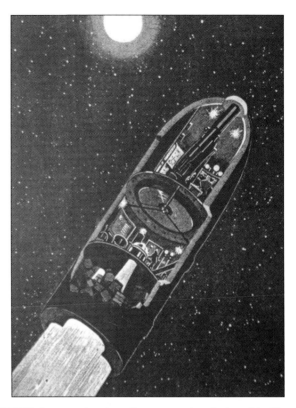

ABOVE Cover of the March 1922 issue of French magazine *la Science et la Vie* (Science and Life) featuring a space rocket on its launch ramp. *Author's collection*

BELOW The slingshot system imagined by Henri de Graffigny. Legend: Pompe à vide = vacuum pump; fractions de tunnel = tunnel sections; injecteur d'huile = oil injector; truck glissant sur rails huilés = truck sliding on oiled tracks; aiguillage = switch. *Author's collection*

ABOVE Cutaway drawing of a space rocket designed by Henri de Graffigny. Note that the liquid-fuel rocket engine is located in the middle section of the spacecraft. *Author's collection*

BELOW A major promoter of space travel was Alexander Ananoff whose book, *l'Astronautique,* published in 1950, greatly influenced the view Frenchmen had of spaceflight. *Author's collection*

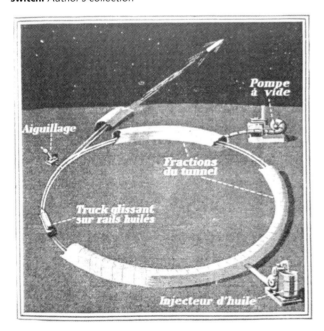

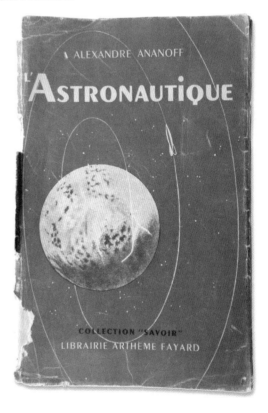

ABOVE In terms of real hardware, the Véronique sounding rocket was instrumental in affirming France's position into space during the 'fifties and the early 'sixties. *SEREB-Aerospatiale*

To assist development of such a vehicle, France, and particularly Nord-Aviation, had an ace up its sleeve: its knowledge of ramjets – especially turbo-ramjets – accumulated during development of the Griffon manned technology demonstrator and CT.41 target drone. SNCASE had also worked on ramjet-powered missiles, and all could get help from ONERA (Office National d'Etudes et de Recherches Aéronautiques: National Office for Studies and Aeronautical Research), the French Aerospace Research Office (Chapter 3).

Taking on Eurospace 'Space Transporter' the Centre de Prospective et d'Evaluations (Centre for Assessment Forward Planning: a research committee set-up in 1964 by the French Defence Ministry to investigate long-term defence and weapon systems) asked Nord and Sud-Aviation to study the feasibility of a reusable, winged first stage (Chapter 4). This in turn led some companies, notably Sud-Aviation to study highly hypersonic vehicles well beyond the Mach 5–Mach 6 speed target of the CPE programme (Chapter 5).

In the meantime, Nord-Aviation tackled the actual construction of a hypersonic space glider VERAS (Véhicule Experimental de Recherches Aérothermodynamiques et Structurales; Experimental Vehicle for Structural and Aerothermodynamical Studies). Although never flown, it gave France a first, hands-on experiment in working on the exotic materials required for such a vehicle (Chapter 6)—after which the whole concept of winged spaceplanes was forgotten until CNES (Centre National d'Etudes Spatiales: National Centre for Space Studies) began thinking about putting a French astronaut into orbit during the 'seventies. Between 1979 and 1984, those studies coalesced into a spaceplane project called Hermes while thoughts were given to a recoverable first stage. (Chapter 7). After a call for proposals in 1985, the detailed design of Hermes was allocated to Aerospatiale (the whole project) and Dassault (the spaceplane, specifically) and progressed amid many re-orientations and political debates (Chapter 8).

ABOVE **The first two French 'astronauts' were Hector the rat and Félicette the cat. Félicette is the female form of Felix, a common name for cats. When the selected tomcat escaped as scientists tried to put him into the spacecraft, the back-up cat took his place—but the back-up was a female, hence the renaming.** *Author's collection*

Another European company had high hopes of skipping the 'rocket + spaceplane' stage and so returned to the Aerospace Transporter concept: MBB with the Sänger II vehicle. BAe was even more ambitious and proposed a winged SSTO, the single-stage-to-orbit HOTOL. Both vied for European financing, concurrently with Hermes, but neither succeeded (Chapter 9). Others approached Aerospatiale with proposals for jointly developing a spaceplane: India, Japan and Russia. Neither of these co-operative initiatives really took off (Chapter 10). Indeed, with budgets being slimmed down, ESA and CNES had difficulties launching Hermes and, even with reduced ambitions, had to cancel the programme in 1992 (Chapter 11). Many projects had been developed concurrently with Hermes, including a space factory, the MTFF (Man-Tended Free-Flyer platform); various fully-reusable launchers; and even the study of …a Moon base (Chapter 12).

Acknowledgements

The author would like to commend Henri Lacaze (formerly of Aerospatiale) for contributions to this book that went far beyond his own reminiscences of the projects he actually worked on, and also for sharing a wealth of documentation on the projects he did not work on. Luc Berger, Michel Rigault and Philippe Coué at Dassault Aviation, Alain de Leffe from CNES, Ron Miller, Dan Sharp, Tony Buttler and Paul Gauge equally contributed help and material for this work.

Philippe Ricco, was invaluable when it came to the various aircraft mentioned here.

And finally, thanks are due to Caroline Thibodeau and Sylvie Voisin for their help.

Chapter One
In the Beginning

ABOVE X-20 Dyna-Soar in orbit with the visor protecting the windscreen just being ejected. *NASA*

A – Pre-War Research

The creator of the spaceplane concept, in Europe and in the World, was a German scientist named Eugen Sänger. In 1970, his wife and collaborator of many years, Irene Bredt, recounted his story in a paper titled *The Silver Bird Story: a Memoir*. This Chapter will draw a lot from that study, with additional notes from Sänger-Bredt's own works and contemporary commentaries.

During his college years, Eugen Sänger was influenced by Kurt Lasswitz's *Auf Zwei Planeten*, a famous German SF novel of the late 19th Century. Lasswitz was a doctor of physics who was a pioneer of science-fiction in Germany. Von Braun claimed he, too, found inspiration in *Auf Zwei Planeten*. On the serious side, Sänger, like most of his contemporaries in Germany, was also influenced by Hermann Oberth, After reading Oberth's book *Die Rakete zu den Planentenraumen* (The Rocket into Interplanetary Space) published in 1923, and revised in 1929, which discussed the feasibility of spaceflight, he switched from civil engineering at the Technische Hochschule of Vienna to aircraft engineering.

However Sänger did not share Oberth's views on the use of missiles (non-winged) to go into space, and was more attuned to the paths advocated by Max Valier (who, during the twenties, experimented with rocket cars and planes and described how he planned to convert a Junkers G23 trimotor into a spaceplane!); and by Franz von Hoefft (who designed a real ancestor of the Transporteur Aérospatial [Aerospace Transporter] in the form of two-stage, rocket-powered spaceplane having a flying boat as its first stage) and favoured the design of a spaceplane. In 1927 Hoefft, working in the aerodynamic department of the Hochschule, needed an assistant for his rocket tests and Sänger applied—but apparently did not follow-up, although he had been accepted.

He passed his doctorate examination on 5 July 1929 in '*the statistics of multi-sparred, parallel-ridged, total or half cantilever, indirectly or directly panelled wings*'. Actually, he had wanted to dissertate on spaceplanes but according to a radio interview he gave in 1964[1] '*I wished to obtain my doctoral degree some years later in the field of spaceflight. But then my good old teacher Katzmayr, with whom I studied aviation told me, "it is much more practical to prepare for your doctoral examination in a classical field—the event will then pass silently across the stage*[2]*. If you try today to take your doctor's degree in spaceflight, you will most probably be an old man with a long beard before you have succeeded in obtaining your doctorate."*'

So Sänger was able to publish his views '*on construction principles and performance of rocketplanes*' in February 1933. In this he proposed a rocket-powered plane able to reach 6,214mph (10,000km/h – about Mach 10) at altitudes between 197,000 and 230,000 feet (60 and 70 km). The engine would use common aviation fuel and liquid oxygen. He envisioned that the final speed of his aircraft would only be limited by the quantity of propellant embarked, which would constrain the duration of the acceleration phase. '*At first sight the limit is given merely by the possible fuel load. It is this restricted fuel load which prevents these planes from increasing their flight velocity up to the orbital velocity of about 18,000mph (29,000km/h) when the centrifugal force of the curved trajectory is equal to the weight of the planes; the wings then don't need to provide lift anymore, and the plane circles the Earth continuously, like a moon in a free inertial orbit without needing any driving power.*'

Sänger finalised his thoughts on the subject in the book *Raketenflugtechnik* (the technics of rocket flight), which was turned down by most of the major German science publishers of the time. Only Oldenbourg, a Munich company that had already published Oberth and Valier's books, agreed to handle the volume—provided Sänger contributed 2,000 Reichmarks to the printing costs *and* bought the first 50 copies. This put his finances in the red for the next four years.

In 1933, Sänger synthesised, for the first time, his views on the rocketplane in a special issue of the journal *Flug*, but that failed to place it at the centre of interest of the various German research institutes. Only on 1 February 1936 did he succeed in winning a contract with the Deutsche Versuchsanstalt für Luftfahrt (DVL: the German Aviation Research Institute). From August 1937 to August 1942 Sänger worked there on developing a large liquid propellant rocket engine. On this occasion, in the autumn of 1937, he met a mathematician who was appointed to his team to do computing: Irene Bredt[3] whom he married, after the war, during his stay in France.

In the meantime he kept an eye on his pet project, the rocketplane. In October 1938, he had a steel model made of his design (now called Silbervogel, the 'Silver Bird' on account of the gleaming appearance of this model), a plano-convex supersonic glider-plane which he tested in wind tunnels. The shape of the model, with its flat undersurface and its dome-shaped fuselage, induced Sänger's assistants to liken it to a flat iron.

In the later Antipodal Bomber report Sänger wrote that he began working on the development of the rocket engine in 1933-34 at the Technische Hochschule

ABOVE Portrait of Eugen Sänger, probably just after the war.
Author's collection

in Vienna. There he made his first experiments with small models with 66lb (30kg) thrust. However, the report continued, '*Experiments with larger models were delayed for several years and could only be resumed after the construction of the Trau Aeronautical Test Station in 1939. The construction of the experimental installation was under the direction of H Zborowski; the construction of the components was directed by H Ziebland; A Hedfeld directed the experimental work.*' The mention of Helmut von Zboroswki is interesting, as he was another of the prominent German scientists who worked for French companies after the war.

Because of the need to accelerate his project plane without having to burden it with immense fuel tanks, Sänger proposed to propel it along a horizontal ramp by a booster to achieve launch speed. He intended the ramp to be a large metal rail, many kilometres long. For this reason, he asked Irene Bredt to research the effect of the friction between the rail and the booster/Silbervogel ensemble at speed of several hundred feet per second. She went to see friction experts Hermann Fottinger and G Vogelpohl at the Kaiser Wilhelm Institute, but they could not help her because they had never studied friction at such velocities.

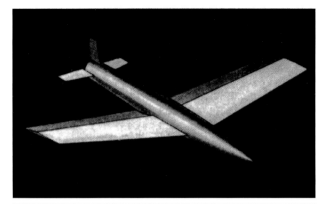

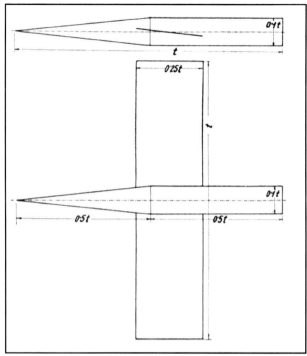

ABOVE Conceptual drawing of the Rocketplane, showing the simple shapes: an elongated cone followed by a cylinder. The wings are just symbolised by lines showing their angle of attack in relation to the flight axis. *Author's collection*

TOP LEFT Artist's impression of the first iteration of the future Silbervogel. This comes from Sänger's initial publication, *On Construction Principles and Performance of Rocketplanes*. At that time it was just called 'rocket plane' and had no intention of going round the Earth. *Author's collection*

LEFT Three-view drawing of the Rocketplane, this time fully rendered as a model. Again, the document comes from *On Construction Principles and Performance of Rocketplanes*, as reprinted in a special issue of the magazine *Flug*. *Author's collection*

Sänger then suggested building a test stand consisting of a military rifle firing into a spiral-shaped lubricated groove. This novel experiment allowed him to test how to overcome the problem induced by having a vehicle running along a track at such high speed. These tests began on 2 June 1939.

B – Wartime and the Silbervogel

Three months later, the Second World War began. Theoretical science research as carried out by Sänger's team was just no longer required. What was needed was technological prowess able to boost Germany's war effort. For this reason, Sänger, recognised as a jet-propulsion specialist, was tasked with designing 'offenrohren'. Offenrohr (literally, open pipe) was the German nickname for ramjets because they were just hollow tubes, without moving parts. Thus, they were perceived as simple, easy-to-design and cheap propulsion systems for high speed aircraft. That was very convenient at this stage of the war.

While working on the ramjet programme (which incidentally, had not produced any operational outcome when the war stopped), Sänger renamed his Silbervogel as a Raketen Bomber (shortened as RaBo[4]) to promote it within the Reich Luftfahrt Ministerium.

An advanced version was designed with the following features:

- Flight velocities at the end of the power flight phase of about 26,250ft/s (8,000m/s), corresponding to the necessary thrust for reaching orbital velocity with a 2,240lb (1 tonne) payload.
- Altitudes in the ballistic section of the flight path up to 186 miles (300km).
- Loading capacities for a transport to the antipodal point of the Earth 12,430 miles (20,000km) up to 7.87 tons (8 tonnes).
- Flight distance to a single Earth orbit with a payload of 3.94 tons (4 tonnes), or up to two and half orbits with a 2,240lb (1 tonne) payload.

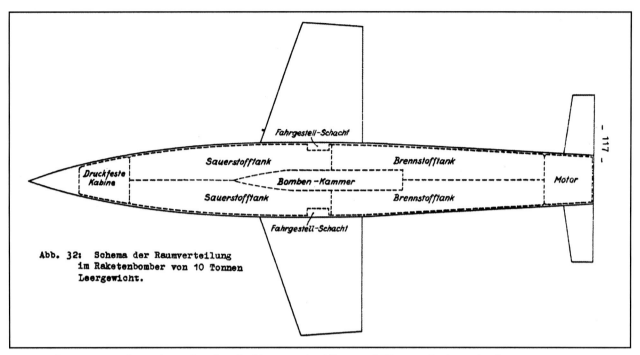

ABOVE About 10 years later, the Rocketplane had become the Silbervogel (Silver Bird). The author has never seen any document with the acronym Ra(kete)Bo(mber) which may have been used in internal memos of DVL. This comes from report UM3538, as later translated by the Allies. *Author's collection*

BELOW From the same source, a drawing depicting the booster-sled on the launch rail. *Author's collection*

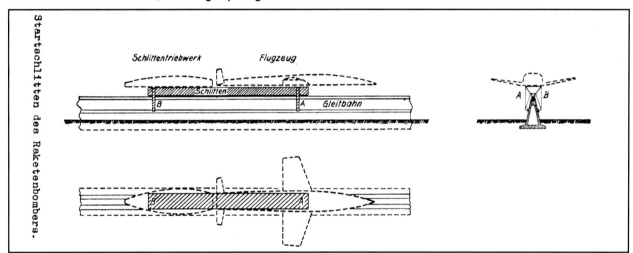

The calculation was based on the following assumptions:

- Take-off (launch) weight of 98.42 tons (100 tonnes) of which 88.58 tons (90 tonnes) was propellant and payload.
- Liquid oxygen rocket engine with a combustion chamber pressure of 100 atmospheres delivering a thrust of 98.42 tons (100 tonnes).

The report, re-tailored for military consumption as *Uber ein Raketenantrieb für Fernbomber* (About the Rocket Propulsion for a Long-Range Bomber) was forwarded to the RLM on 3 December 1941. It was rejected on 17 March 1942. For this reason Sänger was not allowed to continue developing the 100-tonne rocket engine, but was urged to keep working on the ramjet which was urgently needed for high-speed interceptors. On 22 March 1944, through intense lobbying with his friend Pr Walter Georgii, Sänger managed to have his report published under the reference UM3538. Irene Bredt mentioned that, '*The official approval for printing reached Sänger on his 39th birthday, 22 September 1944.*'

Yet the project was not developed further and Sänger continued his work on ramjets until the end of the war, offering a ramjet arrangement to boost Messerschmitt Me 262 fighter in addition to his collaboration with the Czech company Skoda-Kauba in designing a cheap ramjet-powered interceptor.

Here is a summary of the design, as given in the report:

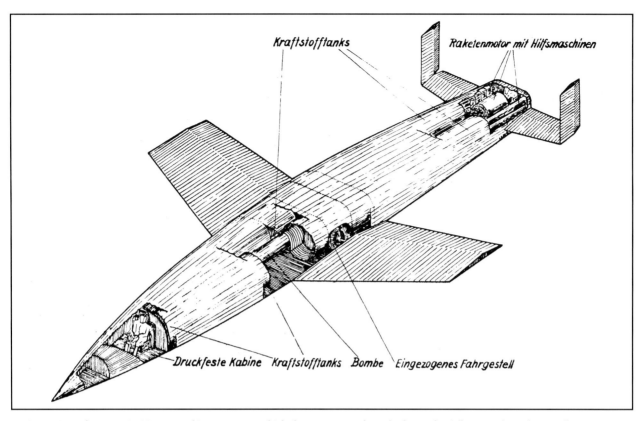

ABOVE Direct from UM3538 comes this cutaway—which does not reveal much about the Silbervogel. *Author's collection*

'The rocket bomber is brought to a speed of about 1,640ft/s (500m/s) at the surface of the Earth by a ground-based rocket drive, covering 1.86 miles (3km) in 11 seconds. It then climbs at full power, to a height of 31–93 miles (50–150 km), along a path which is initially inclined at 30° to the horizon but later becomes flatter; thus it reaches final velocities up to more than twice the exhaust speed. The duration of the climb is 6–8 minutes; during this time the on-board fuel supply will usually be consumed. At the end of the climb the rocket motor is turned off and the aircraft, because of its kinetic and potential energy, continues on its path in a sort of oscillating gliding flight with steadily decreasing amplitude of oscillation. This type of motion is similar to the path of a long-range projectile which from similar heights follows a descending glide-path. Because of its wings the aircraft, descending in its ballistic curve, bounces on the lower layers of the atmosphere and is again flung upwards, like a flat stone ricocheting[5] on the surface of water, though during the entrance into the denser air each time, a fraction of the kinetic energy is consumed, so that the initially big jumps steadily become smaller and finally go over into a steady, gliding flight. At the same time the in-flight speed, over a long glide-path of several thousand miles, decreases from the high initial value to normal landing speed.'

One might note that not a single word is written in all of the above about the intense heat the aircraft (and its pilot) would have to withstand during the ricochet flight. That is all the more remarkable as this would become one of the more difficult problems facing engineers eager to design a spaceplane in the next fifty years.

A large section of the report was devoted to the different ways a target could be bombed from the stratosphere. *'If the descending path (which is within certain limits controllable by the pilot) lies in the direction of the target, the bombs are released at a predetermined moment, and the craft returns to its starting point (or some other landing field) in a wide arc, while the bombs go toward the target along the original direction of flight. Even if the target is very distant from the take-off point, the bombs are only dropped near it, so that the scattering of the bombs can be compensated by a large number of releases on the target which will, in this way, be covered by a Gaussian distribution of hits. The military use is completely independent of weather and time of day at the target, and of enemy counteraction, because of the possibility of using astronomical navigation in the stratosphere and because of the height and speed of flight.'*

In his text, Sänger also evoked the possibility of attack *'with sacrifice of the bomber'* and *'with circumnavigation'*, the later dimension being what was implicit in the whole report: an aircraft able to make a complete, circular flight around the Earth, landing on the very field from which it had taken off.

Sänger is drawn into describing his project as a wonder weapon at its most fantastic, even suggesting targeted assassination from a stratospheric plane:

'From the characteristics given for the rocket bomber, it follows that this is not the development of an improved military craft, which will gradually replace present types, but rather that a problem has been solved for which no solution existed up to now, namely the bombardment and bombing over distances of 620 to 12,425 miles (1,000 to 20,000km). With a single rocket bomber, targeted attacks can be made, for example, from Central Europe, on distant specific targets like a warship on the high seas, or a canal lock; even a single man in the other hemisphere can be fired upon. With a wing of 100 rocket bombers, surfaces of the size of a large city at arbitrary places on the Earth's surface can be completely destroyed in a few days'.

While only very general drawings of the aircraft (with detailed drawing of the rocket motor) were presented, Sänger had obviously given some thoughts to the details of his Silbervogel, for example the cabin for the pilot. *'Behind the bow is the pressure-tight cabin in which the single pilot seats. It is sealed tight to maintain an inside pressure of 0.4–0.5 atmospheres. With a vacuum outside, this should permit rapid exit of the pilot in case of danger (for example, after take-off). Because of the smooth external shape, visibility from the cabin is very poor. In free flight at high velocity, side view slits and optical aids are sufficient. For landing, a kind of detachable windscreen can be used, since the pressurisation of the cabin and maintaining the bullet shape are unimportant. A further essential arrangement for the cabin is that the pilot's seat be so arranged that the pilot can stand the high acceleration along the aircraft's axis in the best possible*

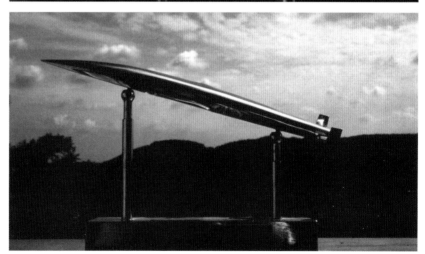

ABOVE The original Silbervogel model has been preserved by the Sänger family. *Ron Miller*

attitude, so that not only body and head, but also feet and arms have good supporting surfaces and at the same position can be shifted. The remaining equipment of the pilot's cabin – instruments, D/F and radio equipment,

1933 Raketen flugzeug	
Status	Concept
Span	–
Length	–
Gross Wing area	–
Gross weight	–
Engine	–
Maximum speed/height	–
Load	None

FRENCH SECRET PROJECTS: FRENCH AND EUROPEAN SPACEPLANE DESIGNS 1964-1994

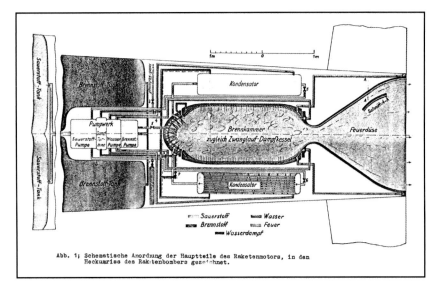

LEFT The large rocket engine intended to power the Silbervogel was presented in more details in UM3538. *Author's collection*

1942-44 Silbervogel	
Status	Project (not very advanced)
Span	49ft 3in (15.00m)
Length	91ft 10in (28.00m)
Gross Wing area	13,632sq ft (1266.5sq m)
Gross weight	220,460lb (100,000kg) (22,046lb (10,000kg) empty weight)
Engine	1 × 100 t thrust rocket engine
Maximum speed/height	–
Load	22,046lb (10,000kg) standard

BELOW Some material was actually built as part of this project—here, one of Sänger's experimental rocket motors is running on a bench. This poor photo comes from the Allied translation of UM3538. *Author's collection*

BELOW Sänger's 1 ton-thrust rocket motor in action. Silbervogel would have required a motor one-hundred times more powerful. *Author's collection*

LEFT Another poor but impressive (note the figure of a man in the lower right corner) photograph of an Oxygen tank in Sänger's laboratory. Capacity 56 tonnes and a daily evaporation of 140kg. *Author's collection*

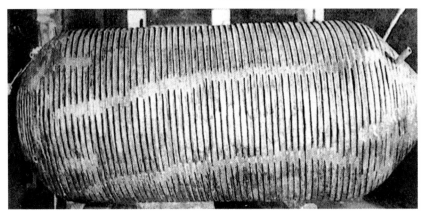

BELOW According to the Allied translation of 3538, this photograph depicts a 2-tonne-thrust rocket engine. Actually it looks more like a fuel tank. Perhaps a test model for the tanks required for Silbervogel? *Author's collection*

ventilation, etc – is not considered further.'

Sänger also commentated about the landing gear. *'The front view of the aircraft does not show the retractable front wheel, which operates in conjunction with a retractable tail skid, and the landing gear which is retractable into the fuselage between the wings. The front wheel serves to prevent dangerous contact of the nose with the ground during the bouncing motion of the aircraft during landing, and to slow it down (with the aid of the landing gear) as quickly as possible; it comes to ground at 93mph (150km/h) and has practically no resistance then because of the small wings [...] Just before the main spar meets the fuselage there are chambers for the extensible landing wheels, these chambers are partly on the vertical sidewalls of the fuselage and partly recessed into the tank cylinders.'*

At landing speed, with fuel and bombs expended, the aircraft would have weighed only 22,046lb (10,000kg), so no difficulties were expected in this area. Again, future engineers who tried to have their space gliders take off from conventional runways at full weight found the question of the landing gear much more complex to solve.

C – Post-War Heritage

After the war, Sänger's UM3538 report was widely circulated among the victors, being translated into English by M Hamermesh of Radio Research Laboratory under the auspices of the Technical Information Branch of the US Navy's Bureau of Aeronautics. Strangely, the author has not found any copy of UM3538 translated into French. The UM3557 report about *l'Accroissement du Rendement du Me 262 par le Moteur Auxiliaire Lorin*[6] (increasing the efficiency of the Me 262 through the use of Lorin – ramjet – auxiliary motors) was the only wartime report by Sänger the author has found translated into French.

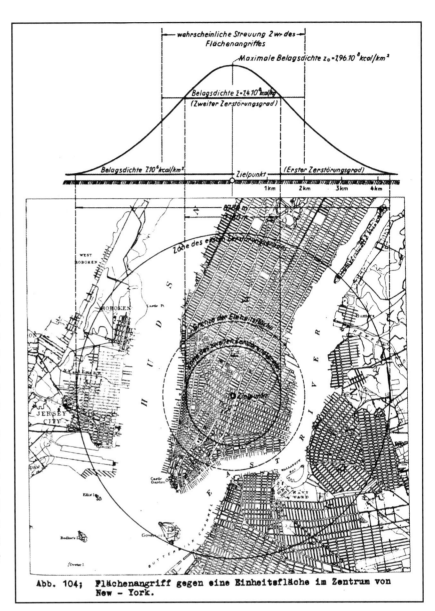

ABOVE To 'sell' the project to the military, Sänger gave-over a large part of his report to bomb trajectories and distribution statistics of bombs launched from the stratosphere. This is the most famous, representing the bombing of Manhattan. *Author's collection*

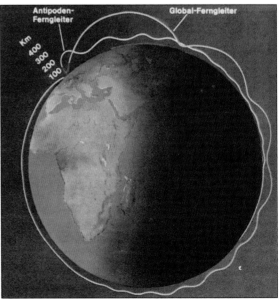

RIGHT But this is the really interesting part: a post-war representation of a round-the-World trip by the Antipodal Bomber, skipping on each atmospheric layer. *Author's collection*

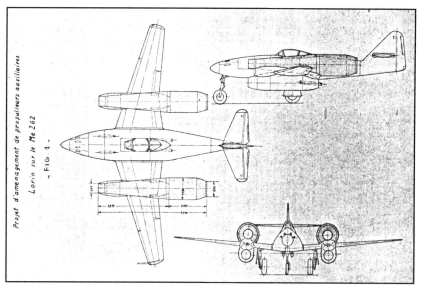

ABOVE The only Sänger-related proposal the author has found in French documents—an Me 262 with ramjets. CAA

i) In France: the Standard-Bearer of Ramjet Research

Sänger decided to go to France after the war, but so far, no translation in French of his work has surfaced.

While in France, Sänger was appointed to the Arsenal de l'Aéronautique. Irene Bredt wrote, '*Subsequent collaboration with French officials, engineers, and workers was agreeable to Sänger in terms of both technical and human relationships, one reason being an early and generous grant of freedom of publishing for Sänger and his collaborators.*' The actual impact of Sänger on French aeronautical research is debated to this day. In *France and the Peenemünde Legacy*, rocket expert and historian Jacques Villain devoted a complete paragraph to '*The Role of Professor Sänger*' but after recognising that, '*[Sänger's] name will remain attached to his rocket-bomber-plane project of 1942, designed to fly by ricochets,*' he concluded, '*in connection with the ramjet, Sänger helped to add to France's existing fund of knowledge. All this clearly underlies the strong position that France has been able to acquire and maintain in the field.*' Indeed, Sänger's work in France appears to have focused on ramjets. '*I, personally, translated the Bredt report regarding the flying testbeds of ramjet in Germany, circa 1943,*' wrote M Beaussard, formerly of the Arsenal de l'Aéronautique, '*but we were lucky, sort of, to obtain a good German expert, Professor Sänger.*'

LEFT While in France, Sänger distributed mementos of his work to his French colleagues. Here is a separate reprint of Sänger's original *Raketen Flugzeug* article in *Flug*. Note the handwritten 'Paris Sept 47, [signed] Sänger' text. *Author's collection*

However, according to Albert Gozlan, Directeur Technique de Nord-Aviation, '*He was mostly used as standard-bearer for our work. We must do ramjets. We have with us a famous German theorist on this subject and so on. He was so needed he did not even belong to our department.*'[7] Again, one should note that only ramjets are referred to here—and, even regarding ramjets, Sänger's input appears to have been minimal. But he was efficient in promoting the Arsenal de l'Aéronautique as a writer in aeronautical journals. He published in 1949 *Kinematics of Spaceflight* in *Interavia* and in 1951 an *Atlas of Selected Trajectories of Rocketplanes to the Spacestation and Back* in the Research Series of the Northwest German Society for Space Research.

In 1954 Eugen Sänger returned to (West) Germany.

ii) In Russia: Stalin Fancies the Antipodal Bomber

It seems the UM3538 report gained some purchase in Soviet Union where Stalin ordered that a Silbervogel be developed.

Within Research Institute No 1 (NII-1), Aleksei Isayev, who had been among the first Russian engineers to enter the Peenemünde experimental station, was tasked to set up a design bureau concentrating on rocket engines specifically suited to a Sänger-type 'antipodal' aircraft. Leonid S Dushkin, another engine expert, was named as the head of a second design bureau attached to NII-1 with the goal of working on rocket engines in general. The third NII-1 design bureau was headed by Mark M Bondaryuk, a ramjet specialist. A copy of the Sänger report was turned over to NII-1 Deputy Director Bolkhovitinov in 1945. Initial studies of the Silbervogel report were not too encouraging: the initial NII-1 evaluation was that it would take at least a decade to develop such design.

In November 1946, the 25-year-old mathematician Mstislav Vsevolodovich Keldysh, was appointed Director of

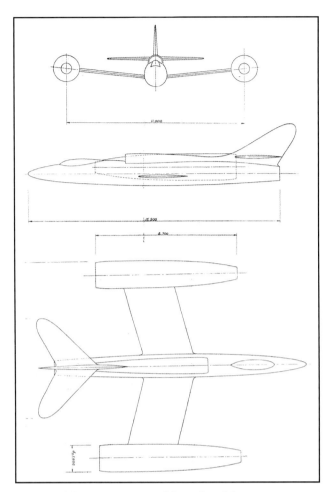

ABOVE While Sänger's Antipodal Bomber did not generate any serious interest in France, his more general ramjet work led to the Arsenal de l'Aéronautique to coming up with various ramjet-powered aircraft: here the ARS 1910. *Philippe Ricco*

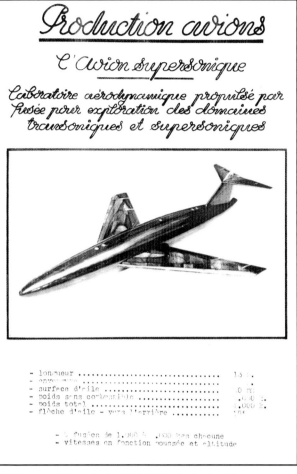

ABOVE —and here is the ARS 2301. *Author's collection*

NII-1 and immediately decided to investigate the Sänger Silbervogel's feasibility with the intention of developing it into a new weapon for the Soviet arsenal. On 3 April of the following year Keldysh synthesised his views on the project for the government, notably the question of the 98.42 ton (100 tonne)-thrust rocket engine. Stalin got wind of the project and became enthusiastic. While Russian scientists were tasked with improving the Antipodal Bomber, Stalin ordered his son Vasily, Colonel-General Serov and Colonel Tokayev (of the Central Aerohydrodynamic Institute [TsAGI] at Zhukovsky) to go out and 'obtain' Sänger for work in the USSR. The Soviet commando explored Western Europe but failed to approach Sänger and Bredt who had been forewarned by the French Intelligence Services. To add insult to injury, Tokayev took the opportunity of this mission to defect to the United Kingdom.

Although tackling the project head-on and displaying great willingness, Keldysh soon discovered that the path was quite difficult and joined those thinking that a decade would be needed to complete it. But Stalin wanted results and not that kind of defeatist ideas. So Keldysh prepared a report *On Power Plants for a Stratospheric High-Speed Aircraft*, which actually described a vehicle largely resembling the Silbervogel. The Institute's scientists doubted Russian ability to develop the required 100-tonne-thrust RKDS-100 rocket engine at a time when the best that had been achieved with comparable powerplants was 24.6 tons (25 tonnes), so they opted for a mixed propulsion system adding ramjet engines to the liquid-propellant main engine. It was intended to use one RKDS-100 engine for the aircraft and five, or even six more for the sled booster. (Keldysh followed Sänger's inspiration by putting his aircraft on a 'catapult' ramp with a booster fitted with the same rocket engines as the aircraft). After 10 to 12 seconds, the aircraft would have achieved a speed of *1,640ft/s (500m/s)*, when the ramjets would take over until it reached 65,600ft (20,000m). At this point, the RKDS-100 rocket engine would have ignited to enable the aircraft to reach the upper tiers of the atmosphere. After stopping the rocket engine, the aircraft would have continued on a suborbital trajectory, bouncing on each atmospheric layer, eventually flying over its target.

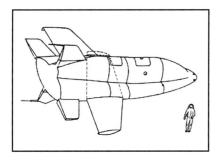

ABOVE Pavel Tsybin's Gliding Space Apparatus (PKA) as rendered by Igor Afanasyev in a NASA report. *NASA*

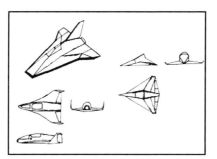

ABOVE Rendering by Asif Siddiqi of Myasishchev's Article 48 spaceplane. *NASA*

Beyond the conceptual stage came a second phase involving detail design of the engines. While Leonid S Dushkin was in charge of developing the RKDS-100, Mark M Bondaryuk was to design the ramjets. The operational range was estimated as 7,460 miles (12,000km). NII-1 specialised in rocket propulsion and the project should have been taken over by a proper OKB (Opytnoye Konstruktorskoye Buro: Experimental Design Bureau), but this was never done. It is known that VIAM (the All-Russian Scientific Research Institute of Aviation Material, under S T Kichkin) would have been involved. It seems Keldysh tried to incorporate 'Germans' in this project, but it is uncertain who those Germans were. What is known is that a copy of the wartime Sänger report was forwarded to them and that their reaction was totally negative— from the idea of using a catapult ramp launch, to stresses affecting the pilot and the airframe during each atmospheric bounce, which had not been investigated by Sänger.

This, together with the inability to 'obtain' Sänger, led to the cancellation of the project.

While the direct Sänger influence in the USSR stopped there, Russian engineers kept on proposing various spaceplanes. *Sovetskaiya Aviatsiya*, a Soviet Air Force daily magazine had published in December 1957 an article about the 'rocketplane, aircraft of the future.' This article did not refer to rocket-boosted interceptors but to a supersonic aircraft, which would accelerate to 9,320 mph (15,000km/h) in the stratosphere and then skip on the various layers of the atmosphere. One can see in this description the shadow of the Silbervogel. This was the overt indication of an interest in spaceplanes by the Soviet Air Force and indeed, in 1958 a secret Soviet Air Force committee recommended some avenues of research:

■ *An early stage with aircraft flying at 3,730 to 4,350mph (6,000 to 7,000km/h) and altitudes of 50 to 62 miles (80 to 100km) for research into aerodynamic heating and flight dynamics at high speeds and altitudes.*

■ *A later stage with velocity and altitude increased to more than Mach 10 and 62 to 93 miles (100 to 150km), respectively.*[8]

In January 1958, a capsule was selected over the space glider for the purpose of recovering a future cosmonaut. Yet work on space gliders was not halted and went on under the auspices of the Soviet Air Force which could sponsor such projects.

Sergei P Korolev's OKB-1 had designed the R-7 Zemiorka rocket and was studying how to use it to put a man into orbit. Meanwhile, aircraft designer Pavel V Tsybin, an old friend of Korolev, was designing a supersonic, long-range reconnaissance aircraft (RS) at OKB-256. Korolev, upon meeting him to discuss ejection seats for the future cosmonauts, asked him if he would be interested in designing a spaceplane that would fit atop the R-7. On 17 May 1959 an early study was delivered to Korolev, describing the Gliding Space Apparatus (PKA). This single-seat vehicle was designed with two large wings which would be folded at launch and would only deploy for re-entry. It was also considered feasible

LEFT Modern model representation of Keldysh's project directly derivated from the Silbervogel. Note the large wingtip ramjet.
Renaud Mangallon

to disengage the PKA from the Zemiorka if a problem occurred during the boost phase. In this situation the cosmonaut would deploy the wings and the PKA would land as a glider. If trouble occurred immediately after launch, at an altitude below 32,800ft (10,000m) the pilot could use his ejection seat.

Two rocket engines, running on kerosene and nitric acid provided propulsion for insertion into orbit and for manoeuvring. Mission duration was expected to be no more than 24 to 26 hours. During re-entry the fuselage would initially provide lift alone due to its shape (Korolev nicknamed it *Lapotov* [sandal] because of its upturned nose). At a height of 65,600ft (20,000m) the wings would be deployed and the aircraft would land conventionally on a runway using two skids in tandem, the rear touching first. Heat stress (the calculated maximum temperature encountered being 1,200° C) was dealt with by the special materials, namely organic silicon covering backed-up by ultra-fine fibre. The nose and other 'hot' areas of the plane were actively cooled by internal circulation of liquid lithium. No windscreen was provided for the pilot, but he had two large side windows and a smaller one in the of cockpit roof for navigation.

Wind tunnel testing was carried out at TsAGI, Zhukovsky, but the project never went beyond that. It was soon realised that surviving the heat of re-entry was much more complex than anticipated—notably, the articulations of the folding wings acted as a heat trap through which heat could penetrate the vehicle.

Reorganisation of the Soviet aerospace industry in 1959 – which resulted in the closing down of OKB-256 – certainly did not help.

But Korolev was not the type to put all his eggs in one basket. So in 1957, he took the same spaceplane idea he had discussed with Tsybin to another of his friends, Vladimir Mikhailovich Myasishchev, the renowned and *avant-garde* aircraft designer, along with whom he had once been a political prisoner. Data about the Zemiorka was transferred to Myasishchev's OKB-23 which began working on the design of a single-seat spaceplane and went on to design 'Article 48'—a space glider to be launched from the nose of an R-7. Gennady Dermichev was project chief with Evgeny Kulaga as deputy. Article 48 was a 9,920-lb (4,400-kg) aircraft capable of a 250-mile (400-km) orbit (the maximum attainable using the R-7 Zemiorka from the initial calculations of OKB-23 engineers).

Unlike Tsybin's PKA, it appears the Article 48 study was officially sanctioned, even though Party Leader Khrushchev had pointed out to Myasishchev that rockets were the responsibility of the government, not of aviation designers. Yet on 8 April 1960, the State Committee for Aviation Technology reviewed the design, but could not agree on any conclusion except that something should be done as a counterpart to the US Dyna-Soar (Dynamic Soaring) project.

During re-entry Article 48 could glide in the upper layers of the atmosphere with a small turbojet giving an extra manoeuvrability to the vehicle. The cosmonaut would eject at about 26,250ft (8,000m) and land by his own

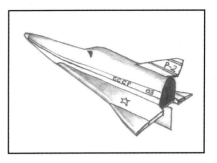

ABOVE Chelomey's Raketoplan R2, rendered by Asif Siddiqi for a NASA report on Soviet space studies. *NASA*

parachute while the glider continued to land automatically on three skids. The heat question was investigated: Niobium would not stand the stress (OKB-23 calculated that heat of up to 1,100°C would be encountered during re-entry). Ablative protection like that used on Vostok capsules was not applicable to a glider which had to maintain a constant aerodynamic surface.

Ceramics were also fraught with problems, developing cracks when submitted to mechanical deformation stress such as the airframe would encounter during flight. In the end, Russian engineers proposed to cut the ceramic revetment in squares, like a mosaic. These squares were tested in the stream of a jet engine and survived. Unfortunately, nothing more is known of the design, as the archive relating to it has been destroyed. Evgeny Kulaga claimed in his memoires that those ceramic squares had anti-radar properties. Indeed, the modern reconstructions of the VKA (aerospace vehicle) by Mark Wade and Asif Siddiqi show a faceted design.

BELOW In 1959, the Soviet press published this somewhat fanciful design of 'the rocket-plane' which will carry the first cosmonaut into space and bring him back safely to Earth. *Author's archives*

NII-1 Soviet hypersonic bomber	
Status	Project
Span	49ft 3in (15.00m)
Length	91ft 10in (28.00m)
Gross Wing area	-
Gross weight	220,460lb (100,000kg) of which 22% would be dry weight
Engine	KDS-100 rocket engine + 2 × ramjets
Maximum speed/height	15,650mph at 161 mi (25,200km/h at 260km)
Load	Up to 13,230lb (6,000kg) of bombs

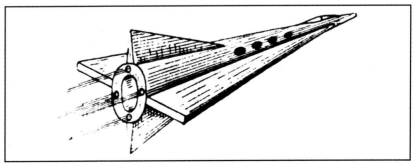

A third designer arrived on the scene with his own ideas about what he termed a Raketoplan or Kosmoplan: Vladimir Nikolayevich Chelomey, head of OKB-52—and lifelong competitor to Korolev. The Kosmoplan was a fantastic design intended for landing cosmonauts on Mars; while the Raketoplan was more akin to an update of the Silbervogel idea. The Raketoplan, which existed in two variants (one with a range of 4,970 miles [8,000km], the other with a range of 24,850 miles [40,000km]) was a fully-recoverable, two-stage design. The combination would be launched vertically and, after separation, the first stage would glide back to land on a conventional runway while the second stage continued the mission toward its target. In April 1960, Chelomey was given the opportunity to present all his grand schemes directly to Khrushchev who quickly disposed of all except for a missile project, which he retained. So ended Chelomey's Raketoplan. During this meeting, Chelomey had been offered OKB-23's facilities to expand his purview. Thereby also ended Myasishchev's own spaceplane project.

Chelomey persisted and managed to have his Raketoplan, now re-labelled an anti-satellite weapon, approved on 23 June 1960. (The idea to fit the machine with a large bay to capture enemy spacecraft '*à la* James Bond's SPECTRE', had been floated to Khrushchev during the April 1960 meeting.) Raketoplans powered by nuclear engines were even considered. In 1961 an AK-4 spaceplane, single-seat and intended for Earth observation, was designed. Both the Raketoplan (approved by the Politburo) and the Kosmoplan (rejected by the Politburo) were tested as small models in wind tunnels at TsAGI.

Even spacesuits for future cosmonauts were designed by this team. A technology demonstrator named MP-1 (MP = Maneuvering, Piloted) was built. On 28 September 1961, MP-1 was launched from an R-12 missile, and landed successfully (by parachute) after a hypersonic re-entry. Even the heat shield survived the experiment intact. Working from these auspicious beginnings, Chelomey's team drafted plans for four variants of the Raketoplan:

■ Single-seat orbital anti-satellite spaceplane.

■ Single-seat orbital bomber against ground targets (direct descendant of Sänger's idea).

■ Seven-passenger ballistic spacecraft for intercontinental ranges.

■ Two-seat scientific spaceship for circumlunar observation (this was a Raketoplan only by name, as it had no wings).

A new hypersonic glider, the M-12– similar in design with the MP-1 but with conventional airbrakes instead of the 'umbrellas' of the MP-1 – was launched atop an R-12 missile on 21 March 1963. Nominally, this launch was part of the programme to design a ballistic warhead, but unofficially it doubled as a testbed for the Raketoplan project. Unfortunately, this test ended with the destruction of the machine during re-entry, apparently due to failure of the heat shield.

But the X-20 Dyna-Soar, which had been the source of the Soviet impetus to work on a spaceplane of their own, was also the cause of its demise because after the US programme was abandoned, the USSR was less keen to finance similar projects and interest in such designs ultimately waned. Besides, the USSR space programme had now another goal: to get to the Moon first, before the Americans could do it. The Spaceplane was just no longer 'in.' Yet, Chelomey managed to keep the flame alive. On 18 June 1964 the new five-year plan for space-based military reconnaissance was published. Spaceplane 'R' was included in this programme. R-1 was to be a technology demonstrator and R-2 a larger, operational version. In 1965 the programme was halted for undetermined reasons (both political and technical causes have been suggested) after a full-size mockup had been built. The R material was then turned over to the Mikoyan design bureau, which was beginning work on a project known as 'Spiral'.

iii) In the USA

In 1952, Walter Dornberger, who had been the head of the Wehrmacht's rocket programme (producing the Aggregat 4 missile, so-called V2) and who was then working for Bell Aircraft, offered Sänger a position in the US firm, which he turned down.

Mark Wade of www.astronautix.com summarised the position of Bell Aircraft at that time. '*Bell was a pioneer in rocket-powered aircraft, having built the X-1, which was the first to break the sound barrier. In 1950 they hired Walter Dornberger, former commander of Peenemuende. He brought on board Krafft Ehricke. They sought to pick up where development at Peenemuende had left off—to pursue development in the United States of rocket-powered, high speed aircraft for both military and civilian purposes. An attempt to recruit Eugen Saenger in France in early 1952 was unsuccessful, but the Saenger Antipodal Bomber became the starting point.*'[9]

Just about every aircraft manufacturer in the USA produced its own project of what a spaceplane would be. So many designs appeared that listing them all would require a book of its own, even though NASA scientists like John V Becker – who actually studied re-entry and hypersonic vehicles from 1945 to his retirement in 1975 – warned, '*Historians and lay readers should not be confused by the multiplicity of aircraft-like vehicle configurations appearing in semi-technical literature of the 1955 to 1965 period under the general subject heading of "Space Transportation System Studies." Vehicles carrying hundreds of passengers and cargo to and from orbit were often depicted in detail by imaginative artists. The main interest of nearly all these studies centred on the propulsion system and the "cost per pound in orbit" of operating the system. The enormous problems of re-entry were not treated except for arbitrary and usually meaningless weight allowance for mythical "guidance and control systems" or equally non-existent "heat protection systems." Obviously none of these studies contributed anything toward solution of the re-entry problem.*'[10]

ABOVE This is Bell's 'Brass Bell' conception of a partially recoverable, global reconnaissance plane. *NASA*

Bell, under the direct influence of Dornberger, designed what were later described as *'greatly advanced and improved Sanger concepts incorporating advanced technology and greater technical depth.'*[11] Under USAF sponsorship, Bell produced designs for 'BoMi', 'Brass Bell' and 'RoBo' and were instrumental in inducing the USAF to launch the X-20 Dyna-Soar programme.

Bomber-Missile (BoMi) was a two-stage vehicle very much in the spirit of the latter-day 'Aerospace Transporter' (see Chapters 2 and 4) proposed in 1952. Both mothership and the actual bomber were manned. A suborbital variant with a range of 2,980 miles (4,800km) and an orbital variant were proposed. Load would have been two nuclear bombs ejected to the rear. The project being unsolicited, the USAF viewed it poorly but thought it could be evolved into a long-range reconnaissance aircraft if Bell studied better the thermal protection (which was thought to be deficient on the original design).

The recoverable first stage was abandoned and replaced by a conventional, vertical, booster launch. Partially-recoverable, the reconnaissance type became 'Brass Bell' in 1956.

The Rocket Bomber (RoBo) programme was a competition to which seven manufacturers replied. It was eventually merged with Dyna-Soar in 1957.

Little hardware was produced but the North-American X-15, if not actually a true re-entry vehicle, was at the very least an aircraft flown on the verges of space. As such, it had to overcome high-Mach aerodynamic problems; to find heat-resistant material; and to obtain an engine able to thrust it to the high stratosphere and smaller engines to control it where aerodynamics could no longer apply.

A major input in spaceplane research was how to make the X-15 stable and controllable during its return through the higher atmosphere. The question of the re-entry of winged vehicles was also tackled for the first time.

Interestingly, at the same moment the Ames *Exploratory Comparative Study of Hypersonic Systems* of late 1955, while assessing mostly suborbital long range vehicles (missiles or reconnaissance aircraft) arrived at the conclusion that a bullet-shaped vehicle was a better option than a winged vehicle. *'For very long ranges, it is better to throw it [the vehicle] than to fly it.'*[12] This had an influence in the choice of cone-shaped vehicles for Project Mercury and its follow up Gemini and Apollo, instead of winged vehicles.[13]

Following on from the X-15, the USAF pushed for the Hyward (Hypersonic Weapon and R&D) system, the objective being to design a vehicle able to reach Mach 15. Actually the scientists working on this project found that Mach 18 was the key number: at this speed, the spaceplane neared its most difficult heating environment. Beyond that, the altitude reached made atmosphere scarce and therefore heat decreased. Here appeared what is today the rather classic shape of the spaceplane: delta wing and flat undersurface with the fuselage above the wing to benefit from a relatively cooler environment in the aerodynamic shadow of the wing. This was also a major difference from the earlier Bell design which had the wing positioned at mid-fuselage. Not all suggestions proved beneficial: Engineer E S Love proposed to attach large cones to the wings' elevons. These were supposed to improve effectiveness in directional and longitudinal stability and control during flight in the upper speed range. They were soon abandoned, because fins and rudders attached to the wing extremities proved lighter, having better heat resistance while performing adequately.

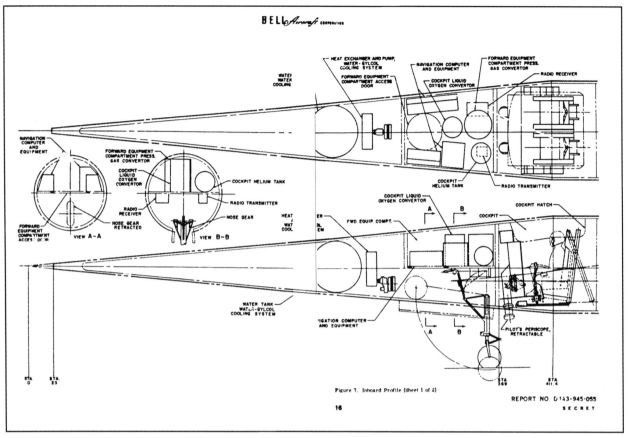
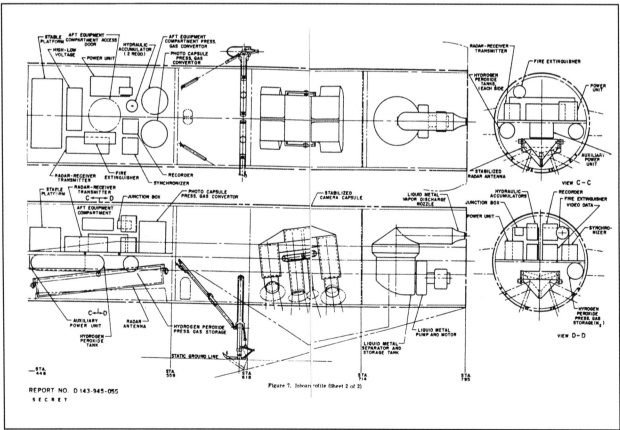

ABOVE **Two detailed cutaway sections of the Bell 'Brass Bell' design.** *NASA*

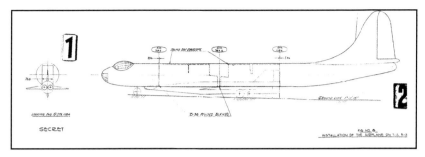

ABOVE **Drawing of the MX-2276 reconnaissance aircraft underneath its Convair B-36 mothership.** *NASA*

The question of heat stress of the leading edge and nose of the spaceplane was a constant difficulty for the teams trying to develop such vehicles at that time (and had been completely overlooked by Sänger in his Silbervogel design). They tried to circumvent the problem by actively cooling those parts through the circulation of a liquid but that was fraught with complications. So the aircraft engineers concluded that their designs were dependant on what the metallurgists could offer—which was quite limited during the mid-'fifties. With some humour, the required but elusive heat-resistant material was nicknamed 'unobtainium.'

The same question would resurface ten years later for French designers at Nord-Aviation: the adequate materials were by now identified, but would it be possible to buy enough to build a vehicle? (see Chapter 6). A contemporary computation was that the mass of coolant liquid required varied wildly for small changes in temperature. If a material able to bear 1,800°F (982°C) was used, a very large weight of coolant would be required. But if a material able to bear 2,000°F (1,093°C) was used, the need for coolant was eliminated except for the sharply curved (small-radii) wing leading edges.

When the X-20 Dyna-Soar's 'call for proposals' was made, no fewer than nine contributions were received. Of these, only Boeing and Bell-Martin were regarded as developed enough to warrant a second round of evaluation. During this second phase, Boeing was selected as the better design but Martin was retained as the provider of the Titan launcher. In the early days of the programme, the USAF was very confident in its project, believing that Mercury would never be ready in time and that Dyna-Soar would put the first American into space. Various operational variants of the Dyna-Soar were also sketched, from reconnaissance aircraft to satellite inspection vehicle—and even nuclear bomber.

ABOVE **Desk model of the Bell 'RoBo' Rocket Bomber, a 'global' bomber echoing Sänger's original ambition, but with a completely different launch strategy.** *Author's archives*

BELOW **Cutaway of the MX-2276.** *NASA*

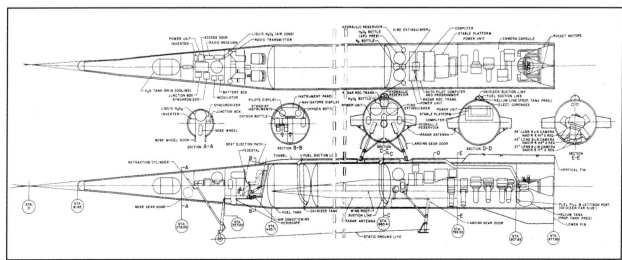

FRENCH SECRET PROJECTS: FRENCH AND EUROPEAN SPACEPLANE DESIGNS 1964-1994

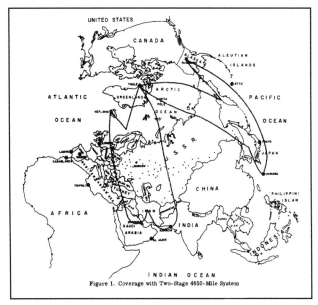

ABOVE Map of the different flight trajectories which MX-2276 could have flown for its reconnaissance missions. *NASA*

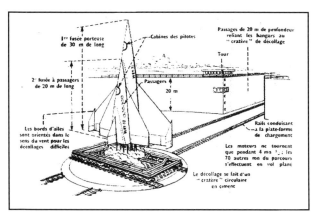

ABOVE Line-drawing of a Bell design for a suborbital transport aircraft. Caption: Passages de 20 m de profondeur reliant les hangars aux 'cratères' de décollage = 20m deep passageways leading from hangars to take-off 'craters'; Rails conduisant à la plateforme de chargement = Rails leading to loading platform; Les moteurs ne tournent que pendant 4mn, les 70 autres mn du parcours s'effectuent en vol plané = Engines are lit for only 4 minutes, the remainder of the flight (70mn) is in gliding mode; le Décollage se fait d'un 'cratère' circulaire en ciment = Take-off is from a circular cement 'crater'; Les bords d'attaque des ailes sont orientés dans le sens du vent pour les décollages difficiles = Wing leading edges are aligned with the wind in case of difficult take-off ; 2ᵉ fusée à passagers de 20m de long = 2nd passenger rocket, 20m long; 1ᵉ fusée porteuse de 30m de long = 1st carrier rocket, 30m long; Cabines des pilotes = Pilots' positions; passagers = passengers; tour = control tower. *Author's archives*

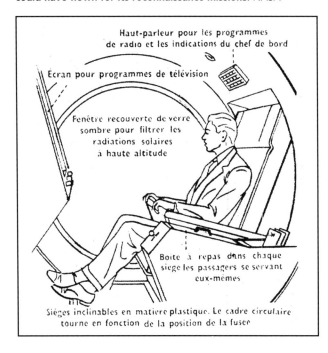

LEFT A passenger's seat in the previously illustrated design. Caption: Ecran pour programmes de television = Screen for TV programmes; Haut-parleur pour les programmes de radio et les indications du chef de bord = Loudspeaker for radio programmes and the chief steward's announcements, Fenêtre recouverte de verre sombre pour filtrer les radiations solaires à haute altitude = Porthole covered in darkened glass to filter solar radiation at high altitude; Boite à repas dans chaque siège, les passagers se servent eux-même = Self-service lunch-box for passengers' meals; Sièges inclinables en matière plastique. Le cadre circulaire tourne en fonction de la position de la fusée = Reclining plastic seats, rotate to follow the attitude of the rocket. *Author's archives*

Bell BoMi (first stage)	
Status	Proposal
Span	59ft 1in (18.00m)
Length	121ft 5in (37.00m)
Gross Wing area	–
Gross weight	793,650lb (360,000kg) with bomber
Engine	5 × rocket engines (fuels: N$_2$O$_4$/UDMH)
Maximum speed/height	Separation occurs at Mach 4 and 19 miles (30km)
Load	Bomber

Bell BoMi (bomber)	
Status	Proposal
Span	36ft 1in (11.00m)
Length	59ft 1in (18.00m)
Gross Wing area	–
Gross weight	–
Engine	3 × rocket engines (fuels: N$_2$O$_4$/UDMH)
Maximum speed/height	–
Load	1 bomb 3,970lb (1,800kg)

Boeing X-20 Dyna-Soar	
Status	Project
Span	20ft 8in (6.30m)
Length	35ft 4in (10.77m)
Gross Wing area	345sq ft (32.1sq m)
Gross weight	11,387lb (5,165kg)
Engine	2 × AJ10-138 rocket engine, 8,000lb st (36kN) each

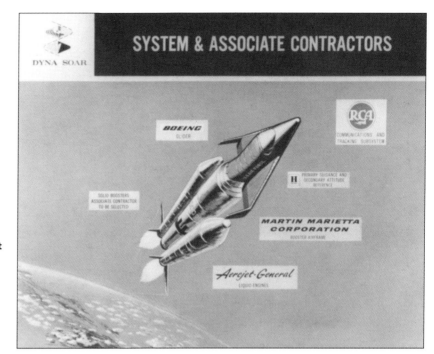

RIGHT The ultimate 'spaceplane' project of the 'sixties was the X-20 Dyna-Soar. All the main actors in the programme are indicated in this picture. *NASA*

BELOW A rather angular wind tunnel model being tested in 1959. Later variants had more rounded shapes. *NASA*

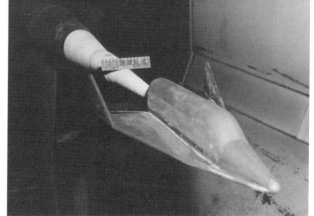

ABOVE RIGHT Many models were tested in wind tunnels. Here is one depicting the spacecraft as of September 1960. *NASA*

RIGHT A nice model of the launch base of the X-20 as it would have appeared in the late 'sixties if the programme had been carried to its conclusion. *USAF archives*

But by being 'in competition' with the Mercury programme, the spaceplane advocates laid themselves open to many criticisms, mostly concerning the additional weight caused by the wings (a ballistic capsule obviously dispensed with the extra weight of the wings which a spaceplane required for re-entry). A half-cone design was also mooted which appeared simpler but presented no specific advantage except it could carry into orbit a larger volume (but not a larger mass). The Aerothermoelastic Structures Systems Environmental Tests (ASSET) programme was initiated in 1960 to assess the thermal protection material imagined for the X-20, but it did not really fly until after the cancellation of the X-20. By December 1961 it was planned to fly two unmanned X-20 suborbital flights (from Cape Canaveral to Edwards AFB), followed by one manned suborbital flight and finally, the maiden orbital flight.

However, after two more years of uncertainties, in December 1963, the whole X-20 programme was cancelled by the USAF. Only the ASSET experimental programme generated hardware that was flown.

iv) In the United Kingdom

It does not appear that Sänger had any direct influence on British designers.

ABOVE Beautiful painting of the X-20 taking off atop a Titan III launcher.
USAF archives

[1] As reported in Irene Bredt's paper.
[2] He probably meant that things will go smoothly and without any fuss.
[3] The film *Hidden Figures* revealed to many the role of women as computers at NASA during the late 'fifties. This situation was obviously not particular to the USA and also existed in other countries like Germany and France.
[4] In fact, the author has never seen a period document, Axis or Allied, in which the nickname 'RaBo' has been used… neither the Silbervogel nickname. Silbervogel seems to have appeared during the 'fifties in reference to a steel wind tunnel model that survived the war. The brilliant appearance of this small (less than one foot?) model generated the name. Being more evocative than RaBo or Antipodal Bomber (another post-war designation), this is the generic term used by the author to describe this aircraft.
[5] The German word *rikoschettier* was used in the original version of the report.
[6] René Lorin was a French scientist holding a diploma from the Ecole Centrale, who is recognised as the inventor of the ramjet. His name was given by the Germans to all ramjet engines.
[7] Both quotes appear in *Mémoire d'Usine,* the history of the Nord-Aviation Chatillon factory.
[8] Quoted in *Challenge to Apollo: the Soviet Union and the Space Race 1945-1975,* by Asif A Siddiqi, NASA report 2000-4408.
[9] http://www.astronautix.com/b/bomi.html.
[10] *The Development of Winged Reentry Vehicles 1952-1963* by John V Becker 23 May 1983.
[11] *The Development of Winged Reentry Vehicles 1952-1963* by John V Becker 23 May 1983.
[12] Quoted in *The Development of Winged Reentry Vehicles 1952-1963* by John V Becker 23 May 1983.
[13] However, winged Apollo spacecraft were considered.

Chapter Two
Eurospace Ideas

ABOVE As late as 1967, Nord-Aviation was still promoting its vehicle through this painting by Paul Lengellé.
Nord-Aviation via author's archives

In 1954 Eugen Sänger left France and returned to Germany.

In June 1960 he was instrumental in setting up the Groupe d'Etudes Européen pour les Recherches Spatiales (GEERS: European Study Group for Space Research); and then events accelerated.

On 1 July 1961, Sänger was recruited as consultant by the German aviation firm Junkers to help design its Raum Transporter (Space Transporter) which followed rather closely the original Silbervogel pattern described in Chapter 1. Following his initiative, a winged aircraft was designed to achieve orbit after a catapult launch. Shortly after, on 21 September, a group of more than 80 aerospace firms joined an entity called Eurospace, which was to promote space industry. On 4 October, the Sänger couple was nominated 'associate members' of Eurospace. The organisation is still active today.

At that point, the need for a Space Transporter (its 'Aero-' prefix mostly discarded) emerged in a context in which rocket launchers were far from perfect. In the USA, the launch of the Explorer 1 satellite had proved quite difficult, with numerous attempts failing, and the ELDO Europa rocket was still in the future. Another argument in favour of an alternative to the vertically-launched rocket was that it took decades for the first two space powers, USA and USSR, to develop a reliable launcher, even with the experience obtained from military ballistic missiles. So Europe was obviously lagging in this area and, instead of arduously following behind USA and USSR, it seemed logical to develop an aircraft-type launcher. And basing this on the work of Eugen Sänger was an obvious step.

This approach was backed by Eugen Sänger in Germany and by former ONERA director Maurice Roy in France. Professor Roy claimed that if the use of missiles as satellite launchers was a quick, short-term solution, then in the long term, only an aircraft-type launcher could be the answer as:

- A 'consumable' launcher was just uneconomic.
- An aircraft-type machine would make a more efficient use of the different types of engines—air breathing versus rocket.
- To broaden access to space for non-astronauts, acceleration should be limited to 2.5G, something which only an aircraft-type launcher would allow.

Eugen Sänger himself pushed for a quick decision on the project. '*Considering that it will take at least ten years to realise such an ambitious project as the Aerospace Transporter, work on it must be started without delay. If development in this field has not yet gathered momentum in the United States, nor possibly in the Soviet Union, this is due to the fact that the entire intellectual and material resources available to space research in those countries are at present being concentrated on pioneer projects, in particular the race to the Moon. As soon as this effort is relaxed, these countries will direct their full capacity toward the Aerospace Transporter phase of spaceflight, as clearly can be seen from the preliminary work of the American aerospace industry in this area. There is therefore at the moment a unique, but only a short-lived opportunity for Europe, with its great intellectual and material resources, to become active in a sector of spaceflight in which the major space powers have not yet achieved an insuperable lead…*'[14]

In practice, the concept remained on the drawing board, governments preferring to finance quick, consumable launchers rather than investing in the long term with a fully re-usable launcher. For example, ELDO, the European Launcher Development Organisation, never investigated this kind of vehicle and concentrated on the Europa Launching Rocket … without success (and Europa would indeed prove impossible to perfect during the 1967–1971 period, with six failures out of six launches). Even the American Space Shuttle proved a failure in this respect despite only the 'drop tank' being consumable (see Chapter 12 regarding the economics of reusability in a space launcher context).

Eurospace was subdivided into five 'working groups' of which Group III was the Technical Studies Group under Professor Baxter, followed by Professor Roy. Group III took a special interest in the Space Transporter studies and on 30 April 1963, Sänger was appointed as project leader at Eurospace. A report, *Proposals for a European Aerospace Programme* was published that year. In 1964, as a follow-on to this, a detailed study focusing on the Space Transporter project was issued at that year's Eurospace symposium organised in Brussels on 23/24 January and presided over by none other than Eugen Sänger and Maurice Roy. Over 200 people from the European space community congregated at this symposium. (Representing France were the companies Sud-Aviation, Nord-Aviation, Dassault, SNECMA, SEPR and Giravions Dorand). That the concept was quite new at the time may be judged by the fact that the French aviation magazine *Air & Cosmos* had to explain what a *Transporteur Aérospatial* was (issue No 41), '*A manned vehicle able to carry a load up to a determined orbit and to come back to Earth to be re-used*'.

Curiously, Eugen Sänger reported in *Preliminary Proposals for the Development of a European Space Vehicle* (completed just a few hours before his death), writing about the Nord-Aviation Transporteur Aéro-Spatial, mentioned an *Eurospace glider model which he compared to a reduced variation of the X-20. (Nord-Aviation/EUROPE).*' The mention of a glider model is intriguing and the reference to the X-20 is quite odd, as this project had nothing to do with the American glider. Maybe there was a mix-up with VERAS, which was another Nord-Aviation project which did lead to actual hardware development (see Chapter 6).

Eurospace's project was defined as follows:

1. Mission: to put into orbit a payload and two men
2. Typical orbit was circular, at a height of 186 miles (300km)
3. The load (including fuel to transfer the satellite to a higher orbit) was to be 5,512 lbs (2,500 kg) and 353cu ft (10cu metres)
4. Take-off was to be from a base located in Europe, with a take-off weight not exceeding 440,925lb (200,000kg)
5. The Transporter was to be a two-stage vehicle, fully recoverable (preferably by return to the take-off base). The life cycle of the vehicle was expected to be fifty flights over two years
6. The fully-atmospheric first stage could use either rocket or jet engines…or a combination of both
7. The second stage would use an oxygen/hydrogen rocket engine
8. Maximum acceleration was to be 2.5G

Upon returning to Earth, the pilot would be able to select from a wide range of airfields and the craft would land like a conventional aircraft.

Eurospace noted that while the Soviets had described 'raketoplans,' it was uncertain if these were propaganda devices, sci-fi designs or actual projects. And the Americans had just abandoned their project Dyna-Soar. Europeans themselves had debated the launch procedure for such a big machine as an Aerospace Transporter: the Germans promoted a horizontal ramp while the French were in favour of a normal take-off on a runway. However, there was an agreement that if the machine was to be (as it most probably would be) a two-stage composite, both stages should be piloted but using two different propulsion

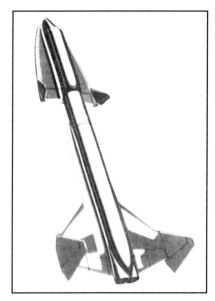

ABOVE A very early design by the German consortium ERNO was this 'shuttle' shown here at take-off. *NASA*

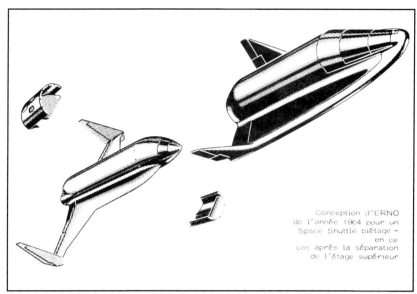

ABOVE This other illustration depicts the separation phase of the two stages of the ERNO Shuttle. *ERNO via Philippe Ricco*

systems: turbo-ramjets for the first stage and rockets for the second stage. Actually the big questions were political and economic. No one doubted the economics of a reusable vehicle, which would not require a special launch infrastructure but there was uncertainty whether the Americans, with their all-powerful industry, would let the Europeans enter this market.

Anecdotally, issue 42 of *Air & Cosmos* was the alpha and omega of this story as it featured (a) the obituary of Eugen Sänger who was the arch-promoter of the concept; (b) one page on the Eurospace symposium; and (c) two pages on the Diamant rocket (which, being the first French satellite launcher in 1965, paved the way for the current Ariane). 1964 was also notable for the signature of a partnership between French companies Nord-Aviation and SNECMA and German company ERNO to collaborate on the design and development of a Space Transporter.

The following year, Eurospace proposed to the European governments that it should produce a feasibility study on this kind of design in two years' time, with a budget of $6 million. There was no direct reply from the governments, although they sent representatives to the industry meetings held during the years following.

In the end, unfortunately, it was concluded that too many parameters were lacking in details for a theoretical research. As pointed out by H Tolle in a summary report, an erroneous estimate of just 20% on the calculation of the structural weight of a 14.8-ton (15-tonne) second stage intended to put into orbit a 2.95-ton (3-tonne) payload, would burn a quantity of fuel equal to the payload.

Interestingly, in the September/October 1985 issue of *Spaceflight*, the magazine of the British Interplanetary Society (BIS), Rex Turner, the Technical Secretary of Eurospace reminded his readers that the Space Transporter had been a key theme at the November 1963 BIS Symposium and that it had fallen into oblivion after the second Eurospace US/European Conference where the concept had been presented to the US Industry and NASA. '*US industry's reaction to this proposal can almost be called derisive; it was suggested that such a project was too advanced, well beyond the capability of European industry and that the Europeans should confine themselves to more modest targets, preferably participating to US projects*'.

This was, indeed, what the German companies chose to do, preferring to work with McDonnell Douglas on its shuttle project.

Rex Turner added, '*Fortunately for the future of European space activities, this view was not shared by other European countries and, as a result, Europe has achieved a respectable degree of independence in space activities which it is hoped will expand as a result of the ESA conference at Ministerial level in Rome during January* [1985] [see Chapter 7].

BELOW A very simple plan view of the ERNO Shuttle. *Via Dan Sharp*

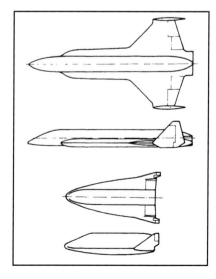

However, would not Europe and the UK in particular, be in an even more advantageous position if the governments had exercised just a little more imagination 20 years ago? Perhaps HOTOL would already be flying?' [see Chapter 9].

A – GERMAN PROJECTS

In West Germany, the government followed Eurospace advice and integrated the Space Transporter research in its National Space Programme under Research Project 623.

i) ERNO study

As part of this Project 623, the Entwicklungsring Nord (ERNO = Northern Development Circle) designed a two-stage, winged 'space shuttle' concept but abandoned it on grounds of cost. ERNO was a 1961 working group between Weserflug, Focke-Wulf and Hamburger Flugzeugbau (HFB) to develop space programmes. In 1964 Weser, Focke-Wulf and Heinkel merged into Vereinigte Flugzeug Werke. During 1967, VFW and HFB (Hamburger Flugzeug Bau) formalised their association by setting up the ERNO Raumfahrttechnik GmbH company with a capital of DM10 million (approximately £900,000), thus emphasising its space activities and objectives (Raumfahrttechnik means spaceflight technology).

The initial ERNO proposal was a vertically-launched 'space shuttle' two-stage vehicle, in which both stages were rocket-propelled, had wings and were manned and recovered. Then the second stage, or rather an enlarged variant of it, was used in the collaborative effort with Nord-Aviation and SNECMA. This design was produced during the period 1963–1964, in time for the 1964 symposium.

In 1965, ERNO associated with Nord-Aviation in the Transporteur Spatial project although its contribution to this project is unclear. ERNO is said to have worked with Dornier and Junkers on Junkers' Raum Transporter proposal.

When, in 1969, the first news of the US Space Shuttle (then labelled by ERNO as a 'Post Apollo Programme – PAP') reached Germany the project was reoriented as participation in the American programme through an association with McDonnell Douglas on the Shuttle MDC A. From 1970, ERNO engineers went to St Louis, Missouri, to work in the American 'space shuttle team.' In Germany, wind tunnel models were tested, notably in the Porz-Wahn establishment of DFVLR (Deutsche Forschungs-und Versuchsanstalt für Luft- und Raumfahrt = German Research & Development Institute for Air & Space Travel). ERNO at that point had high hopes of participating in the design and construction of the American shuttle and therefore invested heavily in various testing programmes.

A model LB-1 (LB for Lifting Body) was tested in wind tunnels. The tests indicated that lateral stability was poor, which by further increments led to the LB-10 model. With LB-10, lateral stability was obtained and the design was further refined until the LB-21 model was considered the final design. LB-21 was then tested in super- and hypersonic wind tunnels. Instability reappeared at low speed, which required the addition of a central fin.

With the programme reoriented, tests were carried out with a 10-foot (3-m) long free-flight model named Bumerang (Boomerang) in 1971. These tests involved dropping the model from a Transall C-160 transport aircraft and recovering it by parachute. The first flight (Flug F1) of Bumerang took place in the German archipelago of Heligoland 12 August 1971. While launch from an underwing pylon (like the US Lifting Bodies) was mocked-up, actual flight test involved dropping Bumerang on a pallet from the rear door of a Luftwaffe C-160D. At 16,400ft (5,000m), and about 250mph (400km/h), the pallet was extracted by a parachute and 2½ seconds later Bumerang separated from the pallet and began its free flight. When an altitude of 1,640ft (500m) was reached, a parachute deployed from the back of the test vehicle for landing. Later flights were carried out from the Salto di Quirra Missile Test Range in southeast Sardinia.

BELOW Meanwhile, in Germany, ERNO began investigating the lifting-body shape. Here is the first model LB-1. *Author's archives*

BELOW Another model, this time of the LB-10. Design would be refined until LB-21 was felt adequate. *Author's archives*

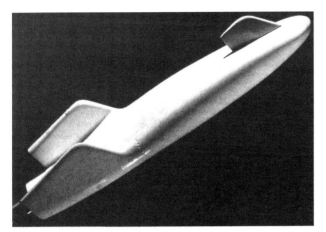

ABOVE When ERNO became involved with the McDonnell Douglas shuttle programme, it took the shape of the LB-21 wind tunnel model and produced a free-flying model named Bumerang (Boomerang). *Author's archives*

ABOVE Here is a low-resolution picture of an early 'rubber' model of Bumerang. *Author's archives*

BELOW Bumerang was reusable and is depicted here being recovered at sea after an airdrop. *Author's archives*

BELOW Rather poor photographs of Bumerang being air dropped from a Transall C-160. *Author's archives*

The vehicle involved was now called Bumerang II because it was radio-controlled while the first had been a free-flyer. Four flights F2, F3, F4 and F5 were carried out there. *Flight* magazine reported that *'two submarines of unknown nationality'* were there to observe those tests. The experiments were regarded as highly successful by ERNO.

Those trials have sometimes been associated with the French Transporteur Aérospatial project but they do not appear related other than being carried out by ERNO, which was partnered with Nord-Aviation in this study.

Using the aerodynamic results of these tests, computer simulations were programmed to evaluate the heat constraints to which such a lifting body would be subjected. On the metallurgy aspect, solutions were investigated to enable the vehicle to support the heat stress. A flying model 10ft 2in (3.10m) long and weighing 1,323lb (600kg), released from atop a Europa launcher, was expected to be the next step in the study (observe the similarity with the French programme VERAS in Chapter 5) but was not proceeded with. After the abandonment of the Europa programme in 1972, ERNO joined ESA in 1975.

BELOW German firm Bölkow also contributed a Raum Transporter design as part of the Eurospace research. *Via Dan Sharp*

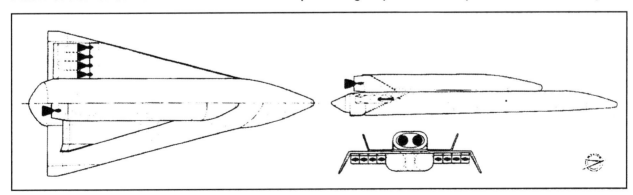

ERNO participated in a further ELDO programme called the 'space tug' which involved designing an unmanned vehicle which could take a satellite from low orbit at 124 miles (200km) up to a geosynchronous orbit at 22,370 miles (36,000km). This was studied during 1971–72 but did not lead to any hardware.

ii) Junkers

In the same time period as the early ERNO study (circa 1961–63), Junkers

Lambrecht SSTO early	
Status	Study
Span	–
Length	84ft 0in (25.60m)
Gross Wing area	–
Gross weight	–
Engine	Rocket O_2/H_2 fuel
Maximum speed/height	–
Load	1,102lb (500kg)
Lambrecht SSTO late	
Status	Study
Span	–
Length	59ft 9in (18.20m)
Gross Wing area	–
Gross weight	–
Engine	Rocket F_2/N_2H_4 fuel
Maximum speed/height	–
Load	1,102lb (500kg)

engineer Jurgen Lambrecht, in *Design Problems for a One-Stage Transporter* written in November 1963, described a single-stage-to-orbit vehicle, launched by a 'steam catapult'. The device appears to refer to a concept originated by Eugen Sänger in which the actual spaceplane would be boosted along a horizontal rail 10,500ft (3,200m) long. It was not a steam catapult comparable to those used aboard aircraft carriers. The booster (which remained attached

BELOW Rare picture of Eugen Sänger's own view of what a Raum Transporter would be. *Author's archives*

to the rail) used pressurised steam for propulsion. Although his initial design weighed 147.6 tons (150 tonnes), Lambrecht was able to reduce the all-up weight to 83.7 tons (85 tonnes) for a design using as fuels H_2/O_2 (the most favourable mix between the two fuels being 7:1) and 49.2 tons (50 tonnes) for the F_2/N_2H_4 design (2.2:1 mix) which he naturally favoured as both had the same ability: 1,102lb (500kg) of payload in a 186-mile (300-km) circular orbit.

The vehicle would be launched in an elliptical trajectory, with the main engine being shut down when an altitude of 31 miles (50km) was reached (about 150 seconds after take-off), the vehicle continuing on its trajectory until it achieved its apogee at 186 miles (300km). There it would manoeuvre to place its payload in orbit, or dock with an orbiting space station (for example to bring back crew members to Earth). Because of these rendezvous manoeuvres Lambrecht indicated that '*it* [is] *desirable that the Space Transporter be manned*'. The landing speed on return was expected to be 171mph (275 km/h). Rocket engine specifics were not discussed.

Lambrecht's studies were not really oriented toward the full design of a Space Transporter, but rather to identify

BELOW A major actor in Germany's RaumTransporter project was Junkers. Here is one of its lesser-known propositions: a lifting body-type orbiter, comprising a delta-shaped first stage with a non-rigid wing (*membranflügel*), all this being carried by a specially adapted B-52. *Author's archives*

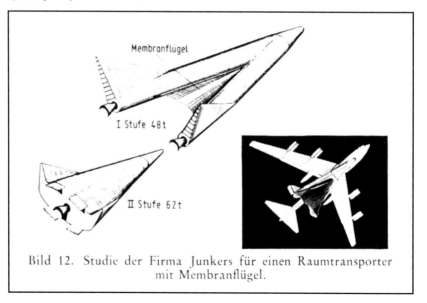

the key parameters in designing one. In this respect he demonstrated the importance of selecting the densest fuels and that the take-off weight grew much more quickly than the payload weight. In December 1964, he applied for a patent describing an H_2/O_2-powered spaceplane in which he referred specifically to a Eugen Sänger patent registered one month before (which the author has not been able to trace). An original feature of this design was that the combustion chamber of the rocket engine was created by the undersurface of the wing and the rear, tapered, section of the fuselage.

In 1965 there appeared the two-stage RT-8-01(RT-8 because it was probably the eighth iteration of the basic project) a machine with a large, high-mounted wing, and a rocket-propelled first stage topped by a low-wing, also rocket-propelled orbiter. Both designs had delta wings with up- (orbiter) and down- (mothercraft) turned tips. Take-off weight was 184 tons (187 tonnes), of which 2.70 tons (2.75 tonnes) was payload. As in the wartime project, the two aircraft, docked together, were placed on a rocket-propelled launch booster running on 10,500ft (3,200m) of track. Take-off (separation from the booster) would occur at 560mph

ABOVE Junkers, with Eugen Sänger acting as consultant, designed the Raum Transporter, of which the RT-8 (presumed to be the final iteration) is illustrated here. *Author's archives*

ABOVE Another view of the RT-8 model, with what appears to be a mockup of GRS-1 Azur – the first German satellite to be launched by an US Scout rocket in 1969 – on the right and an unidentified satellite mockup on the left. *Author's archives*

ABOVE RIGHT This illustration depicts the two launch options imagined by Junkers engineers: either under the wing of a Boeing B-52, or along a ramp. *MBB via author's archives*

RIGHT In this illustration the RT-8-1 stages, still docked together, climb through the atmosphere.
MBB via author's archives

Junkers RT-08 (stage1)	
Status	Study
Span	-
Length	144ft (44.00m) (approx)
Gross Wing area	-
Gross weight	412,265lb (187,000kg)
Engine	Rockets LOX/LH$_2$
Maximum speed/height	-
Load	66,140lb (30,000kg) (2nd stage)
Junkers RT-08 (stage2)	
Status	Study
Span	101ft 8in (31.00m) (upper stage)
Length	39ft 4in (12.00m) (upper stage)
Gross Wing area	-
Gross weight	66,140lb (30,000kg)
Engine	Rockets LOX/LH$_2$
Maximum speed/height	-
Load	2.7 tons (2.75 tonnes) in 186 mile (300km) orbit

ABOVE Illustration of the separation of the two stages of RT-8.
MBB via author's archives

(900km/h), the two vehicles parting in flight, the lower mothership returning to Earth while the orbiter, with a crew of two astronauts, continued its flight until the orbit altitude of 186 miles (300km) was reached.

On the return trip, the orbiter was to land on three retractable skids. The Junkers effort culminated in the RT-8-02 'Sänger' but difference between the dash 1 and dash 2 variant is unknown. It was mentioned in press reports of the time that the launch from a Boeing B-52 Stratofortress was also investigated. Considering that the B-52 payload is usually given at 70,000lb (31,750kg) it is unclear how Junkers expected to do it (notwithstanding the question of acquiring a B-52—perhaps leasing NASA's assigned B-52B?)

iii) Bolköw/MBB

In 1964–65 a two-stage, rocket-powered vehicle was designed. Both stages used the same engines (eight in the first stage, two on the orbiter) but the first stage incorporated air augmentation to boost engine performance. 'Air augmentation'[15] was to have the combustion chamber of the rocket engines only partly shrouded. It was hoped to obtain a significant gain in specific impulse without adding much weight for shrouds. Take-off was horizontal and was to be assisted by undefined means. In 1966, research focused on re-entry vehicles along the 'lifting body' design methodology.

On the whole, when promoting the Sänger II project in 1983, MBB claimed that 12 million Deutsche Marks and 220,000 man-hours had been allocated to the project from 1962 to 1968.

BELOW Detailed drawing of Junkers RT-8. *MBB via author's archives*

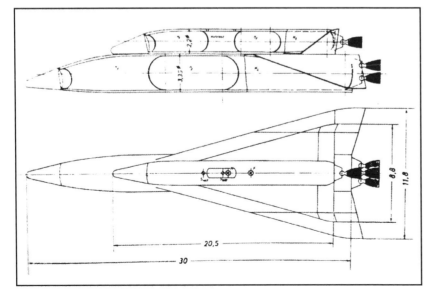

Bumerang	
Status	Test model
Span	6ft 3in (1.90m)
Length	10ft 2in (3.10m)
Gross Wing area	–
Gross weight	772lb (350kg)
Engine	–
Maximum speed/height	Up to 249mph (400 km/h) at 16,400ft (5,000m)
Load	none

Bolköw/MBB 1 stage	
Status	Study
Span	–
Length	–
Gross Wing area	–
Gross weight	396,825lb (180,000kg)
Engine	8 × rockets (LOX/LH$_2$) with air augmentation
Maximum speed/height	–
Load	114,640lb (52,000kg)

Bolköw/MBB 2 stage	
Status	Study
Span	–
Length	–
Gross Wing area	–
Gross weight	114,640lb (52,000kg)
Engine	2 × rockets (LOX/LH$_2$)
Maximum speed/height	6,614lb (3,000kg)
Load	–

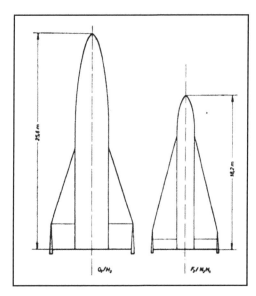

ABOVE From a Junkers report written by Jurgen Lambrecht, comparison of the size of O$_2$/H$_2$ fuelled vehicle (left) with a F$_2$/N$_2$H$_4$ fuelled vehicle on the right. This second option was obviously better in terms of size. *MBB via author's archives*

BELOW Afterwards, ERNO worked with Nord-Aviation on a first iteration of the Transporteur Aérospatial. This design was promoted through a leaflet distributed at the 1965 Paris Air Show. *Author's collection*

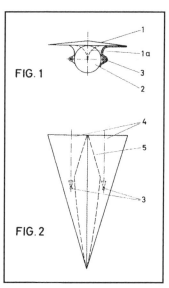

ABOVE In 1964 Jurgen Lambrecht of Junkers patented this launcher in which the ramjet combustion chamber was delimited by the wing undersurfaces and fuselage sides. *Author's archives*

B – FRANCO-GERMAN (NORD/SNECMA-ERNO) CO-OPERATION

Following the 1964 co-operation agreement between Nord-Aviation, SNECMA and ERNO, a pamphlet describing a Transporteur Aérospatial collectively branded to the three companies was made available at the Paris Air Show 1965. '*The goal of such a vehicle would be, first and foremost, the construction and resupply of the future space stations.*' At this time the project was still conceived as a two-stage vehicle. '*The first-stage plane is propelled by four turbo-ramjet combined engines. It can reach an altitude of 114,800ft (35,000 metres) and a speed of Mach 7, then, after a ballistic phase, get back to its base. The second stage, set underneath the engines of the carrier-plane, is mostly geared toward atmospheric re-entry, thanks to its six rocket engines using*

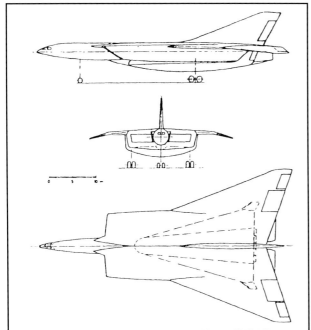

JOINT FRANCO-GERMAN COOPERATIVE SHUTTLE DESIGN STUDY BY NORD, SNECMA, AND ENTWICKLUNGSRING NORD (ERNO) FOR AN "AEROSPACE TRANSPORTER" KNOWN AS THE MISTRAL, CONSISTING OF A WINGED JET-PROPELLED LAUNCH AIRCRAFT AND A LIFTING BODY ORBITER NESTLED BENEATH IT. LATER VERSIONS OF THE MISTRAL WERE LESS ELEGANT IN APPEARANCE, AND CONFIGURED TO LAUNCH CONVENTIONAL UNMANNED UPPER STAGES.

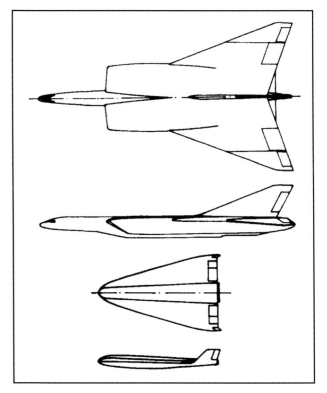

ABOVE Another drawing of the same design showing detail differences to the previously illustrated version (intakes, undersides, etc). *Author's archives*

ABOVE LEFT Three-view drawing of the Nord-ERNO design. This document is from a 1970s NASA report which mixed-up the Nord-ERNO Transporteur Aerospatial and the later Nord Mistral in its captioning. The drawing refers only to the Nord-ERNO vehicle. *NASA*

hydrogen and liquid oxygen. This second stage, once on the selected orbit, can manoeuvre and put on orbit up to 8 tons of payload. Upon completion of the mission, its pilot can select a landing field and land in the usual way. The operational deployment of this machine could be expected to be around 1980.' The six engines of the second stage were actually four propulsion engines and two vernier engines.

A scale model of the Nord TAS was also exhibited at this show.

On 25 June, just five days after the the Paris Air Show closed, the Nord model was again exhibited at the

LEFT To promote its design, Nord-Aviation had a large display model built by La Maquette d'Etudes et d'Exposition. *Copyright AIRBUS/AIRITAGE/DR*

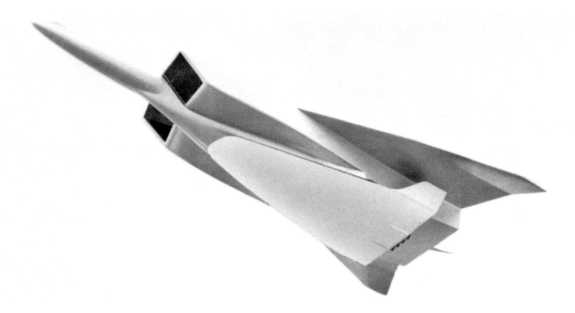

ABOVE **Another view of this superb model.** *Via Philippe Ricco*

BELOW **Ink illustration of the Transporteur Aérospatial designed jointly by Nord and ERNO in a Nord newsletter. Note the downward turned winglets. Too little is known about this project to tell if that change had technical reasons or was just the (unknown) artist's fancy. Note, however, the dotted outline of an intermediate stage, thus recognizing that a purely two-stage, fully recoverable design, was not feasible.** *Nord-Aviation via author's archives*

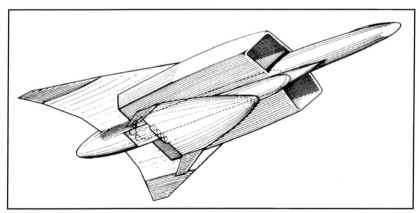

Nord Spaceplane (1st stage) (1965)

Status	Study
Span	-
Length	169ft 11in (51.80m)
Gross Wing area	-
Gross weight	264,550lb (120,000kg)
Engine	4 × turbo-ramjets
Maximum speed/height	Mach 7 at 114,830ft (35,000m) separation speed and altitude
Load	176,370lb (80,000kg) load

Nord Spaceplane (2nd stage)

Status	Study
Span	-
Length	85ft 0in (25.90m)
Gross Wing area	-
Gross weight	176,370lb (80,000kg)
Engine	6 × LOX/LH$_2$ rockets
Maximum speed/height	-
Load	6,614lb (3,000kg)

ERNO vertical launch vehicle S1

Status	Study
Span	-
Length	-
Gross Wing area	-
Gross weight	403,450lb (183,000kg)
Engine	Rockets LOX/LH$_2$
Maximum speed/height	-
Load	108,025lb (49,000kg) (orbiter)

ERNO vertical launch vehicle S2

Status	Study
Span	-
Length	-
Gross Wing area	-
Gross weight	108,025lb (49,000kg)
Engine	Rockets LOX/LH$_2$
Maximum speed/height	-
Load	6,614lb (3,000kg)

International Communications and Transport Exhibition in Munich.

The Nord-Aviation/SNECMA/ERNO project gained a short-lived celebrity before falling into oblivion and the dustbin of history.

According to a German ERNO source, the cost of building a Transporteur Aérospatial was estimated at that time at 20 billion Deutschmarks (about €32 billion) while the study of the orbiter (this source indicated a goal of putting a 2.95-ton [3-tonne] payload in a 310-mile [500-km] orbit which is significantly more ambitious than the original Eurospace project) revealed a lot of uncertainties regarding structure, type of engines, and thermal insulation. For this reason, the orbiter part of the studies was dropped and the two parties separated.

Probably because this fully-recoverable design was too uncertain, it was not among those presented in the report made to CPE in September 1966 (see Chapter 4) but curiously was still being publicised in early 1967 press releases of the Direction des Recherches et des Moyens d'Essais of the Ministère des Armées…for disinformation?

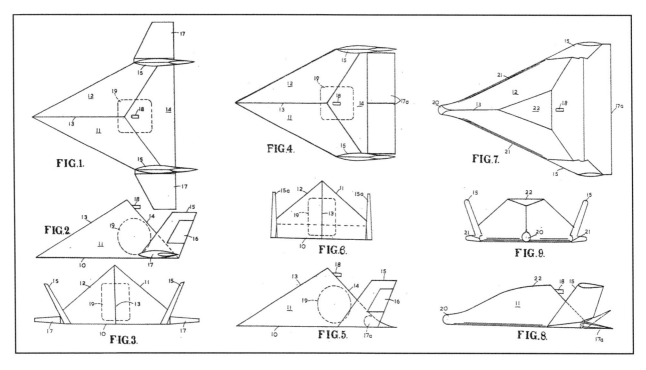

ABOVE Three drawings from the Armstrong Whitworth patent showing variants of the so-called 'Pyramid' spacecraft. In all three versions, the 'fuselage' was basically the same pyramidal structure; what differed greatly was the location of the fins and flaps. *Author's archives*

In a similar approach, the Transporteur Aérospatial theme was selected as the All Fools' Day spoof for the 1 April 1968 edition of *Aviation Magazine*. The joke was based around the fact that two men named René Leduc had contributed to French aviation: one had studied ramjets since before the Second World War, while the other was a designer of small private aircraft.

Dassault's response to the Eurospace call for feasibility studies is described in Chapter 4.

C –UK CONTRIBUTIONS

Beside the German and Franco-German initiatives described above, Eurospace also received contributions from UK.

i) The 'Pyramid'

This design came about in 1960, so was well before the Eurospace initiative, but it is still very interesting, because it was seemingly the first British project to tackle the question of bringing back astronauts from orbit using gliding re-entry. Armstrong Whitworth, in partnership with Hawker Siddeley, proposed a two-man Waverider to be launched using a Blue Streak medium-range ballistic missile. The Waverider concept was of direct Sänger descent, even if it had been renamed. The Waverider itself was designed by Terence R F Nonweiler in 1951 but the whole concept was later taken over by H R Watson at Armstrong Whitworth and Dr W F Hilton at Hawker Siddeley. It seems the original design had a more conventional, aircraft shape but the necessary 'compacity' (to allow it to be put it on top of a rocket) led to the more compact but unusual design. The spacecraft soon gained fame because of its unusual pyramidial shape, although it seems the 'pyramid' word was never used officially.

This configuration meant that for launch, a symmetrical shape was installed back-to-back (or more exactly bottom-to-bottom) with the spaceplane (that 'image' device could have been used as an extra fuel tank for the first

BELOW The two main designs are illustrated side by side in *RAF Flying Review*, August 1958 edition. *RAF Flying Review via author's archives*

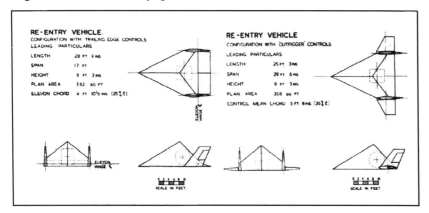

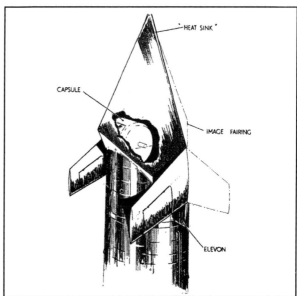

ABOVE H R Watson, technical director of Armstrong Whitworth, with a model of the 'Pyramid' spacecraft. This is the second version, with the fins on the sides and the flaps positioned at the trailing edge of the fuselage. This configuration, called 'body flap' would be used on the Rockwell orbiter and on late French hypersonic designs.
Author's archives

ABOVE RIGHT Artist's drawing of the launch configuration, illustrating the cylindrical crew capsule inside the 'Pyramid' and the need for an 'image' fairing atop the launcher.
RAF Flying Review via author's archives

AW 'Pyramid'*	
Status	Project
Span	29 ft 6 in (9.00m)
Length	25ft 3in (7.70m)
Gross Wing area	3,658sq ft (340sq m)
Gross weight	4,143lb (1,879kg)
Engine	Rocket
Maximum speed/height	80-mile (129-km) orbit
Load	2 × astronauts

stage and be dropped at an altitude of 40 miles [64 km]). This resulted in a marked diamond shape for the launcher head. A patent was submitted on 13 April 1960 which described three variants of the basic 'pyramid' design. A curious feature was a 'controllable vent pipe' installed in the rear wall which was intended to *'reduce, by ejector action, the air density within the vehicle and around the [crew] compartment and thereby reduce the transfer, by convection or by conduction of the heat generated by atmospheric friction with the interior of the vehicle.'* This appears to be an exceedingly optimistic view of re-entry heat stress control.

Its crew compartment was equipped with a parachute and could be ejected in case of emergency. The patent does not describe any rocket motor for orbital changes or de-orbiting but explains that attitude control jets would be used for manoeuvring in space; and fin, rudders and flaps for manoeuvring in the atmosphere. Where to put those rudder and flaps is a question that remained unsolved. Aerodynamically, the best solution appeared to put them as wing extensions (bringing the span to 29ft 6in [9.00m]), but it also meant a less compact vehicle during the launch sequence and more leading edge which had to be protected from heat stress during re-entry. Another option was to place them behind the vehicle, between the rear fins.

RIGHT Another artist's impression attempting to depict the complete launch configuration. The drawing is based on a photograph of an American Atlas ICBM, while the intended launcher was a derivative of the British Blue Streak.
RAF Flying Review via author's archives

* Data from *RAF Flying Review* 1959. Ron Miller's *The Dream Machines* indicates length: 25ft, width 18ft, weight 'about 2 tons'. Booster first stage weight 133.5 tons; second stage 22 tons; third stage weight unspecified.

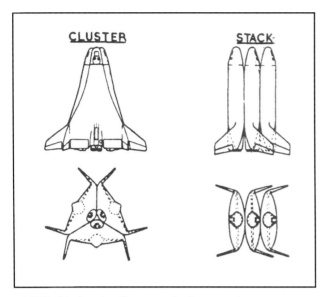

ABOVE This drawing illustrates the famous 'triamese' configuration of the BAC Mustard concept which was also presented at an Eurospace seminar. *NASA*

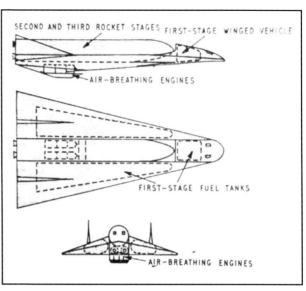

ABOVE Three-view illustration of the Bristol Siddeley space launcher announced in 1964. *Author's archives*

After lift-off, the Blue Streak (or a similar launcher) with a weight of 133.5 tons[16], a length of 41ft (12.5m) and a diameter of 13ft (3.95m), would ascend to an altitude of 40 miles (64km), at an angle of 40°. At this point, separation from the second stage would occur, the latter then climbing to 75 miles (121 km), where the third stage would separate. At 80 miles (129 km), the third stage engine would fire to assure insertion into a 700 × 80-mile (1,126 × 129-km) orbit. It was expected that the mission, encompassing two complete orbits, would have duration of 6¾ hours. Landing on a conventional undercarriage would take place at 80mph (129km/h). To investigate low-speed performance of the design, small 1ft (30cm) wide balsa models were catapulted across a field. Several successful flights were achieved, but it was also plainly obvious that success was much dependant on the model being launched at the right angle. If not, it would just 'tumble and crash.' This showed that correct recovery of the design would require precise manoeuvring.

LEFT Lifted from a contemporary publication, these 1964 illustrations depict the Eurospace Bristol Siddeley design. The half-buried position of the second stage is well shown here. *Author's archives*

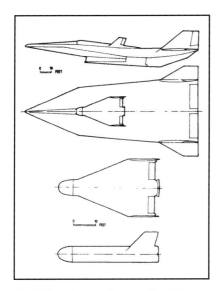

ABOVE Evolution of research at Bristol Siddeley led to this later (1966) design in which both stages were winged. *Author's archives*

ABOVE The ultimate incarnation of this train of thought appears to be this design, published in 1970 as a Hawker Siddeley Dynamics project. *NASA*

The launcher component of this project is ill-defined. It seems it was not really a Blue Streak missile but an 'enlarged' Blue Streak with four engines and a second stage based on the Black Knight missile, probably modified as well.[17] This debate is not helped by the fact that the famous art depicting the Waverider on its launcher is actually drawn from a picture of an American Atlas ICBM. Further vagueness surrounds the 'third stage' engine. Nothing is illustrated or even mentioned in the patent. Description of the 'Pyramid' only refers to 'retrorockets' of unspecified type, which weighed 200lb (91kg) including the fuel tanks. It was estimated that only 17lb (7.7kg) of fuel was needed to de-orbit the vehicle and bring it down to an altitude of 6.5 miles (10km) where it was expected ('hoped' might be more accurate) enough atmosphere would be present to facilitate aerodynamic manoeuvring.

Return to the ground from orbit was expected to take 35 minutes out of a total flight time of nearly 7 hours.

Hawker Siddeley aerodynamicist Dr W F Hilton admitted at the time that extensive wind tunnel tests were required to solve the problems associated with the controlled re-entry. Even after that it was accepted that choosing the actual landing area was going to be complicated. Armstrong Whitworth's Technical Director H R Watson could only say, '*in the early space flights, it will represent a considerable achievement to land away from inhospitable areas such as sea, mountain or jungle and a touchdown probably hundreds of miles from the target point must be expected. At best we must expect to land on rough ground*'[18]. At that point, the ejection of the crew compartment under a parachute, described as an emergency solution in the patent, appears to have been considered the normal landing method for the astronauts.

The whole project is described here to illustrate the quantum leap that was demanded by Eurospace to spacecraft designers when approaching the Aerospace Transporter, just three years later.

ii) British Aircraft Corporation 'Mustard'

BAC proposed a stack of three near-identical lifting-body vehicles, of which only one was destined to reach orbit. The three were manned and entirely recoverable.

For a detailed view of this project see, from the same publisher: *British Secret Projects No5 'Britain's Space Shuttle'* by Dan Sharp.

iii) Bristol Siddeley Engines Ltd

The British company contributed to the Eurospace studies a two- or three-stage vehicle which used a turbo-ramjet powered delta-shaped aircraft as its first stage. Turbojets accelerated the machine to Mach 2 with ramjets taking over to bring it up to Mach 7 (some sources say Mach 12) when separation occurred. Designer R J Lane admitted, '*Obviously, the air-breathing stage is a complex and expensive vehicle…*'[19] A large, rocket-propelled second stage was nested in a recess on top of the delta plane. The missile-shaped second stage was not recoverable. The third stage was designed specifically for orbit insertion/manoeuvring. Estimates were that a payload of up to 12% of the take-off weight could be put into orbit. This vehicle was designed during 1963-64 and was presented at the 1964 symposium.

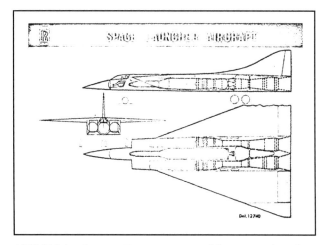

ABOVE John Gregory Keenan's proposal for a space launcher propelled by turbo-rockets generated so much interest at French engine manufacturer SNECMA that it had the original Rolls-Royce brochure re-labelled a SNECMA brochure for internal circulation. *SNECMA via H Lacaze*

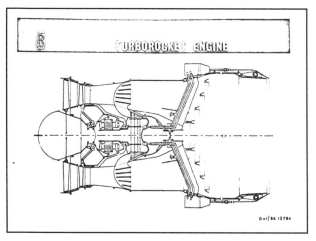

ABOVE From the same Rolls-Royce brochure is this illustration of the turbo-rocket engine. *SNECMA via H Lacaze*

Bristol Siddeley Engines Ltd 1 stage	
Status	Study
Span	-
Length	-
Gross Wing area	-
Gross weight	-
Engine	Rockets LOX/LH$_2$
Maximum speed/height	M 7 (separation speed)
Load	-
Bristol Siddeley Engines Ltd 2 and 3 stage	
Status	Study
Span	-
Length	-
Gross Wing area	-
Gross weight	-
Engine	Rockets LOX/LH$_2$
Maximum speed/height	-
Load	-

iv) Hawker Siddeley

In 1966, the Hawker Advanced Project Group, led by Robert Hugh Francis, in response to the ELDO Europa vertical-rocket launcher, proposed a horizontal take-off machine, along the line of Eurospace Aerospace Transporter ideas. The Hawker design was only attempting to prove that another option to the vertical rocket existed. It was a two-stage vehicle: an air-breathing, ramjet-powered first stage able to reach Mach 12, and a 19.7-ton (20-tonne) second-stage. Both were recoverable. In-orbit payload was to be 2.26 tons (2.3 tonnes). For take-off, a launch trolley was to be used—a solution that would be resurrected twenty years later in HOTOL (see Chapter 9) and which echoed the landing gear questions which the French Transporteur Aérospatial studies raised (see Chapter 4).

R H Francis' main idea was a propulsion system with the air-augmentation concept, also used by ERNO. For take-off, the rockets were ignited and projected the machine up to Mach 5. The rocket thrust flow would mix with air entering the front intakes.

BELOW John Keenan patented this annular-wing launcher in which the second stage (of which only the lower section is shown here) was positioned in the flight axis of the launcher, thus paralleling a French Zborowski proposal. *Author's archive*

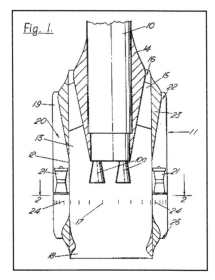

In this way it was hoped that the overall thrust of the rocket would be greatly increased. After Mach 5, the ramjet cut in and powered the first stage up to Mach 12 when separation occurred at an altitude of 25 miles (40km). The supersonic ramjets would be kept in operation until the final approach when small turbojets would be used for landing. The second stage orbiter employed standard rocket engines.

BELOW For larger vehicles, John Keenan suggested using a battery of his annular-wing launchers. Here, three are attached to the lower section of the second stage. *Author's archive*

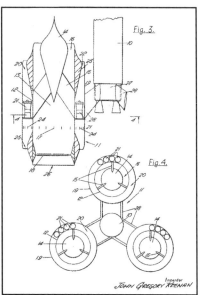

Hawker Siddeley 1 stage	
Status	Study
Span	200ft (61.00m)
Length	121ft 5in (37.00m)
Gross Wing area	-
Gross weight	374,775lb (170,000kg)
Engine	Ducted LOX/LH$_2$ rockets and supersonic ramjets. Auxiliary turbojets
Maximum speed/height	Mach 12.5 (separation speed)
Load	43,430lb (19,700 kg) (2nd stage)

v) Rolls-Royce Space Launcher System

In 1964, long-term project engineer John Gregory Keenan, of Rolls-Royce Ltd, proposed the 'turbo-rocket' for propelling Space Launchers. Turbo-rocket function was explained in J G Keenan's brochure. '*The turbo-rocket operates by burning oxygen and fuel in a rocket-type combustion chamber. The efflux from this chamber goes through a turbine which drives a compressor taking the air from the atmosphere. The turbine exhaust is mixed with air ducted from the compressor and additional fuel is burned in the tail pipe in a similar way to a ramjet or a reheated turbojet.*'[20] Keenan estimated that a turbo-rocket would be about one-third the weight of a turbojet designed to operate at Mach 4 and while fuel mass was about 1.5 times the consumption of a reheated turbojet; '*it was three times better than the rocket.*'

To illustrate the potential of the turbo-rocket, Keenan proposed a Space Launcher System consisting of a large manned delta aircraft powered by two turbo-rocket engines fuelled by kerosene/oxygen, carrying inside its fuselage a two-stage rocket. The take-off weight of the ensemble was 100,000lb (45,360kg) for an orbiting payload of 500lb (227kg). The second- and third-stage rocket weighed 27,000lb (12,250kg). Separation occurred at Mach 4.5 at 74,000ft (22,550m) after an 8-minute climb. Following separation, the manned first stage returned to its base.

Consideration was given to design of the air intakes and integration of the engine(s) in the airframe with the conclusion that '*powerplant installation […] may have features suited for a particular aircraft mission.*'

While the aircraft launcher depicted in the turbo-rocket brochure may appear rather conventional, much more original was Keenan's other launcher proposal which he patented that same year.

Here we see an 'annular-wing wingless vehicle' used, alone or in combination, to launch a centrally-mounted rocket missile. At take-off the machine is propelled by 'any air-breathing engine' which could be turbo-rockets. The said engines are fitted inside the annular wing, at the centre of which is a ramjet which

Rolls-Royce Turbo-rocket Launcher 1 stage	
Status	Study
Span	-
Length	-
Gross Wing area	-
Gross weight	100,000lb (45,360kg)
Engine	2 × turbo-rockets
Maximum speed/height	Mach 4.5 at 74,000ft (22,555m)
Load	27,000lb (12,247kg)

Rolls-Royce Turbo-rocket Launcher 2/3 stage	
Status	Study
Span	-
Length	-
Gross Wing area	-
Gross weight	27,000lb (12,245kg)
Engine	2 × rockets
Maximum speed/height	-
Load	500lb (227kg) in low Earth orbit

comes into operation at higher speeds. Various configurations were proposed, ranging from a single 'wingless vehicle' in which the rocket second stage is placed centrally (this is close to an earlier von Zborowski 'Coléoptère' patent to which, correctly, Keenan's patent refers) to an arrangement of multiple 'wingless vehicles' attached by short arms to the rocket's second stage.

The machine described in this patent, while designed concurrently with the horizontal take-off launcher, was not presented in any Eurospace meeting.

---

[14] Quoted in *Dream Machines* by Ron Miller.
[15] *West German Approach to Reusable Launch Vehicles* by Jürgen Lambrecht and Edwin Schäfer, Junkers Flugzeug und Motorenwerk, 1967.
[16] In a 1959 Conference, H R Watson talked about '160 tons.'
[17] Some sources indicate a size of 100ft (30.5m) and three stages.
[18] *RAF Flying Review*, Vol XV No 2.
[19] Reported in *The Dream Machines* by Ron Miller, Krieger 1993.
[20] *Turbo-fusée pour Véhicules Hypersoniques*; A A Lombard et J G Keenan, Rolls-Royce conference WGLE, SNECMA.

Chapter Three
The Way of the Ramjet

As we have seen in the previous chapters, the Silbervogel heritage took two separate paths. The first was the development of an aircraft, winged but able to reach very high speeds and very high altitudes, which could be used as the first (ideally the only) stage of a satellite launcher; the second would be a real spaceplane, which would re-enter the Earth's atmosphere the same way the Silbervogel bounced on it.

ABOVE **Prototype No 10 is seen here at the 1961 Paris Air Show.** *Philippe Ricco*

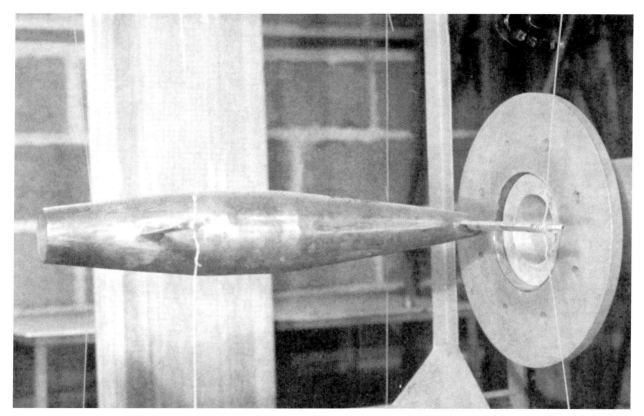

ABOVE A poor, but interesting, photograph of the first ramjet experiment by ONERA on a workbench at Chalais-Meudon in 1947. *Author's collection*

In France, the first path, which was to lead to the Transporteur Aérospatial concept, is very much associated with the development of ramjets. And in this work, culminating in supersonic ramjets, the three major players were:

- Nord-Aviation, having developed the turbo-ramjet for its Griffon experimental aircraft, wanted to experiment with this new type of propulsion for all types of aircraft.
- SNCASE, which became part of Sud-Aviation in 1958, had a Bureau d'Etudes in Cannes (project number range: 4000+, hence all their designations began with 4) that had earlier specialised in missiles and rockets under Jean Béteille.
- In 1948, ONERA began researching very high speeds with experimental missiles and sounding rockets.

A leader in the field of the ramjet (*tuyère thermopropulsive*) was René Leduc who began researching before the Second World War. However, he called it quits for lack of financing in 1958 with his Leduc 022 having only achieved 1,200km/h (745mph).

Ramjet Research at ONERA

1A – Early Ramjet Work

In 1954[21] test flying of ramjets began with the Fusée (Rocket) 2140 two-stage missile (2140 was the identification of the study at ONERA). The first stage was a rocket booster which separated at Mach 1.4. The second stage, built around a 12.6-in (320-mm) ramjet continued its flight, reaching Mach 3. Use of a rocket booster was the inspiration of Paul Lygrisse of ONERA's OP section (OP = physics department). '*The idea occurred to me to develop small experimental rocket models, named Opd, which enabled us to test in real condition of launch and flight our experiments,*' he recalled. It was reported that those rocket boosters were initially more akin to hobbyist models than scientific vehicles. Ten Fusée 2140s were test flown in 1955 from the île du Levant range. However, they proved unstable during flight, so a variant with two identical ramjets was test flown in 1958 as the VD-2120. Although intended to reach Mach 3, it never went beyond Mach 2.5. A later three-stage variant (the first two stages were solid-rockets) flew on 28 November 1957 as the **Ardaltex**. This research programme was intended to demonstrate how a ramjet worked at high altitudes.

In parallel, as part of research to develop the PARCA missile (PARCA = Projectile Autopropulsé Radioguidé Contre Avion: Self-propelled Anti-aircraft Radio-controlled Projectile), the DEFA[22] had built and successfully tested an experimental missile named **NA-250** which used a 10-in (250-mm) ramjet boosted by four solid rockets of the same type used on the PARCA. This missile, fired from the île du Levant test range, reached Mach 3 circa 1954.

ABOVE A ramjet rocket alleged to have been launched by ONERA from the Camp de Mailly in 1951. *Author's collection*

ABOVE The NA250 was a 250mm (slightly over 10in) ramjet which began testing in 1953. *Author's collection*

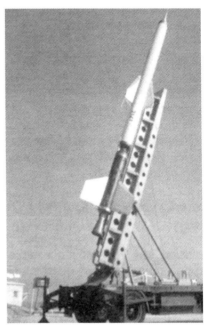

ABOVE A fusée 2141, two-stage ramjet test missile ready for launch in 1956 on the Ile du Levant. *Author's collection*

However, what DEFA required was a 20-in (500-mm[23]) ramjet for its Super-PARCA project. For this it approached ONERA on the basis of the experience of this laboratory with Ramjets. An alternative version of this affair, recounted by Philippe Varnoteaux in *l'Aventure Spatiale Française* was that, as part of a reshuffling of the programmes required by the French Government for budgetary reasons, DEFA was required to transfer its ramjet research to ONERA. So in 1957, the ramjet expert at DEFA, Roger Marguet, moved to ONERA, bringing with him his then-current work, the NA-250.

In the end, ONERA designed the **Statex** 20-in (500-mm) ramjet with Marguet as the programme head. Statex followed the usual configuration of a two-stage machine, ramp launched (it is possible ramp launch technology also came from DEFA). About ten Statexes were launched, reaching Mach 3 before the programme was terminated: France had opted to buy Raytheon Hawk SAMs for its defence. Statex was followed by **Stataltex**, a rather simple machine, with a fixed-geometry ramjet having an annular intake placed at the tip of the nose and a simplified fuel flow regulation which allowed the mixture to remain compatible with the location of the combustion point during the whole duration of the flight. Stataltex took-off from an inclined ramp and was accelerated to Mach 2 with a solid rocket booster. In 1962, it reached a height of 127,950ft (39,000m) and a speed above Mach 5, which made it the fastest ramjet-powered missile in the World for many years.

BELOW A picture of the twin-ramjet variant of 2140. *Author's collection*

BELOW Statex 500 mm ramjet stage in close-up. *Author's collection*

BELOW Statex No 10 being readied for launch. *Author's collection*

CHAPTER THREE
THE WAY OF THE RAMJET

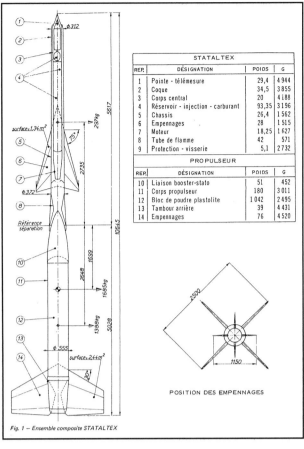

ABOVE Plan view of Stataltex. The experimental ramjet Stataltex is represented by the upper part of this drawing, above the section '10'. *ONERA*

LEFT The Stataltex experimental missile No 7, ready for launch at Hammaguir in 1962. *ONERA*

ONERA Statex
Status	Flown
Span	-
Length	-
Gross Wing area	-
Gross weight	-
Engine	1 × ramjet + rocket booster
Maximum speed/height	-
Load	N/A

ONERA Stataltex
Status	Flown
Span	8ft 2in (2.50m)
Length	34ft 11in (10.645m)
Gross Wing area	-
Gross weight	654lb (297kg) (2nd stage) 3,060lb (1,388kg) (booster) = 3,715lb (1,685kg)
Engine	1 × ramjet + rocket booster
Maximum speed/height	Mach 5
Load	N/A

ONERA Scorpion
Status	Wind tunnel model
Span	28ft 10in (8.80m)
Length	-
Gross Wing area	-
Gross weight	-
Engine	1 × ramjet
Maximum speed/height	up to Mach 6
Load	N/A

1B – Supersonic Ramjets

After **Stataltex**, ONERA under M Marguet and C Huet went on studying ramjets, notably supersonic combustion ramjets and for this purpose built specific test-benches at Palaiseau and Modane.

Research began in 1966 on supersonic ramjets. Ramjets had proved their worth, in term of efficiency, for speeds from Mach 6 and above. So ONERA first evaluated a subsonic ramjet for the sequence between Mach 3 and Mach 6. That project was called Scorpion. A later design, Esope was a ramjet in which combustion could go from sub- to supersonic and reach Mach 7.

Scorpion was the design for an air interception missile able to be launched at Mach 3; accelerate to Mach 6 at a height of 98,425ft (30,000m); and manoeuvre to engage its target. This research should be viewed in the context of the previously-mentioned studies being carried out at nearly the same time for a *Police de l'Air* Mach 3 interceptor able to tackle an SST (or high speed reconnaissance aircraft). Scorpion, while still a two-stage missile, introduced the side intakes which were later used on the single-stage missile.

ABOVE ONERA also flew – under the name 'Bérénice' – the SNCASE SE 4400 missile. *Author's collection*

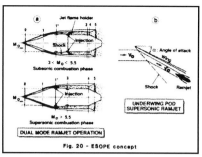

ABOVE Sketches of the later ESOPE ramjet. *ONERA*

configuration underneath VERAS. This project was called **Esope**. Early on, it was proved that a fixed-geometry ramjet was possible if the injection nozzles could be moved inside the combustion chamber. But this solution also presented risks: around Mach 5.5 difficulties were encountered to ignite the fuel. The project was then reorganised and downsized so only the study of the intake in wind tunnels and research into the combustion chamber during the transition phase between sub- and supersonic airflow were carried out.

The experiment involved a 16.5-in (420-mm) ramjet model being 'flown' at Mach 5.5 to Mach 6 in the S4 wind tunnel at simulated heights of 82,000 to 95,100ft (25,000 to 29,000m). The conclusion was that this type of ramjet had performance approaching that of a subsonic ramjet fitted with a variable-geometry intake. An interesting experiment was to compare the results of a ramjet flying alone with the ramjet fitted underneath a wing. It was found that this configuration offered the following benefits:

- The wing acts as a ramp, leading air into the ramjet intake so that the air feed is guaranteed at all attitudes.
- For a given thrust, the diameter of the ramjet pod under a wing is two-to-three times smaller than the diameter of a ramjet flying solo.
- If the ramjet is used as a booster, its thrust is improved and the increase in thrust is proportional to the increase in speed.
- If the ramjet is used as a cruising engine, then the ramjet increases the glide ratio of the complete aircraft and thus increases the range of the machine.

The case of the ramjet attached to the side of the aircraft was also studied and it was found that the overall efficiency of such a ramjet was lower than that of a ramjet under a wing.

Tests on a 7.2-in (200-mm) model positioned inside a free-flow conduit of the Modane S3 wind tunnel were carried out at a simulated airspeed of Mach 2 at 3,280ft (1,000m) height. The high-performance Modane S4 wind tunnel was also used for these tests after its opening in 1969. These proved the validity of the concept with the new intake architecture ('inverted bi-dimensional intakes') and new exhaust nozzle. Another series of tests on a Palaiseau test bench resulted in the adoption of a mechanical flame-holder which was more efficient when the airflow became distorted as the speed increased.

In this research, the heat question was wantonly ignored. Since the operational outcome of this work was an air-to-air missile, the duration of the flight was expected to be brief – in the order of 200 seconds – and for this short duration, materials with high thermal resistance were not required. Covering the ramjet surfaces with ablative material was deemed sufficient. Therefore the use of fuel as coolant was not considered. We will see in the next Chapter that for the Transporteur Aérospatial, which was to fly at speeds over Mach 5 for tens of minutes, this consideration resurfaced. For fuel hydrogen, pentaborane and kerosene were considered but, in the end, gaseous hydrogen was selected for its good energetic properties (although early experiments appeared to have been carried out using kerosene).

Later, ONERA proposed, as part of the project VERAS (see Chapter 5) to develop a ramjet able to operate successively in sub and supersonic modes which could be attached in pod

ONERA Esope	
Status	Wind tunnel model
Span	2ft 0in (0.60m)
Length	8ft 2in (2.50m)
Gross Wing area	–
Gross weight	352lb (160 kg)
Engine	1 × ramjet
Maximum speed/height	up to Mach 6
Load	N/A

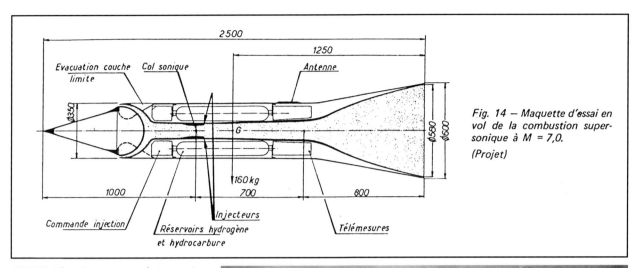

ABOVE A late 'sixties or early 'seventies design for a Mach 7 ramjet. ONERA

RIGHT Wind tunnel model of the Scorpion ramjet-powered air-to-air missile from the later part of the 'sixties. ONERA

Sud SO 9050 Trident III	
Status	Built
Span	21ft 10in (6.95m)
Length	43ft 6in (13.26m)
Gross Wing area	156sq ft (14.50sq m)
Gross weight	13,007lb (5,900kg)
Engine	2 × Turbomeca Gabizo 3,375lb (15kN) and two chambers SEPR 63 3,375lb (15kN)
Maximum speed/height	Mach 2 at 68,900ft (21,000m)
Load	1 × MATRA R511 missile

1C – Later Research During the 'Seventies

Later, in 1972, ONERA proposed combining the rocket for the booster and the ramjet into a single unit. The nozzle was common to both engines: the rocket fired through an inner section while the ramjet used the outer, annular section of the nozzle. This architecture was much more compact than the earlier rocket booster (first stage) + ramjet (second stage). The intakes for the ramjet were on the side of the airframe, which enabled the front section to be used in an operational role. The system was also improved by a gas generator supplying fuel for the ramjet. Using gaseous fuel did not require pressurisation, injection or regulation and was

Sud EM 3	
Status	Project
Span	29ft 6in (9.00m)
Length	59ft 1in (18.00m)
Gross Wing area	311sq ft (29sq m)
Gross weight	24,059lb (10,913 kg)
Engine	Super ATAR and 2 × ramjets
Maximum speed/height	Mach 3
Load	?

therefore simpler than the earlier liquid (kerosene) fuel. Blanks blocking the air intakes during the rocket phase of the flight were ejected when the ramjet was put into action. This compact engine was named *stato-fusée* (ram-rocket).

By 1977 the research had reached a point at which new wind tunnels, able to test well above Mach 7 were required …and were just unavailable.

ONERA also experimented with the OPD-320 test rocket at even higher speeds – up to Mach 10 – as part of studies into re-entry of warheads for France's *Force de Dissuasion* missiles.

Ramjet Research at SNCASE and Other Manufacturers

Early ramjet experiments concerned the development of a surface-to-surface missile named **SE-4200** produced by SNCASE. The first ramjet-powered machine was the SE-4204 01 flown on 3 October 1950. A speed of Mach 0.6 was reached. In 1955, the missile became operational as the SE-4263. About 600 were produced up to 1963.

ABOVE A SNCASE SE 4200 missile on its launchpad. Although the SE 4200 became operational, its performance was quite poor compared with other vehicles in this book, at just Mach 0.9. The machine is seen here during tests in French Algeria. *Author's collection*

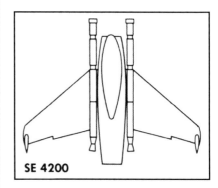

ABOVE Drawing of SE 4200 from above. *Author's collection*

LEFT An SE 4400 missile on its launch pad, location unknown. The SE 4400 reached Mach 3.2 in 1957. *Author's collection*

ABOVE The SE 4500 was an enlarged variant of SE 4200 intended as a carrier for the French nuclear bomb. The programme was abandoned when the decision was taken to use a manned carrier, the Dassault Mirage IV. *Author's collection*

SNCASE continued with the surface-to-air missile **SE-4400**. Like most other designs, it comprised a first stage solid rocket booster with a ramjet-powered second stage. The first stage was fitted with a cruciform wing attached to a fuselage 12ft 5in (380mm) in length, housing an SEPR 505 rocket engine delivering a thrust of 24,250lb (108kN). The second stage consisted of a 1ft 8-in (500-mm) ramjet in a 16ft 3-in (4.96-m) long fuselage with four wings with a span of 5ft 11in (1.80m). A central body contained the avionics and the fuel (kerosene) tank.

The first two launches at the Hammaguir test range were failures but after many flights, the SE-4400 managed to reach a height of 58,100ft (17,700m) and a speed of Mach 2.7 on 5 October 1955, more than one year after the first (failed) flight of 10 April 1954. Mach 3.2 was achieved on 15 February 1957. Separation between the first and second stages occurred around Mach 1. The SE-4400 lost out to the Raytheon Hawk in the competition to supply the French Army with a surface-to-air missile but it managed to achieve new speed records with Mach 3.7 on 18 April 1958. It reached a height of 219,800ft (67,000m) on 10 March 1961. In all, 92 missiles were fired.

A direct evolution of the SE-4200 was the **SE-4500**. Planned to carry the French atomic bomb, it was basically a SE-4200 enlarged by 30%. Its ramjet was intended to propel it at about Mach1. However. the machine was initially unsuccessful and was finally abandoned when the decision was taken to use a manned aircraft (Mirage IV) as the carrier of the A-bomb.

During the sixties, ramjets resurfaced at the Cannes Bureau d'Etudes (now operating under Sud-Aviation), as part of the cruise missile research initiated for the Minerve military programme. (Minerve was a planned replacement of the Dassault Mirage IV nuclear bomber

SE 4200	
Status	Series production
Span	7ft 1in (2. 15 m)
Length	6ft 5in (1.95 m)
Gross Wing area	–
Gross weight	660lb (300kg)
Engine	1 × ramjet
Maximum speed/height	Mach 0.9
Load	Surface-to-surface artillery missile

SE 4400	
Status	Flown
Span	–
Length	26ft 7in (8.10m)
Gross Wing area	–
Gross weight	2,645lb (1,200,kg)
Engine	1 × ramjet
Maximum speed/height	Mach3.5
Load	Experimental anti-aircraft missile

SE 4500	
Status	–
Span	–
Length	–
Gross Wing area	–
Gross weight	–
Engine	–
Maximum speed/height	–
Load	–

by a longer-range weapon system.) The Minerve carrier aircraft was to be armed with a large cruise missile which would penetrate enemy territory while the carrier would return to base. Various missile designs were considered but it turned out that the turbojet intended for the X 417 missile was difficult to design. Therefore on 31 October 1963 the Sud-Aviation design team met with ONERA people to discuss the possibility of using the ramjet propelling the Stataltex experimental machine. The new design was renamed X 417 S (S for *statoréacteur*). The ONERA ramjet was not retained, Sud selecting the SE 4400 in January 1965. The X 422 became the demonstration model of the X 417. Three X 422s flew on 29 January 1967, 16 March and 20 June. After that, the programme was cancelled, Minerve itself having been abandoned around 1964. The **X 422** was able to demonstrate Mach 2 during a 43-mile (70-km) flight (about 2 minutes).

In 1953, another aircraft manufacturer, **Latécoère**, began working on an anti-aircraft missile: the MASALCA (Marine Salmon Contre Avions: Naval Anti-Aircraft, Max Salmon [who was the designer]) but its ramjet was never perfected and the missile abandoned. About 100 MASALCAs were tested, some reaching Mach 1.3.

MATRA was another missile-specialist manufacturer, which attempted to produce a surface-to-air missile during the 'fifties. Its design, the R-431 (R = Roger Robert, the head of the Bureau d'Etudes, 4 = missile, 3 = third type, 1= first variant) followed the usual configuration with a rocket booster and a ramjet. The 2ft 2-in (614-mm) ramjet was provided by Nord-Aviation for this project. The R-431 flew successfully up to Mach 1.7 but was not pursued, due to the political decision to select the Hawk.

Hypersonic Research at Nord-Aviation: Beyond Griffon

Nord-Aviation had its own high-performance machine, the Nord 1500 Griffon which was to demonstrate the effectiveness of ramjets to power manned aircraft. But even higher performance was obtained during development of missiles and target-drones.

3A – Griffon and Developments

Research had begun in the immediate post-war period with the strange-looking **Arsenal Ars 1910** which was expected to reach Mach 1, but remained as no more than a wind tunnel model. The **Nord 1500 Griffon** (which had begun life as the SFECMAS 1500 Guépard) made its maiden flight on 20 September 1956 and after some reworking of the second airframe (1500-02 Griffon II) actually reached Mach 2.19. Flight testing of the Griffon II continued until 1960, some flights being funded by the US. However, from 1957, the French Defence Ministry had requested Nord-Aviation to begin working on a Mach 3-class aircraft named the **Griffon III**.

Two variants were considered. One was a direct derivative of the Nord 1500 and had a large ramjet coupled to a Super ATAR turbojet to achieve a turbo-ramjet. On this was attached the cockpit pod (reminiscent of the earlier Ars 1404 study), the delta wings and the tail assembly. This may have been the **Nord 1510**. The second configuration (**Nord 1520**) used the architecture of the missile 'SSS' (according to the brochure; nothing is known about this project but in appearance it was quite similar to the Nord CT.41 target, having a long fuselage and straight wings with ramjets at the tips). Again, according to the brochure, the advantages of this architecture were:

- Better aerodynamic finesse 'at all Mach numbers' than the delta wing of the Griffon.
- Ability to replace the ramjets by 1,100-gallon (5,000-l) ferry tanks for long-range missions, for example reconnaissance.

Two variations of this Nord 1520 appear to have been designed: one (probably the earliest version) used a rather straight

BELOW **The Nord 1500 Griffon is shown being tested in the Modane wind tunnel. The width of the tunnel enabled testing of complete aircraft.** *Authors' collection*

Ars 1910	
Status	Project
Span	36ft 1in (11.00m)
Length	50ft 10in (15.50m)
Gross Wing area	323sq ft (30sq m)
Gross weight	21,275lb (9,650kg)
Engine	SNECMA ATAR 101 and 2 × Ars ramjets
Maximum speed/height	Mach 1.5 at 49,215ft (15,000m)
Load	40 × 8.8lb (4kg) rockets

ABOVE Nord's 1500-02 Griffon II is seen here as part of an exhibit of company aircraft in 1960. *Nord-Aviation*

derivative of the Nord 1500 fuselage now housing only an ATAR turbojet of unspecified type, the ramjets of 12.9sq ft (1.2sq m) being repositioned to the squared tips of the wings. This version was calculated to be able to reach above Mach 2.5 at around 82,000ft (25,000m). Described as an 'experimental airplane' it probably would have paved the way for the second version which was to be a fully operational aircraft with military equipment. This second version had a completely new fuselage housing a Super ATAR while the wingtip ramjets had their section reduced to 6.78sq ft (0.63 m²). The pilot gained an escape capsule in the process. This second version was expected to reach Mach 3 at 55,800ft (17,000m). Armament would have been two SNCAN AA25 missiles.

It was noted that the twin-ramjet, straight-wing variant would present more risk than the single-ramjet, delta-wing variant which was a direct extrapolation of the Griffon II. For this reason the Nord CT.41 Narval trisonic target drone, with smaller ramjets at the tips of the straight wing, was to contribute to the development of this design (the CT.41 having not yet flown at that time).

Neither variant was built.

RIGHT A glimpse into the cockpit of the Griffon II. This photograph was probably taken during the same exhibit of Nord-Aviation aircraft as the previous picture. The large white panel lists safety instructions for opening the cockpit from outside in case of an accident. *Nord-Aviation*

BELOW Front view of the Griffon II showing the large intake, common to both the turbojet and the ramjet. *Aerospatiale*

Nord 1500 Griffon II	
Status	Flown
Span	26ft 7in (8.10m)
Length	47ft 8in (14.54m)
Gross Wing area	344sq ft (32sq m)
Gross weight	15,575lb (7,065kg)
Engine	ATAR 101E3 7,875lb (35kN) and Nord ramjet
Maximum speed/height	Mach 2.19
Load	N/A

Nord 1520 Griffon III (ATAR)	
Status	Project
Span	25ft 7in (7.80m)
Length	55ft 1in (16.80m)
Gross Wing area	–
Gross weight	19,840lb (9,000kg)
Engine	ATAR (unspecified)
Maximum speed/height	Mach 2.5 at 65,600ft (20,000m)
Load	N/A

Nord 1510 Griffon III (single ramjet)	
Status	Project
Span	32ft 10in (10.00m)
Length	70ft 8in (21.55m)
Gross Wing area	517sq ft (48sq m)
Gross weight	27,094lb (12,290kg)
Engine	Super ATAR 20,250lb (90kN) and Nord ramjet
Maximum speed/height	Mach 3
Load	2 × missiles

Nord 1520 Griffon III (Super ATAR)	
Status	Project
Span	23ft 7in (7.20m)
Length	76ft 5in (23.3m)
Gross Wing area	517sq ft (48sq m)
Gross weight	27,890lb (12,650kg)
Engine	Super ATAR 20,250lb (90kN) and 2 × Nord ramjet
Maximum speed/height	Mach 3
Load	2 × AA25 missiles

ABOVE André Turcat, pilot of the Griffon II, gained fame as the pilot of the Concorde a few years later. *Aerospatiale*

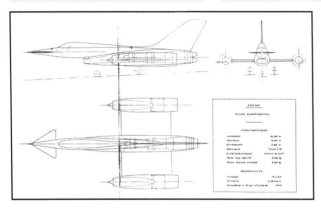

ABOVE Another Nord study, based on the original Griffon with a slightly enlarged fuselage, new rectangular wings and ramjets at the wingtips. Had it been built, it might have borne the moniker 'Griffon III'. *Philippe Ricco*

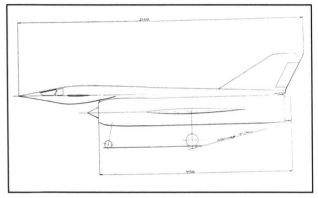

ABOVE Side view of another aircraft project which could have been the Griffon III. *Philippe Ricco*

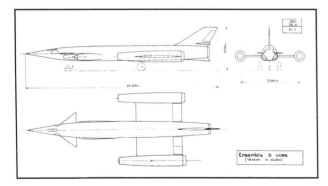

ABOVE A further Griffon evolution was this 'Super Griffon' (and possibly Nord 1520), based on the architecture of the CT.41 target drone. *Philippe Ricco*

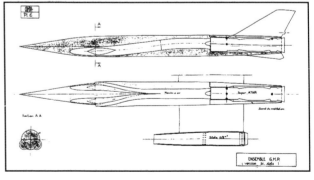

ABOVE Sketch indicating the engine configuration of the Nord 1520 with a Super ATAR turbojet in the fuselage and one 0.9 m ramjet at each wingtip. *Philippe Ricco*

That did not prevent the Nord Bureau d'Études from going one step further with the Nord **M4 Super Griffon**. The 'M4' label was the target speed. This aircraft was designed in 1965 as a two-seat, hypersonic *Police de l'Air* fighter (but armament was never specified, possibly because air-to-air missiles flying above Mach 4 did not exist). For propulsion, combined ram- and turbojets called 'turbo-ramjets', which had been pioneered on the

ABOVE Model of twin-ramjet Griffon III. *George Cox via Tony Buttler*

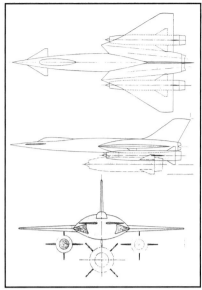

ABOVE Nothing is known of this design beside its use of large boosters (or missiles?) which are obviously ramjet-powered. *Philippe Ricco*

Griffon, were to be used. Turbo-ramjets were also intended as prime drivers for other Nord projects: their own vision of the Super-Caravelle (Mach 3), the Transporteur Aérospatial (see Chapters 2 and 4). ONERA assisted greatly with this project with tests of functioning ramjets in the Modane wind tunnel and later with the Scorpion free-flight sounding missile. Apparently, funding for this project also came from the US.

A **Griffon Mach 5** is rumoured to have been studied but has not yet been declassified.

Nord M4 Super Griffon	
Status	Project
Span	36ft 1in (11.00m)
Length	75ft 2in (22.90m)
Gross Wing area	646sq ft (60.00sq m)
Gross weight	?
Engine	2 × Nord turbo-ramjets
Maximum speed/height	Mach 4.5
Load	?

BELOW The next step for Nord-Aviation was this 'Super Griffon M4', M4 being the intended speed. *Nord-Aviation*

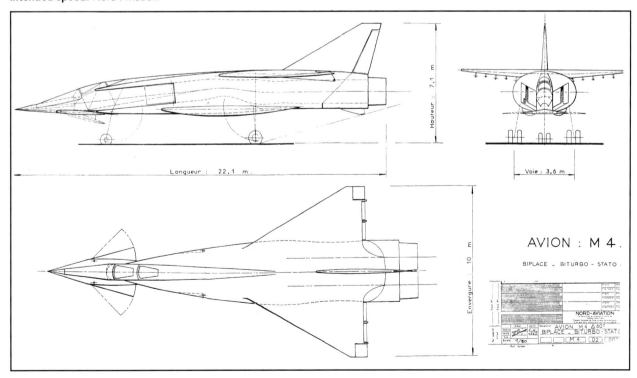

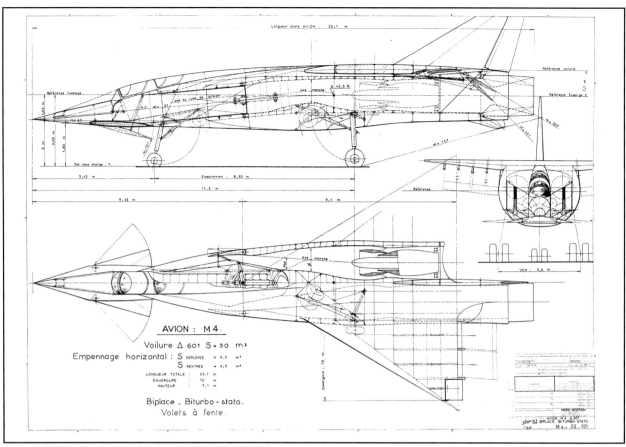
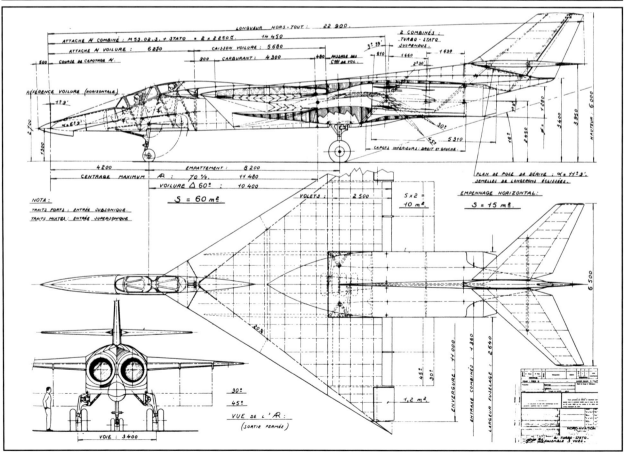

ABOVE Véga test vehicle for a 645mm ramjet. The fins are marked '01', suggesting this is the prototype. *Philippe Ricco*

3B – CT.41 Narval (Narwhal)

Besides the Nord 1500, the Bureau d'Etudes studied other applications of the ramjet. One of them was for a missile. It seems the impetus for this project came from the US Navy. M Alaplantive, an engineer in the ramjet department of Nord-Aviation, related in *Mémoire d'Usine*, 'We studied a ground-to-air missile for the Americans. I think they mainly wanted to know how advanced we were on ramjet propulsion and obtain the results of our work for free.' Work on this subject began in 1957. Ramjets of a diameter of 23.6in (60cm) were first tried out on a bench at Saclay (Centre d'Etudes des Propulseurs CEPr, the organisation, which is today under the umbrella of the DGA Direction Générale de l'Armement, was set up in 1946 at Saclay, in the western suburbs of Paris), then later in free flight from the French Sahara testing range at Colomb-Béchar (the base was returned to Algeria in 1967). Eighteen test missiles, named **Véga** were produced and most were flown. Véga was first accelerated to Mach 2 by a solid-propellant booster, then the booster was dropped and the ramjet took over. On the 10 November 1961, a speed of Mach 4.15 and an altitude of 77,100ft (23,500m) were obtained. Four variants of Véga were designed: experimental (actually flown), target drone, surface to air missile and powerplant pod for larger machines. The Véga ramjet had a diameter of 25.4in (645mm).

The next step was to design an operational machine. The **CT.41** Narval (the name appears sometimes in the period documentation but it is uncertain if it was an official Nord-Aviation designation) was designed as a drone target, propelled by two ramjets. Different problems had to be conquered before the CT.41 could be flown: vibrations during the launch sequence, separation of the booster 'chariot,' regulating the ramjets, and recovery of the missiles after testing. The overall architecture of the CT.41 was unusual with a long fuselage, rectangular wings at the rear supporting two 2ft 6-in (625-mm) ramjets named Sirius. To test Sirius in flight, 15 experimental vehicles called **Nord 625** (for obvious reasons) were fired successfully.

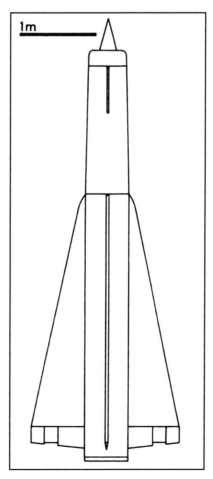

OPPOSITE TOP Nord-Aviation also studied this other Mach 4 design which used a more conventional tail than the canard controls used in other schemes, but featured an unusual retractable cockpit fairing which pivoted downward for landing (the illustrated configuration). *Nord-Aviation*

BOTTOM Detailed drawing of the same M4 aircraft as in the previous illustration. *Philippe Ricco*

LEFT Side view of the Nord Véga vehicle. *Author's collection*

ABOVE A Vega on its launch pad in 1965. The lower section is a solid booster which accelerated the second, ramjet-powered stage to Mach 2, at which speed separation occurred and the ramjet took over. *Philippe Ricco*

When launching the CT.41, two rocket boosters were attached astride the nose of the craft and dropped immediately after the craft had left its ramp.

In 1958 a model of the CT.41 was shown to General de Gaulle. (It seems that immediately after his return to power in June, he was presented the prototypes of French combat aircraft;

Nord Vega

Status	Flown
Span	6 ft 3in (1.9m)
Length	19ft 8in (6.00m)
Gross Wing area	–
Gross weight	1,500lb (670kg)
Engine	1 × ramjet
Maximum speed/height	Up to Mach 5 up to 115,000ft (35,000m)
Load	N/A

Nord ST-450

Status	Flown
Span	–
Length	15 ft 5in (4.7m)
Gross Wing area	–
Gross weight	1,380lb (626kg)
Engine	1 × ramjet
Maximum speed/height	Up to Mach 4
Load	400lb (181kg) (equipment)

Nord 625

Status	Flown
Span	–
Length	–
Gross Wing area	–
Gross weight	–
Engine	Sirius ramjet
Maximum speed/height	Mach 3
Load	Ramjet flying test bed

Nord CT.41

Status	Series production
Span	11ft 9in (3.59m)
Length	32ft 7in (9.93m)
Gross Wing area	33sq ft (3.09sq m)
Gross weight	6,500lb (3,010kg) with boosters. 3,300lb (1,500kg) without boosters
Engine	Solid-propellant booster and two Sirius II ramjet
Maximum speed/height	Mach 2.7 (nominal) over Mach 3 (actual observed speed)
Load	N/A

this may have been the occasion.) The 01 prototype was shown to SHAPE authorities in early 1959; 03 was discovered by the public at large during the Paris Air Show in June of the same year; 05 appeared at the Farnborough Air Show the following year.

On the CT.41, the ramjets were located to the rear for weight considerations. The CT.41s were tested at Colomb-Béchar from 1959 until 1962. Instability observed during the first flights required the addition of canard surfaces. After these initial troubles, speeds exceeding Mach 3 and altitudes of over 82,000ft (25,000m) were attained, but it was the recovery of the airframes after flight which proved the most difficult. '*Recovery was complicated: explosive bolts separated the*

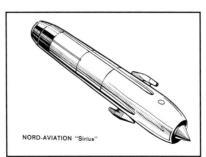

LEFT Line drawing of the Sirius ramjet which was tested solo as the Nord 625. *Nord-Aviation*

boosters. Then after the flight and the ramjets being stopped, the main body with the ramjets descended under parachutes. The fin was then ejected. The nose cone then separated and was recovered under its own parachute. The main body then pivoted and the recovery parachutes were deployed. The 'skirts' of the ramjet, in inox steel, were to cushion the landing and ended-up concertinaing. The fin, when it separated, tumbled in the sky like a kid's paper airplane. We replaced the skirts and put the machine back on its launch ramp.' Missile No 15 was the most successful in this respect and completed three sorties. '*We called it the* Captain*; it had three rank bars [painted on it] because it had flown three times.*' Sixty-two CT.41s were assembled and 26 fired from Colomb-Béchar.

In November 1962, Bell Aerosystems bought six and acquired a production licence for $1.7 million. In July 1963, three of these were test-flown from the Pacific Missile Range at Point Mugu, in California, under contract to the US Navy. Unfortunately the US Navy tests

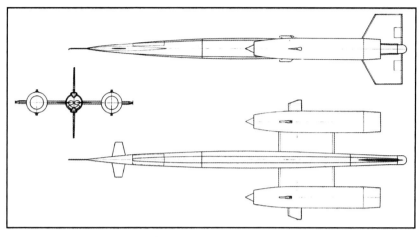

ABOVE Three-view drawing of the CT.41 target missile. The similarity in appearance to later 'Griffon III' designs is obvious. *Nord-Aviation*

were not fully successful: the intake for the Sirius ramjets had a central cone, which in the US version, were painted. At Mach 3, and 400°C the paint just melted and blocked the sensors controlling the functioning of the ramjet—which kept accelerating. In the UK, Hawker Siddeley also bought a production licence but never did anything with it. The CT.41 was

ABOVE Prototype No 05 was exhibited at the 1960 Farnborough SBAC display. *Author's collection*

BELOW Nord CT.41 prototype No 03 being readied on its launchpad—for demonstration purposes only, considering the proximity of the tracking radar array and of the launch control truck. *Author's collection*

LEFT **Possibly taken at the Nord factory at Bagneux, a display of Nord-designed missiles is dominated by the prototype No 20 of the CT.41.** *Author's collection*

cancelled in 1965. The problem with the target drone was that at Mach 3, it was just too speedy to be intercepted by fighters of the 'sixties. '*I remember the tests we did with the CT.41 at Colomb-Béchar. It was the CT.41 which intercepted the interceptor! We had the fighter taking-off and we timed it with our chronometer, we fired the CT.41 when it had a chance to get face-to-face with the fighter. We never succeeded,*' wrote M Malaval of Nord-Aviation in *Mémoire d'Usine*.

The French government, which had funded the flight tests in the Sahara, decided to pull the plug and the programme was cancelled in 1965, curtailing all high-speed ramjet research. '*Usually technologies die because they are obsolete, in this case it died because it was too advanced,*' said M Ravel of Nord-Aviation when interviewed during the 'eighties for *Mémoire d'Usine*.

---

[21] The year of those tests is also given as 1951 or 1956, depending on the sources.
[22] Direction des Études et Fabrications d'Armement: Directorate for Study and Production of Weapons. A department of the French Defence Ministry in charge of research and development. It was created in 1933 as the Direction des Fabrications d'Armement, renamed as the DEFA in 1946.
[23] The original document indicated the diameter of the ramjet. However, to simplify comparison between the two variants, the author has converted this dimension into the section of the combustion chamber.

Chapter Four
Going into Space on a Set of Wings

ABOVE **Near the end of its return flight, the orbiter extends its wings prior to the final braking and landing.**
Dassault Aviation/Direction des Relations Extérieures et de la Communication

1 – The Official Programme

In 1965, the Centre de Prospective et d'Evaluations initiated a wide-ranging project called the Transporteur Aérospatial (TAS). Three companies were asked to produce reports evaluating the opportunity and feasibility of a space launcher using an air-breathing first stage and taking-off 'tangentially' (meaning tangentially to the curvature of Earth, hence from a runway not a launch-pad). The mission was summarised thus: '*Put into a circular orbit at a height of about 186 miles (300km), a 13,230lb (6,000kg) payload while recovering the first stage, the other stages being used-up*.' Nord-Aviation, partnered with SNECMA under contract 340/64/DRME/CPE, supplied its conclusions in September 1966, promoting what they called the 'Mistral' concept. Sud-Aviation, department PLT (long-term projects) received four different contracts: 336/64/DRME/EPO; 909/65/CPE, 906/66/CPE, 902/67/CPE and sent back four conclusive reports from October 1967 to April 1968. Dassault later contributed willingly, outside any government financing.

It should be noted that, contrary to the work done in Germany and the USA (see Chapter 2), the French studies had no grounding in Sänger's earlier works. This is very surprising, considering that Eugene Sänger and

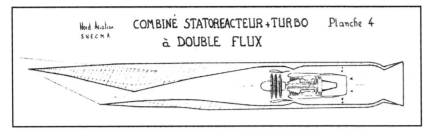

ABOVE Schematic of the Nord-Aviation/SNECMA turbo-ramjet. This design evolved from the studies carried out with the Nord 1500 Griffon. *Nord-Aviation/SNECMA via H Lacaze*

BELOW Close-up schematic of the Nord-Aviation/SNECMA turbo-ramjet. Legend: Compresseur BP = LP Compressor; Mélangeur = Mixer; Compresseur HP = HP Compressor; Chambre de combustion du turboréacteur = Combustion chamber of the turbojet; Turbine HP = HP Turbine; Turbine BP = LP Turbine; Chambre de combustion = Combustion chamber (of the ramjet); Tuyère de sortie variable = Variable (geometry) exhaust nozzle. *Nord-Aviation/SNECMA via H Lacaze*

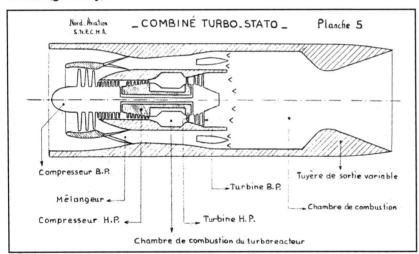

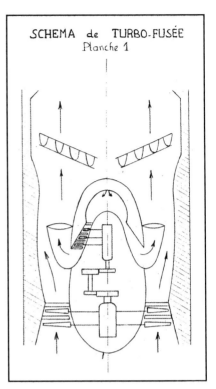

ABOVE Schematic of the turbo-rocket proposed by Rolls-Royce. This is a Nord-Aviation/SNECMA drawing—for the original R-R drawing see Chapter 2. *Nord-Aviation/SNECMA via H Lacaze*

Irene Bredt were under contract by the Arsenal Aéronautique. Yet it seems he never developed his Silbervogel while in France: '*From 1946 to 1954, when he returned to Germany, Sänger contributed the following:*

- Development of liquid- and solid-propellant engines
- Designing of the SS.10 anti-tank missile
- Studies for the [Nord 1500] Griffon turbo-ramjet experimental aircraft
- Development of the RO10 ram-jet missile

This summary of Sänger's work while in France made by Jacques Villain of Société Européenne de Propulsion at the 1992 World Space Congress[24] could be discussed (most people associated with the design and development of the Nord Griffon don't acknowledge much more than a highly theoretical input by Sänger[25]) but does not list any space launcher or antipodal aircraft.

The studies led to a report with the title *Etude Préliminaire sur les Lanceurs Spatiaux à Propulsion Atmosphérique* (Preliminary Study of Space Launchers Using Air-Breathing Propulsion) issued on 13 September 1968. Research went on at ONERA until 1970.

By asking for studies for a Transporteur Aérospatial, CPE (*Centre de Perspective et d'Évaluation*: Centre for Evaluation and [long term] Perspective) thought to '*examine if there were no better designs for a space launcher than those directly extrapolated from ballistic launchers.*' Firstly it was thought that 're-usability' should be investigated, as space launches were expected to become '*weekly or maybe daily*' occurrences in the near future. For this, two tactics could be applied: either to modify existing rocket launchers to be recoverable, or if no pre-existing ballistic launcher was available, then designing an all-new launcher. In the report, US studies were quoted: 'ballute' [balloon-parachute] '*such a system if applied only to the recovery of Saturn V first stage would reduce the payload by about 4%*'; parachutes; gliders for re-entry launched vertically atop a ballistic launcher; and finally gyrogliders. Strangely enough, the BAC Mustard project was mentioned, but not the Boeing X-20 Dyna-Soar and earlier US projects. In France, re-entry gliders had been studied by ONERA and gyrogliders by Giravions Dorand (see next chapter).

This part of the report, accompanied by illustrations of the main French projects, was published in the magazine *Icare*, written by Ingénieur en Chef de l'Armement Thierry Moulin, who had also signed the original report.

So what did this report say? First of all it announced that 'aircraft-type' space launchers would be a reality by 1985, although it distinguished between the machines propelled by turbo-ramjets like the Nord-SNECMA Mistral which appeared 'realistic' (even though the calculations presented by the manufacturers were deemed 'optimistic') and the full-rocket and full-ramjet proposals from Sud-Aviation. *'In the current state of our knowledge, their feasibility at the horizon 1985 does not appear certain.'* As for designs in which both first and second stage were recoverable, they were deemed *'of rather monstrous appearance'* and probably unrealistic in the mind of the writer of the report.

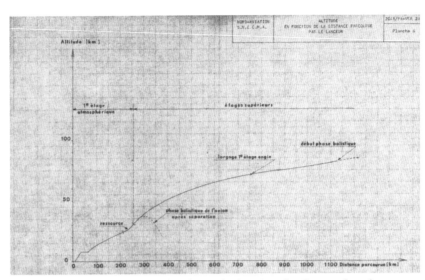

ABOVE Trajectory graph of the height reached against the distance from base. *Nord-Aviation via H Lacaze*

The whole report did not push the development of such advanced concepts: *'By 1980, most of the first stage of classic satellite launchers will be recovered. Rudimentary devices like ballutes, parachutes, parasail wings or more advanced like extendable wings, and hypersonic rotors will be developed by those countries which have today a large choice of launcher, and by secondary powers which cannot afford to develop the new concepts alone. The new concepts, among them launchers with aerodynamic, horizontal landing and take-off first stages, are only of worth to the major powers who launch at least one satellite every eight days, like USA and USSR. By 1980-1985, those new concepts may be of interest to Europe, but for France's space needs, the rudimentary devices previously mentioned appear more appropriate.'*

BELOW The later Nord-Aviation/SNECMA Transporter Aérospatial – now called Mistral (here the smaller variant carrying a high-density propellants rocket) – compared with the Concorde (in dark silhouette). *Nord-Aviation via H Lacaze*

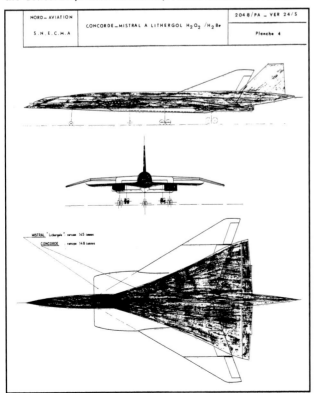

BELOW Mistral variant with the smaller, high-density propellant rocket (in black). *Nord-Aviation via H Lacaze*

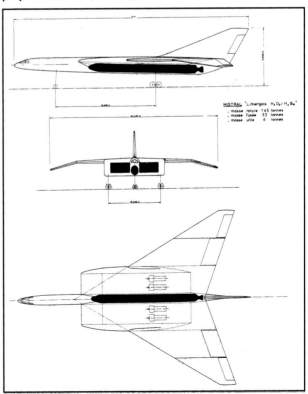

FRENCH SECRET PROJECTS: FRENCH AND EUROPEAN SPACEPLANE DESIGNS 1964-1994

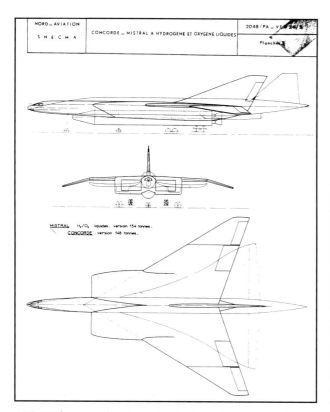

ABOVE The Mistral variant with the larger H₂/O₂ propellant rocket. *Nord-Aviation via H Lacaze*

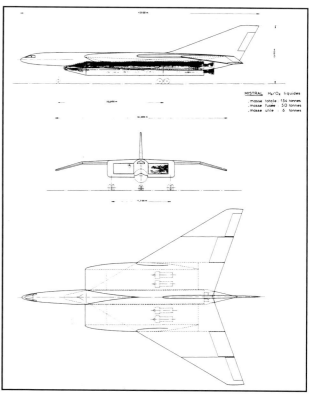

ABOVE The larger H₂/O₂ rocket is here emphasised in black. *Nord-Aviation via H Lacaze*

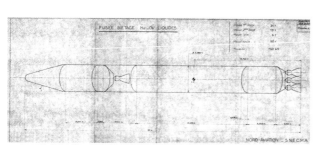

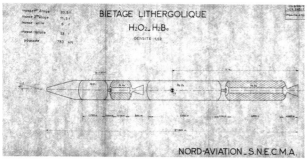

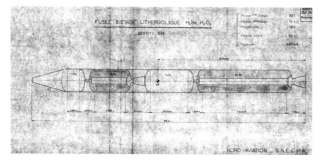

TOP Cutaway drawing of the two-stage H₂/O₂ rocket. *Nord-Aviation/ SNECMA via H Lacaze*

MIDDLE Cutaway drawing of the two-stage, high-density propellant H₂O₂/H₂Be rocket. *Nord-Aviation/ SNECMA via H Lacaze*

BOTTOM Cutaway drawing of another version of the two-stage, high-density propellant rocket. It seems the respective propellants of the two stages have been inverted (H₂Be/H₂O₂) when compared with the previous document. *Nord-Aviation/ SNECMA via H Lacaze*

Various designs had been proposed by the manufacturers; they will be detailed later. One aspect to keep in mind is that these designs were only 'parametric studies' aimed at defining the general size, the aerodynamic shape and the engine requirements to accomplish the job. They were never fully-detailed aircraft studies as would be, for example, Hermès. *'Considering the limited financing of the Mistral contract, the Nord-SNECMA team could not produce a detailed study of the structure of the aircraft stage and only supplied the weight tables and the brackets to be imposed to the possible trajectories. [Sud] had a larger contract which enabled it to go deeper into the structural problems and offered rather precise solutions'.*

The report also pointed out that most of the questions which required answers to develop such machines, were also relevant to other programmes and projects like the high-Mach Sky Police interceptor aircraft, reconnaissance aircraft, air-to-air and air-to-surface missiles, and air defence missiles intended to guard against hypersonic missiles.

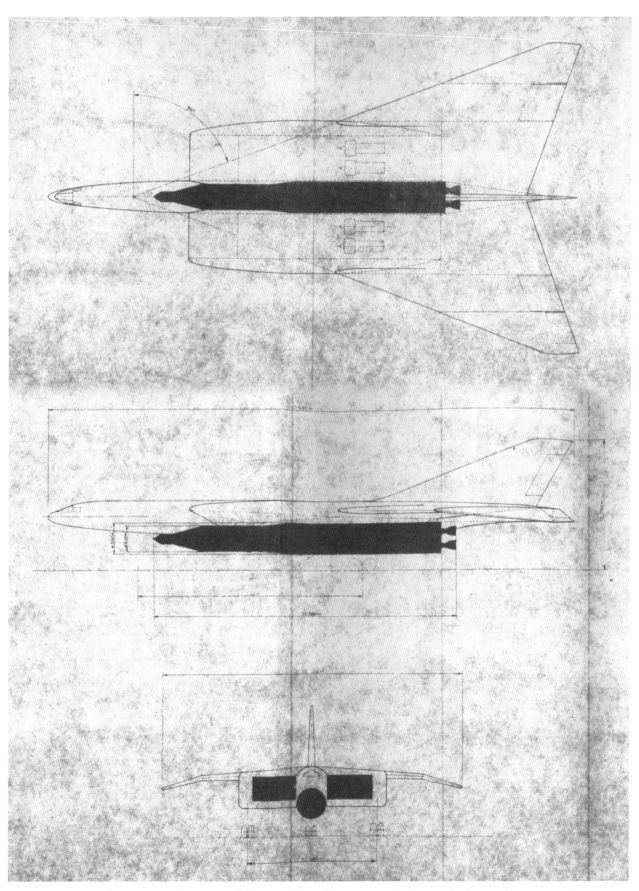

ABOVE There were also variants in the sweep of the wing. This is the version with 70° sweepback... *Nord-Aviation via H Lacaze*

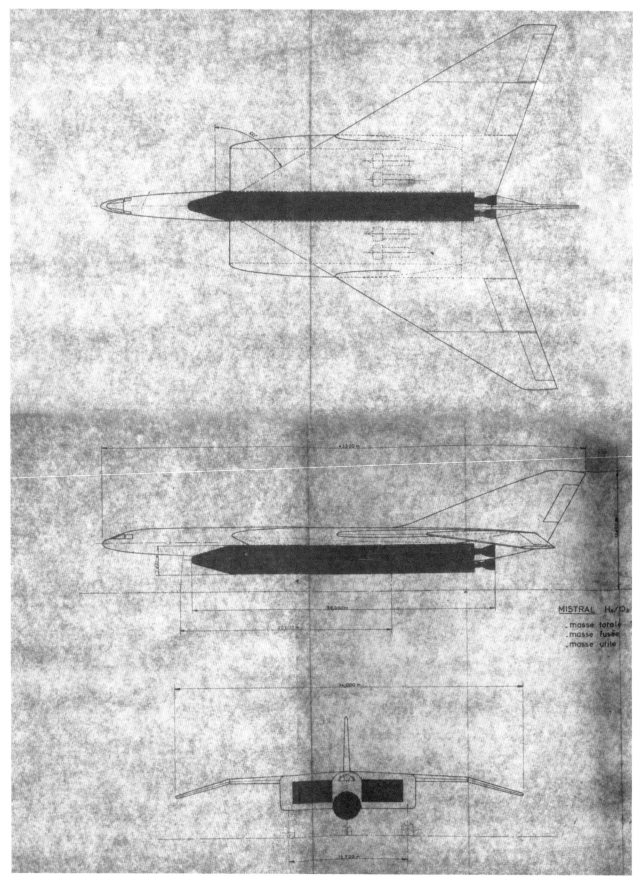

ABOVE ..and here the variant with the 60° swept wing. *Nord-Aviation via H Lacaze*

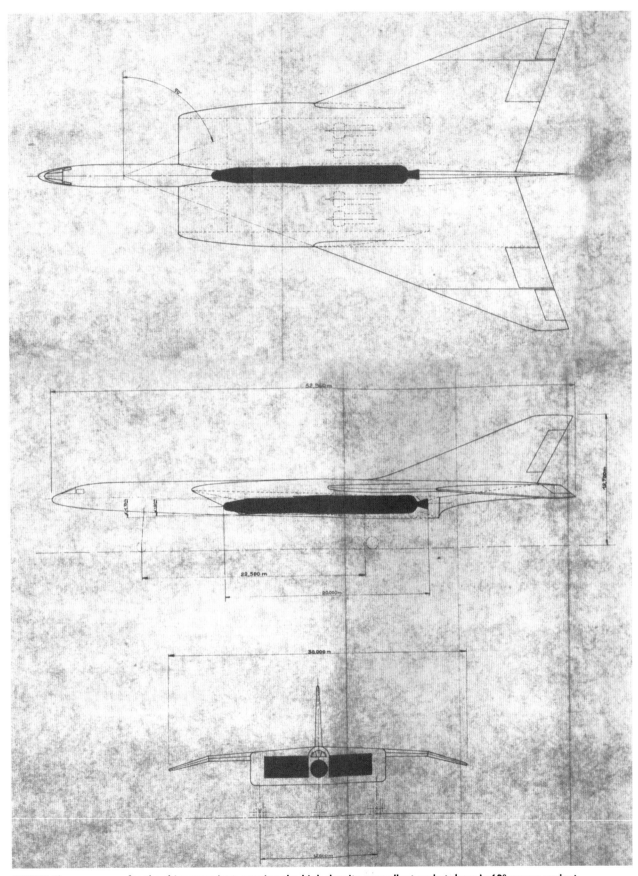

ABOVE The same goes for the thinner variant carrying the high-density propellant rocket: here in 60° sweep variant.
Nord-Aviation via H Lacaze

Among the new designs, the turbo-ramjet-powered first stage flying at Mach 5/6 was favoured as it could be moved by its own means (like an aeroplane) from one area of the World to another, thereby allowing launch on an equatorial orbit after taking-off from any place; it could even be adapted as a hypersonic transport. This referred, obviously, to the SNECMA-Nord Mistral. The Sud designs were regarded as not agile enough or too futuristic. *'Among the new designs, there is the launcher propelled by rockets but fitted with turbojets to assure the flight back to base, which is quite simple [to build]. It would certainly be able to launch over 1.5% of its launch mass. But this concept does not allow it to be ferried [from its base to another], the launch on an equatorial orbit, neither can it be adapted for hypersonic transport [...]* this *kind of launcher could be adapted for a large country wishing to accomplish only a limited number of launches, about 10 yearly'.*

As for more advanced designs flying at Mach 12 or Mach 15 and possibly fully recoverable, they certainly could be designed but not in the intended time target. *'They may be possible around 1990; it is doubtful they could be for 1980. Anyway, it does not seem advisable to use air-breathing engines above Mach 7 for this kind of machine.'*

CPE had put the maximum separation speed at Mach 7 because, *'Mach 7 is a limit after which structures made out of nickel-chromium alloys must be replaced by structures and alloys much more heat-resistant or protected, whose technology is still poorly understood. Mach 7 is also the point beyond which subsonic combustion ramjets are no longer efficient.'*

Viewing this programme with fifty-year hindsight, Henri Lacaze who was among the CPE team thought: *'The very notion of what Dassault called the Transporteur Aérospatial – even though other manufacturers later used it – was a bad idea; while it is in the atmosphere, a vehicle must fight against its drag. Getting out of the atmosphere as soon as possible is a better idea, hence the vertical launching rockets…ideally a SSTO (single-stage-to-orbit) would be the perfect vehicle.'*[26]

Nord Spaceplane (1st stage) (1965)	
Status	Project
Span	-
Length	169ft 11in (51.80 m)
Gross Wing area	-
Gross weight	264,550lb (120,000kg)
Engine	4 × turbo-ramjets
Maximum speed/height	-
Load	176,370lb (80.000kg)

Nord Spaceplane (2nd stage)	
Status	Project
Span	-
Length	85ft 0in (25.90 m)
Gross Wing area	-
Gross weight	176,370lb (80,000kg)
Engine	6 × LH_2 LO_2 rocket engines
Maximum speed/height	-
Load	6,614lb (3,000kg)

Nord Mistral for H2/02 second stage	
Status	Project
Span	111ft 7in (34.00m)
Length	155ft 10in (47.50m)
Gross Wing area	-
Gross weight	-
Engine	4 × turbo-ramjet developed from SNECMA TF106: 54,000Da N (w/oAB), 97,000Da (w/AB)
Maximum speed/height	-
Load	98 × 9ft (30 × 2.80m) diameter second stage 110,230lb (50,000kg)

Nord Mistral for high density fuels second stage	
Status	Project
Span	106ft 4in (32.40m)
Length	147ft 8in (45.00m)
Gross Wing area	-
Gross weight	-
Engine	4 × turbo-ramjet developed from SNECMA TF106: 43,000Da N (w/oAB), 81,000Da (w/AB)
Maximum speed/height	-
Load	72.2 × 5.2ft (22 × 1.60m) diameter second stage 116,845lb (53,000kg)

2 – The Aircraft

2A Nord Transporteur Aérospatial

Nord had amassed a great deal of knowledge on ramjets through the Griffon programme and building on this experience and the studies made under the auspices of EUROSPACE (see Chapter 2) proposed the Mistral launcher.

2A1 – The 1966 Mistral

For the CPE contract, two different iterations of Mistral were presented. (The name evokes the violent, cold, north or northwest wind flowing from the Rhône valley into the Mediterranean, but it could be an acronym, although no explanation was ever given.)

The first variant featured upper stages with low-density propellants (liquid oxygen and hydrogen), partially integrated into the first stage. The second variant featured high-density propellants which could therefore be fully inserted into the first stage airframe. From the CPE final report it appears that many doubts existed as to the actual performance of the high-density propellant rocket. CPE appears highly dubious of the claims made by Nord-SNECMA regarding this mode of propulsion, but recognised it presented a great interest for military variants and thus needed to be investigated.

According to the Nord-SNECMA's *rapport de synthèse* dated September 1966, the launch performance was claimed to be a payload (satellite) of 4% the take-off weight, *'as compared with 2.8% for the Atlas-Centaur rocket.'* This claim was discussed by CPE which estimated the launch performance of the Mistral as 2.4% of its take-off weight.

In those days, computers were not common and in its report Nord-Aviation mentioned its use of an IBM 7040 machine. (However, lower down the report refers to *'the IBM 7040 belonging to SNECMA'*—so it is uncertain whether the two partners each had their own computer or whether they shared one) *'A calculation programme on an IBM 7040 computer has been developed at Nord-Aviation for the computation of dimensions, cycles and performances. To this main programme are added a series of sub-programmes performing thermodynamic calculations. This makes a very important ensemble of over 3,000 [Hollerith] punch cards.'* The main programme, written in Fortran language, was aimed at defining the turbo-ramjet engine required for the Mistral, including air intakes and exhaust nozzles, while secondary programs computed the heating of various points on the aircraft, based on equations supplied by ONERA. The specific turbo-ramjet used in the Mistral was to be based on the SNECMA TF106 for which technical data was summarised as follows: *'Axial double-body, mixed double-flow turbojet with a moderate dilution ratio. Its nine-stage, low-pressure compressor includes a three-stage blower driven by a three-stage turbine. The seven-stage, high pressure compressor is driven by a single-step turbine. The global compression ratio is in the order of 16;1. The length from the front of the engine to the turbine flange is 7.55ft (2.30m), the dry weight is 3,483lb (1,580kg).'* The TF-106-based turbo-ramjet had been studied using the Griffon under an US Air Force contract.

Mistral's size was similar to Concorde with a weight of 339,500lb (154,000kg) (H_2/O_2 second stage) or 319,675lb (145,000kg) (high-density propellants) versus 326,275lb (148,000kg) for the airliner. That's where the lack of structural calculation was found wanting: CPE declared it 'strange' that the wing of the Mistral weighed less per square metre than that of Concorde, an aircraft of similar size but flying at Mach 2.2 not Mach 5. It noticed that the structural weight computations used were based on those of the Concorde, built out of light alloys and without the 'cut frames' as used in rocket launchers. But *'recent information appears to indicate that a steel structure is 15% heavier than a light alloy structure. One is therefore to assume a basic weight of 440,925lb (200,000kg) for the Mistral [...]*. For the Mistral, one should also add, as PLT [Sud] did, a margin of 10% for the fuel consumption, therefore an overall mass increase of 10% more. In the end the CPE estimates that it is more reasonable to evaluate at 485,000lb (220,000kg) the mass of a launcher able to put 13,228lb (6,000kg) in low orbit, with a separation speed of Mach 4.5.'

Mistral was to take-off at a speed of 219mph (352km/h) which – by comparison to Concorde (224mph [360km/h]), Convair B-58 Hustler (236mph [380km/h]) and North American B-70 Valkyrie (205mph [(330km/h)] – Nord found to be acceptable, but noted that it depended on wing loading which was estimated as 61lb/sq ft (300kg/sq m) for the Mistral.

Wing loading was briefly discussed by CPE which approved the choice of Nord-SNECMA. *'There is no need to have a lower loading as the weight of the wings penalise serious the performance of the aircraft.'*

The first stage flew up to a height of 114,825ft (35,000m). Separation occurred once this altitude was reached at a speed of about Mach 5. The question of separation could have required a whole study by itself which, as far as the author knows, was never carried out. CPE recognised that *'the launch of the rocket is a remarkably arduous problem [...] if the rocket is attached on the back of the launcher, separation would occur after a negative pull-out of the launcher and the rocket begins its flight with a negative incidence [...] this method impacts [negatively] the orbited mass and extends the duration of the atmospheric flight of the rocket.'* The second option then was to put the rocket underneath the carrier, solution adopted by Nord. But then two configurations could be built: either with the rocket half-inserted into a recessed bay in the carrier, either with the rocket inside a fully enclosed bay. *'Finally, it is the carrying [of the rocket] in a closed bay – with the separation being achieved by dropping the rocket after opening of the bay doors – which has been selected by [Sud] even though it adds mechanical complications due to the need to design opening doors and the need to stabilise the rocket. Nord-Aviation/SNECMA also reached the same conclusion for the high-density propellants rocket.'*

CPE noted that questions regarding *'thermal flows and efforts inside the bay before it could be closed, effects of aerodynamic interactions [between the first and second stage] which could gravely perturb the launch of the second stage'* should be further explored. It also added consideration regarding the equilibrium of the rocket immediately after launching. Control through jets appeared dangerous, and nozzle swivelling insufficient. So CPE recommended adding fins or skirts to the second stage. That was something which was obviously not studied by Nord/SNECMA.

Interestingly, CPE digressed in a revealing way, showing it had things in mind other than civilian satellite launchers. *'An experimental study would deal, not only with the present problem [rocket launcher separation], but with all launches of air-to-air or air-to-surface missiles from high-Mach interceptor- or bomber aircraft. At speeds above Mach 4, it appears unimaginable to hang missiles under the wings as is currently done with Mach 2 fighter aircraft. Thermal flows would damage propulsion fuels, warloads and avionics after only a few minutes of flight; hence the requirement to carry the missile in a closed, possibly air-conditioned bay.'*

The second (rocket) stage then ignited and brought the third stage (satellite) to 590,550ft (180,000m) at which altitude the satellite separated to enter into a low orbit. Up to 13,228lb (6,000kg) of payload could be put into a low orbit (the commonly accepted definition for LEO is 100 to 1,240 miles [160 to 2,000km] above the Earth's surface).

Again, the choice of a rather low separation speed was criticised by CPE, arguing that the whole point of the aircraft launch was to economise on the recoverable parts of the launcher and the lower the separation speed, the bigger and therefore the more costly the second rocket stage would be. So in the views of CPE: either the operator/country launched a lot of satellites, in which case an aircraft-style first stage was economically justified but should be associated with a smaller second stage separating at high Mach or, if the operator launched few satellites, then a rocket first stage recovered via ballutes or a rotor was sufficient.

This approach did not take into account the flexibility brought by an aircraft-style first stage which could be beneficial by itself.

Nord-Aviation indicated the following advantages for this design:

- Flexibility in terms of achievable orbit through the ability to select the launch trajectory.
- Possibility of ferry flights between bases.
- Ease of return flight, since the spaceplane is able to fly at subsonic speeds.
- Simplified launching equipment similar to that needed for large aircraft.

For ferry flights (but carrying the second stage rocket under its belly), the Mistral could obviously fly at either subsonic or supersonic speeds. Nord determined the optimal range for a ferry mission to be 1,150 miles (1,850 km) in subsonic flight but only 547 miles (880km) in supersonic mode. Therefore it suggested fitting the Mistral with an air refuelling ability.

A quick calculation was also made by Nord of what the performance of a Mistral-derived SST would be. (It should be remembered that Nord-Aviation had been invited to make a proposal for the government-sponsored competition that was to lead to Concorde. However, it had backed the wrong horse, favouring the high-performance Mach 3 design when the 'simpler' Mach 2 design was favoured in this competition). For its Mistral SST, Nord took as basis the slimmer 'high-performance rocket' carrier. It computed that the Mistral SST could have a range of over 3,105 miles (5,000km) if the full 'rocket bay' were given to fuel, but only 2,175 miles (3,500km) if the rocket bay were shared between fuel and payload. No real detail was given; not even the cruising speed intended—Mach 5? Actually, the whole calculation does not make much sense; what use would be a SST in which all available space were given to fuel? And the range appears very limited, considering that the Mistral is described elsewhere in the report as 'over-motorised.'

Another criticism levelled by CPE at the Nord/SNECMA project regarded the variant using high-density propellants ('lithergols': hydride of beryllium and hydrogen peroxide). While recognizing such propellants presented advantages for military applications because they were easily stored, CPE thought that experience was lacking for performance estimates. *'The CPE considers the performance data stated by Nord/SNECMA are not realistic; the practical specific impulse will certainly be smaller than the calculated specific impulse. One must therefore take into account that the replacement of a rocket using O_2/H_2 as propellants by a rocket using H_2Be/O_2H_2, will reduce, if everything else is kept equal, the orbited load which would go down from 5.9 to 4.4 tons (6 to 4.5 tonnes).'*

2A2 – Further Variants

In a later proposal, Nord departed even further from the apparent simplicity of the early projects by launching the whole combination using a sled. It seems Nord-Aviation ran into difficulties when trying to design an undercarriage that could support the whole machine and have clearance from the two upper stages nested underneath the carrier. Interestingly in the 1966 report, it was not take-off but landing which worried Nord engineers. *'The most critical case would be a forced landing with the upper stage(s) [still attached] even admitting a quick flushing out of the fuel of the first stage.*

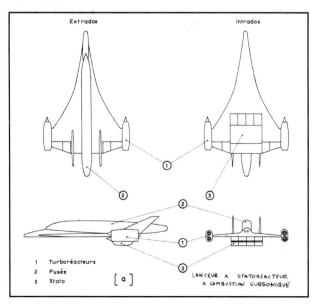

BELOW Four-view of the initial PLT180 study (turbojets and ramjets separated). This is an original Sud-Aviation drawing. *Henri Lacaze*

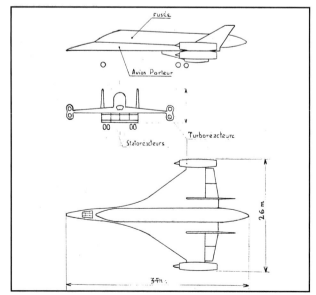

BELOW You might think this is the same drawing as above, but there are differences, the most notable being the addition of the position of the wheels when the composite is on the ground. This drawing was included in the Centre de Perspective et d'Evaluation final report. *Henri Lacaze*

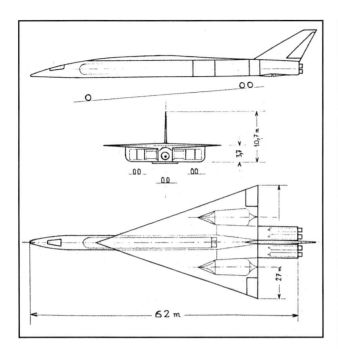

ABOVE **The PLT 186 study, as presented in the Centre de Perspective et d'Evaluation final report. Note, again, addition of the landing gear wheels. The question of the landing gear (or more exactly, the take-off gear) was a serious concern in this report.** *Henri Lacaze*

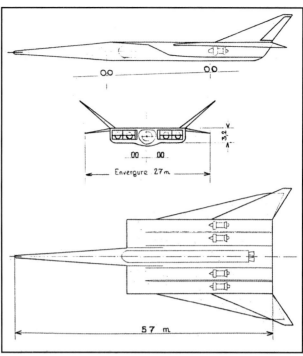

ABOVE **The PLT 219 study, as presented in the Centre de Perspective et d'Evaluation final report. This was the first Sud design with turbo-ramjets.** *Henri Lacaze*

The landing weight is then 75% of take-off weight. The length of the landing strip must therefore be homogenous to that of the take-off track. Therefore, Mistral could land on airfield with standard runaways of 10,500ft (3,200m) length.' But the 1968 CPE report did raise the question *'a core problem for this kind of launcher bloated with fuel, is to have a take-off gear able to support high speeds, the take-off run by itself remains short (about 3,280ft [1,000 m]) because the aircraft is over-motorised.'*

On the whole the following difficulties were expected: heavy structural weight; difficulty in integrating the upper stages into a first stage of aeroplane type; and separation of the components of the composite.

It soon became apparent that heating was to be a major problem. Re-entry was planned at around Mach 10, which generated surface heating which would exclude the use of common aeronautical materials. Hence a new material, called Norsial, was developed. This was a 'metallic sandwich' able to support temperature over 1,000° C and was actually flown on the Diamant and Europa/Coralie[27] rockets.

Finally the project evolved toward a three-stage variant:

i First stage was to be a 148-ton (150-tonne) aircraft propelled by four turbo-ramjets with a total thrust of 71 tons (72 tonnes) at launch.

ii Second stage was a 49-ton (50-tonne) liquid hydrogen/oxygen rocket.

iii Third stage was to be a manned spaceplane weighing about 29.5 tons (30 tonnes) with the ability to place a 7.9-ton (8-tonne) payload into low orbit.

The addition of a consumable second stage was for economic reasons. It was actually easier to develop a smaller recoverable spaceplane than the larger version originally planned. Considering that the ambition to put 5.9 tons (6 tonnes) into orbit was severely criticised, aiming at 7.9 tons (8 tonnes) was most probably more of a marketing exercise than a serious proposal.

Nord's project was seemingly abandoned around 1966.

2B – Sud-Aviation PLT Proposals

Sud-Aviation, which received more money than Nord-Aviation, was able to continue studies until April 1968 and came up with five different designs. The PLT acronym stood for Projets à Long Terme (Long-Term Projects), a special design office within Sud-Aviation, in charge of strategic studies and reflections. The number following PLT does not refer to a specific aircraft design/concept but is the sequential number of the report.

Here is the list of the designs, followed by a translation of the report title:

1 Launcher aircraft with separate turbo- and ramjet propulsion/second stage carried on the back: 'PLT180 preliminary note regarding the study of an air-breathing launcher, version L1' issued 7 January 1965.

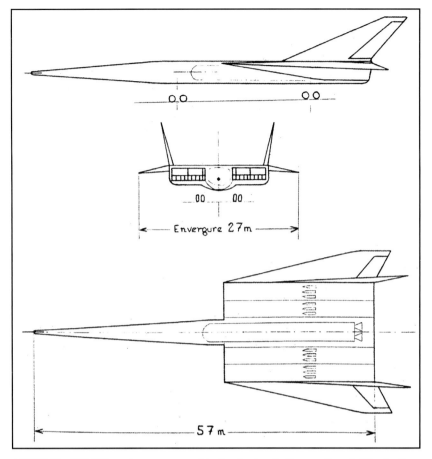

ABOVE The PLT 220 study as presented in the Centre de Perspective et d'Evaluation final report. This was Sud proposal fitted with turbo-rockets. *Henri Lacaze*

SUD PLT 180	
Status	Project
Span	85ft 4in (26.00m)
Length	111ft 6in (34.00m)
Gross Wing area	–
Gross weight	–
Engine	4 × turbojets + 4 × ramjets
Maximum speed/height	–
Load	98 x 9.4ft (30 × 2.85m) diameter second stage 81,571lb (37,000kg)

2 Launcher aircraft with separate turbo and ramjet propulsion/second stage carried recessed in bay: 'PLT186 Study for a Mach 7 launcher propelled by a subsonic-combustion ramjet' issued 25 July 1965 with a complement on 2 February 1966.

3 Launcher aircraft with turbo-ramjets/second stage carried recessed in bay: 'PLT219 Aerospace launcher propelled by turbo-ramjets' issued 20 October 1967.

4 Launcher aircraft with ramjets and rocket propulsion/second stage carried recessed in bay: 'PLT220 Aerospace launcher propelled by rockets and ramjets' issued 5 February 1968.

5 Launcher aircraft with rocket propulsion/second stage carried recessed in bay: 'PLT221 Aerospace launcher with rocket propulsion. Parametric studies. Optimisation studies' issued 15 March 1968.

Actually, Sud-Aviation PLT Office published another report 'PLT 222 Aerospace launcher with supersonic turbo-ramjets' which was issued 22 April 1968. The designs covered in this report investigated a much higher separation speed (up to Mach 12). These were considered as *'more futuristic and therefore lacking in realism'* by CPE. For this reason, and because they were less detailed than the previous ones, these designs will be dealt with in the next chapter.

2B1

Launcher Aircraft With Separate Turbo- and Ramjet Propulsion/Second Stage Carried On The Back: PLT180

This 1965 design was the only proposal to discuss the opportunity to launch the ballistic second stage from the top of the aircraft first stage. PLT decided quickly that this was not the preferred option. Placing the second stage in a closed bay inside the fuselage of the mothership was a much better option in terms of heat control.

This first design was constructed around a set of parameters which were constant for all the other PLT projects covered in this Chapter:

■ Dimensions: the wing surface was set at 5,380sq ft (500sq m) with the wing having a sweep of 70°. From this and the general shape of the aircraft, drag could be estimated. PLT asked ONERA for help in calculating the drag of their projects, notably in the transitional phase into supersonic. The first estimations indicated a Cx of 0.35 which PLT tried to reduce by redesigning the leading edges (thinned) and the air intakes (the ramjets are notorious for generating high drag). CPE was dubious about this low value and asked ONERA to study the drag at Mach 7 of the PLT project. These studies showed that leading edge drag of the *auvents* ('awnings') leading toward the air intakes of the engines was inappropriate: the 'awnings' had been designed with only thermal resistance in mind, leading to a too-thick leading edge which generated drag. Basically, compared with the results of ONERA's studies, CPE thought that PLT had underestimated the drag value of its design(s) and re-evaluated the Cx at 0.67 *'without taking into account drag generated by the interactions of the different elements, the "traps" …'*

- Separation speed: PLT selected Mach 7 as the separation speed because it thought *'considering the launch trajectories, thermal constraints are no more critical at Mach 7 than at Mach 5. Structural mass could therefore be estimated as constant between Mach 5 and Mach 7.'* CPE doubted the validity of this point: it thought that, yes, the outer skin of the vehicle would not suffer more at Mach 7 but, no, it could not apply to air intake, ducts, ramjet combustion chamber and exhaust nozzle. CPE then agreed with Nord-Aviation which had selected the lower speed of Mach 4.5 specifically because it estimated it could design a ramjet operating at this speed but, beyond that, more studies would be required regarding heat constraint.

2B2

Launcher Aircraft With Separate Turbo and Ramjet Propulsion/Second Stage Carried Recessed In Bay: PLT186

Carrying the second stage rocket in an enclosed bay was the solution selected by PLT for all further studies. This added many levels of complexity as (a) the doors needed to be mechanically opened; and (b) when the doors were opened, the air flow was disrupted causing aerodynamic problems and the rocket needed to be stabilized while it crossed this disturbed airflow (its own rocket engine could not be fired until it was separated enough from the mothership). For launching the rocket PLT only studied one procedure: the aircraft mothership takes-off, carrying in its bay the second stage; it increases its speed and altitude until it has reached Mach 7 and 98,425ft (30,000m); launches, turns 180° and returns to its base with the engine on economy power. Viewed by CPE it could be applicable to both civil satellite and military ICBM launchings (*'this [flight plan] is appropriate for a launch toward the East, during a war.'*) PLT estimated that turning as it cut down the engines, while keeping a load factor at 2.3, the speed would be reduced to Mach 2.6 at the end of the turn. Based on that, it elected to descend in spiral, with the engine running [but probably at a lower power than during the ascending phase] to arrive at the end of the turn at an appropriate supersonic speed to come back to the base on economy power.

This procedure was criticised by CPE, stating, *'The intended manoeuvre requires a very delicate piloting of the aircraft; it would be much simpler first to stabilise the aircraft, then decelerate and finally to turn back at low supersonic speed. But this procedure is not applicable to PLT's proposals as it considerably extends the distance away from base.'*

Besides, there was the question of the exterior and interior heat to which the machine would be subjected. For the 'heated load bearing structure' it was proposed to use super-alloys like René 41. For hotspots, PLT recommended use of ceramics or ablative materials. It thought the leading edges would be exchanged during regular maintenance work, as would be tyres. Fuel cooling was also expected to be used for leading edges and intake lips. For hotspots inside ramjets and nozzles, PLT suggested 'reinforced ceramics' but CPE did not agree and thought that cooling via fuel circulation would be required for a machine approaching Mach 7. Due to the debate about heating tolerance of the aircraft, PLT reworked its return trajectory, the spiral descent appearing to inflict too much heating on the machine.

SUD PLT 186

Status	Project
Span	88ft 7in (27.00m)
Length	203ft 5in (62.00m)
Gross Wing area	-
Gross weight	-
Engine	4 × turbojets + 4 × ramjets
Maximum speed/height	-
Load	98 x 9.4ft (30 × 2.85m) diameter second stage 81,571lb (37,000kg)

2B3

Launcher Aircraft With Turbo-Ramjets/Second Stage Carried Recessed In Bay: PLT219

For this and the next proposal, PLT was unable to propose a satisfying estimation of the optimal climb trajectory. It tried a 'quasi-optimal' method with which CPE did not agree.

Initially another type of engine had been investigated—the 'turbo-rocket', proposed by Rolls-Royce, which appeared able to work efficiently up to Mach 5. The idea was to drive the compressor rotor with a rocket and burn the excess of fuel in the air thus compressed. A quick study of this engine was made by Sud-Aviation as report PLT 185, which proved the performance of a launcher fitted with such motors would be as efficient as those of a machine propelled by turbo-ramjets. However, CPE decided to focus instead on turbo-ramjets for two reasons:

- *'This is an all-new type of turbomachine, so a lengthy and costly development must be expected. Mainly there is the question of designing a drive to carry the power from the rotor to the compressor while working in a very difficult caloric environment*

- *The specific consumption at low speed is much higher than for a turbojet, being a value of about 5 (units of fuel weight per hour to propel one unit weight of rocket). Therefore this engine is not fit for loitering at low altitude and its subsequent use on either military of civil aircraft is unlikely'*

Therefore, the turbo-ramjet was preferred by CPE ... and this was the forte of Nord-Aviation. Sud tried to investigate a turbo-ramjet built around special configuration called the 'single-flow' turbojet. This kind of turbojet, which had a good performance up to Mach 3, was then being investigated in the USA as part of the SST programme. Unfortunately, here again, CPE did not agree with the conclusions of PLT as its major hypothesis: that a supersonic

SUD PLT 219	
Status	Project
Span	88ft 7in (27.00m)
Length	187ft 0in (57.00m)
Gross Wing area	–
Gross weight	–
Engine	4 × turbo-ramjets based on turbojets of the 'Pratt and Whitney JTF17 A 21 class'
Maximum speed/height	–
Load	98 x 9.4ft (30 × 2.85m) diameter second stage 81,571lb (37,000kg)

and a subsonic flow could be separated had been proved false during the tests of the Nord Griffon. However, because turbo-ramjet was the engine type favoured by CPE, this was the design which was developed further.

The 187ft (57 m)-long fuselage contained the second stage in a fully enclosed 98-ft (30-m) bay. Engines were grouped in pairs underneath the wings to benefit from the 'awning' effect. They were fed by very large intakes of 215sq ft (20sq m) which represented the largest part of the machine at its widest section. The delta wings generated most of the lift and carried the 646-sq ft (60-sq m) fins which were angled from the vertical to stay efficient at hypersonic speeds and improve lateral and longitudinal stability.

Detailed technical data were given (see table).

Separation was estimated to occur about 140 miles (225km) from base. This is interesting as at the same time, for Concorde, the effect of the Mach 1 booms on the ground-dwellers was carefully studied and here we read of launching a Mach 7 machine from mainland France.

This design appears to have been widely studied in ONERA wind tunnels with different degrees of sweep to the wings and different configurations of fins. Those tests demonstrated the influence of the engine nacelles in the stability of design—the large nacelles required by turbo-ramjets destroyed axial stability at speeds above Mach 5 while smaller pods, as would be used for supersonic ramjet, had a positive influence on the stability between Mach 5 and Mach 7. Another conclusion was the difficulty in designing the rear of the fuselage when a nozzle was integrated with it. Violent manoeuvres could generate asymmetrical effects and could have negative effects on the longitudinal stability of the aircraft, as could have a change in engine speed.

2B4

Launcher Aircraft With Ramjets and Rocket Propulsion/Second Stage Carried Recessed In Bay: PLT220

The propulsion system here comprised four elements:

i Bi-dimensional intake followed by a diffuser placed underneath the large delta wing and therefore benefiting from the 'awning' compression effect of the wing.

ii Divergent duct containing 48 liquid-fuel (JP4 + O_2) rockets in bundles of four of a sufficient length to ensure a good mix of air and rocket-generated gas, and then their combustion (those gases still contain fuel as the rockets work at a richness above 1).

iii Reheat chamber were those gases are finally burnt and where fuel specific to the ramjet is injected and burnt.

iv Convergent-divergent nozzle.

PLT found that this type of engine increased the performance of the rockets (to simplify things PLT studied the functioning of the engine in either full-rocket/no-ramjet mode and no-rocket/full-ramjet mode). The thrust of the rocket could, for example, be doubled by ram effect between Mach 2 and Mach 3. However, the mixer, the intakes and the nozzles needed to be designed with care to avoid 'clogging' [the wording is from the original French text; one may assume it means flaming-out]. However, it also proved that increases in thrust and specific impulsion were greater at low altitude—not exactly what was required here.

SUD PLT 219 Technical Data			
Fuselage length	255ft 11in (78.00m)		
Max height of the fuselage	14ft 1in (4.30m)		
Max width of the fuselage	19ft 0in (5.80m)		
Diameter of the internal bay	10ft 10in (3.30m)		
Length of the bay	114ft 10in (35.00m)		
Width of the central body	53ft 10in (16.40m)		
Span	90ft 7in (27.60m)		
Wing sweep	70°		
Wetted surface	7,136sq ft (663sq m)		
Surface of the fin	861sq ft (80sq m)		
Static thrust of the four rocket engines	595,750lb st (265,000da N)		
Static thrust of the two turbojets	40,465lb st (18,000da N)		
Weights:			
Rocket engines	10,406lb (4,720kg)		
Turbojets	6,393lb (2,900kg)		
Equipment	22,046lb (10,000kg)		
Fuel	577,600lb (262,000kg)		
Total fuel	169,755lb (77,000kg)		
2nd and 3rd stage + 13,228lb (6,000kg) payload	80,028lb (36,300kg)		
Take-off weight	833,350lb (378,000kg)		
Performance:			
Take-off (run 3,280ft [1,000m])	duration: 19 s	Fuel consumed: 3,086lb (1,400kg)	
Acceleration to Mach 1.5	duration 325 s	Fuel consumed: 50,706lb (23,000kg)	
Acceleration from Mach 1.5 to Mach 7	duration 161 s	Fuel consumed: 68,343lb (31,000kg)	
Separation	duration 12 s	Fuel consumed: 2,205lb (1,000kg)	
Return to the base		Fuel consumed: 28,660lb (13,000kg)	

SUD PLT 220	
Status	Project
Span	88ft 7in (27.00m)
Length	187ft 0in (57.00m)
Gross Wing area	–
Gross weight	–
Engine	4 × ramjets + 48 liquid-fuel rockets
Maximum speed/height	–
Load	98 x 9.4ft (30 × 2.85m) diameter second stage 83,776lb (38,000kg)

It was also found that while the 48 small rockets were slightly cheaper than four turbojets, overall, the financial advantage thus gained was negated by the larger size of the aircraft.

The final nail in the coffin of this variant was the realisation that '*subsonic performance of this design is catastrophic. A round trip uses up to 30,865lb (14,000kg) of fuel. This is therefore an unsure launcher which would require a lot of guidance devices. This solution must be avoided in favour of the turbo-ramjet launcher.*'

2B5

Launcher Aircraft With Rocket Propulsion/Second Stage Carried Recessed In Bay: PLT221

In designing this project, the interest was twofold:

i To have a reference against which to compare the air-breathing designs.

ii To have a simple design to test the modelling methods for assessing the parameters of an aerospace vehicle.

This project, which was designed after the others seen previously, followed closely the parameters of those earlier designs:

- Take-off and landing were effected 'tangentially'.
- The LH$_2$-LO$_2$ second- (and third-) stage rocket was carried internally in a closed bay.
- Wing was of delta shape and fitted with elevons.
- Large vertical fin ensured yaw stability.
- Main rocket engines were fuelled by kerosene and liquid oxygen.

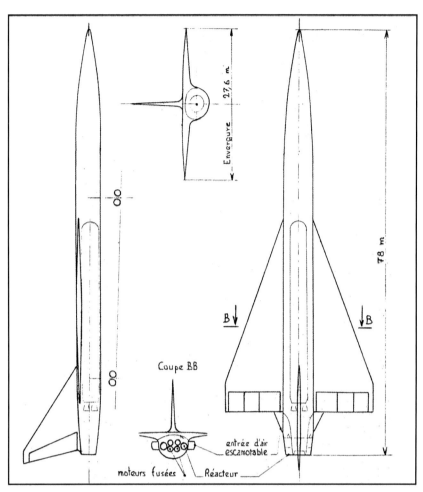

ABOVE This is the PLT221 design, which only used rockets—a solution that was not deemed practical by Sud. *Henri Lacaze*

- As the return flight was a costly phase in fuel if done only on rocket power, small turbojets were designed into the rear fuselage to fly back home at subsonic speed. They were fed with air through retractable intakes (thus avoiding extra drag during the supersonic/hypersonic phase of the flight).

The end result was a very big machine with a length of 256ft (78m) and a take-off weight nearing 837,755ft (380,000kg). Actually, it was found that the greater the overall size of the machine, the better the ratio orbital payload weight on overall mass was. It was also found that the denser the upper stages, the better, which validated the research by Nord-Aviation on the second stage using high-density propellants.

Another study which confirmed Nord-Aviation's opinion was on the question of take-off gear. Take-off runs were short, about 3,280ft (1,000m), because the aircraft was in some ways 'over-engined.' The unfortunate consequence was that it required an undercarriage able to support a very heavy machine at high speed (estimated at 459ft/s (140m/s), about 311mph [500km/h]). Both Nord and Sud research proved this could be the Achilles' heel of the concept, but where Nord suggested the use of a launching 'chariot' (something which had already been studied during the 'fifties by René Leduc, SNCASE, DOP, and Jacques Gérin[28]) Sud went further and proposed (but as far as known, did not design) a 'Hovercraft chariot.'

CPE thought that, for once, separation speeds above Mach 7 could be worth studying, as only the leading edge would be subjected to high heat.

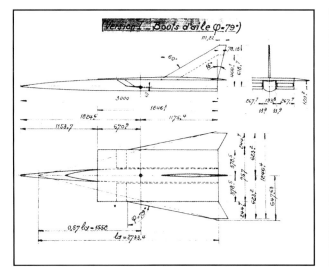
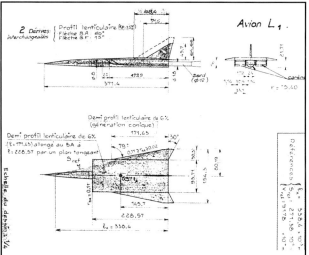

ABOVE ONERA intensively tested the shape of LSM3 in its wind tunnel as the avion L1 (aircraft L1, L probably meaning 'launcher') during October 1968. The two drawings illustrate the kind of step-by-step studies undertaken with different nosecones and exterior wings swept at 79° (drawing 1) and different tails (drawing 2). *ONERA, author's collection*

SUD PLT 221 Technical Data	
Fuselage length	255ft 11in (78.00m)
Fuselage length	255ft 11in (78.00m)
Max height of fuselage	14ft 1in (4.30m)
Max width of fuselage	19ft 0in (5.80m)
Diameter of internal bay	10ft 10in (3.30m)
Length of bay	114ft 10in (35.00m)
Width of central body	53ft 10in (16.40m)
Span	90ft 7in (27.60 m)
Wing sweep	70°
Wetted surface	7,136sq ft (663sq m)
Surface of fin	861sq ft (80sq m)
Static thrust of four rocket engines	595,750lb st (265,000da N)
Static thrust of two turbojets	40,465lb st (18,000da N)
Weight table:	
Rocket engines	10,406lb (4,720kg)
Turbojets	6,393lb (2,900kg)
Equipment	22,046lb (10,000kg)
Fuel	577,600lb (262,000kg)
Total fuel	169,755lb (77,000kg)
2nd and 3rd stage + 13,228lb (6,000kg) payload	80,028lb (36,300kg)
Take-off weight	833,350lb (378,000kg)

Weight table for Vehicle 2 (only the orbiter is recovered)	
Cabin with pilot	4,410lb (2,000kg)
Payload	2,205lb (1,000kg)
Fuel tanks/engine	2,866lb (1,300kg)
Propellant: Liquid H_2 (LH_2) and O_2	3,968lb (1,800kg)
Re-entry and landing system	2,646lb (1,200kg)
Weight table for Vehicle 2 (full recovery)	
Cabin with pilot	4,410lb (2,000kg)
Payload	2,205lb (1,000kg)
Fuel tanks/engine	12,125lb (5,500kg)
Propellant: Liquid H_2 (LH_2) and O_2	7,055lb (3,200kg)
Re-entry and landing system	3,968lb (1,800kg)

In the end, while the turbo-ramjet solution was still favoured, the overall assessment of the full rocket version was that it was *'appealing in its simplicity.'* There are fewer uncertainties than in the other designs, no hypersonic engine to design, no exotic fuel and thus *'it is reasonable to think this launcher would indeed weight slightly under 881,850lb (400,000kg) and be able to put 13,228lb (6,000 kg) in orbit.'* It was also noted that the two turbojets gave it good return and landing abilities. But *'it has no versatility and the high take-off speed is the source of problems which would need to be solved.'* So in the end, CP concluded *'this kind of launcher is appealing, especially if the launch frequency is moderate.'*

But the report-writer also indicated that due to the question of the undercarriage, a few PLT engineers were wondering if it was really worth bothering with 'tangential' take-off,

SUD PLT 221	
Status	Project
Span	90ft 7in (27.60m)
Length	255ft 11in (78.00m)
Gross Wing area	–
Gross weight	–
Engine	2 × turbojets + 4 × rockets
Maximum speed/height	–
Load	98 x 9.4ft (30 × 2.85m) diameter second stage 81,571lb (37,000kg)

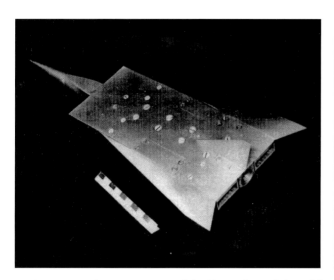
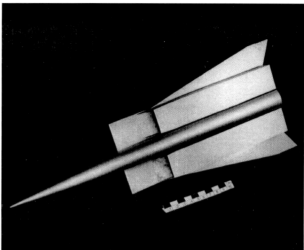

ABOVE Two (rather poor) views of the ¹/₁₇₅ scale model used for these tests. ONERA, via H Lacaze

ABOVE Pictured is a model with twin, angled outside, fins. Note the model is fixed on its belly while the model with a single fin was to be fitted on a 'dard' (a spindle-support) attached to the rear fuselage (see the upper right picture on page 78 where the 'dard' attachment is specifically mentioned).
ONERA, via H Lacaze

if the common though more fuel-consuming, vertical take-off was not a solution. The phrasing of the report does not indicate what method of recovery should be used in this case.

Showing that interest in this version was high, detailed technical data were given (see table).

2C – Avions Marcel Dassault

Dassault, as a private venture, without an official contract, studied two aircraft-type launchers.

In a report to the Journées Techniques [Technical Days] of Eurospace (where Nord-Aviation also presented their views on the Transporteur Aérospatial), Henri Deplante and Pierre Perrier of Générale Aéronautique Marcel Dassault (GAMD) gave their opinions on the TAS.

Michel Rigault, who later studied the TAS archives in preparation of his work on STAR-H (see Chapter 12) pointed out, '*The two authors of this presentation were the director-general in charge of technical design, Henri Deplante (he had been at this position since foundation of the company, just after the Second World War) and the head of aerodynamic studies, Pierre Perrier (who went on to develop all the software pertaining to aerodynamics*

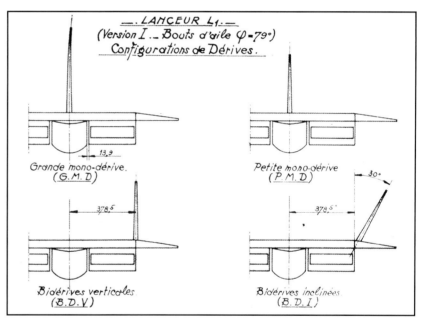

ABOVE Here, the exterior wings are still swept at 79°, but different styles of fins have been studied: single large fin; two smaller vertical fins; two smaller reclined fins. ONERA, author's collection

until his retirement). Perrier's later rôle on Hermes was critical: he was the one person who selected its aerodynamic configuration in 1985.'[29]

In the TAS presentation they explained why they had opted for a horizontal take-off, aircraft-type first stage. '*We think horizontal take-off is more appealing than vertical take-off for many reasons:*

■ *The obvious interest of being able to use any airfield. There is no more any need for a special launch site.*

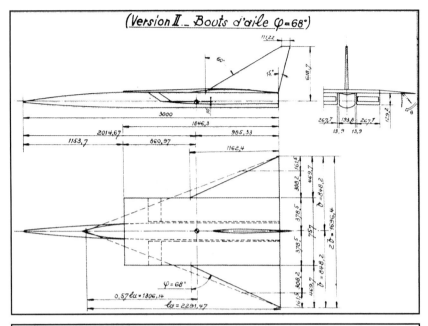

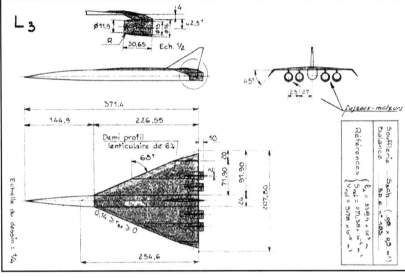

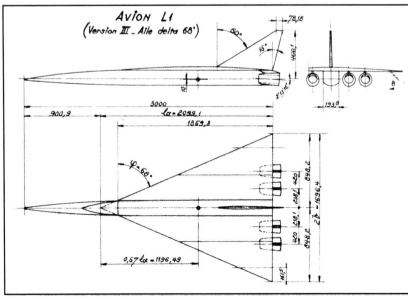

ABOVE This illustration is rather poor, but is the best available of the test model with its downturned wingtips. *ONERA, author's collection*

LEFT Two months later, in December, a new version, with the wings swept at 68°, was tested. The extremities of the wing could be folded down. *ONERA, author's collection*

- A better accessibility for air- and ground crew.
- The ability to control and stabilise the vehicle No1 [the carrier] with aerodynamic controls, taking off occurring at about 220 knots'.

Dassault's design took as its hypothesis that the role of the Transporteur Aérospatial was to transfer a load of 2,205lb (1,000kg) to a space station orbiting at 200 miles (320km). For this goal a composite design using two different vehicles was conceived:

Vehicle 1 was a large delta-winged aircraft, manned and fully recoverable. It was to be powered by turbo-rockets with the addition of liquid-fuel rocket engines.

Vehicle 2 was described as *'a long cigar...of a shape resembling the Falcon missile,'* carried semi-recessed in the belly of Vehicle 1. It was fitted with short wings to be *'totally or partially'* recoverable. This last sentence could be explained by the fact that 'Vehicle 2' was actually a smaller, manned spaceplane which was boosted into orbit by a consumable rocket. Landing was to be assisted by two sustaining engines at both extremities of the vehicle, or by variable-geometry wings. All the documents available describe a variable-geometry wing arrangements; nothing in them implies lift engines.

ABOVE MIDDLE In February 1969, a new model was to be tested with podded engines. *ONERA, author's collection*

LEFT June 1969: a further new model was to be tested with podded engines. *ONERA, author's collection*

CHAPTER FOUR

GOING INTO SPACE ON A SET OF WINGS

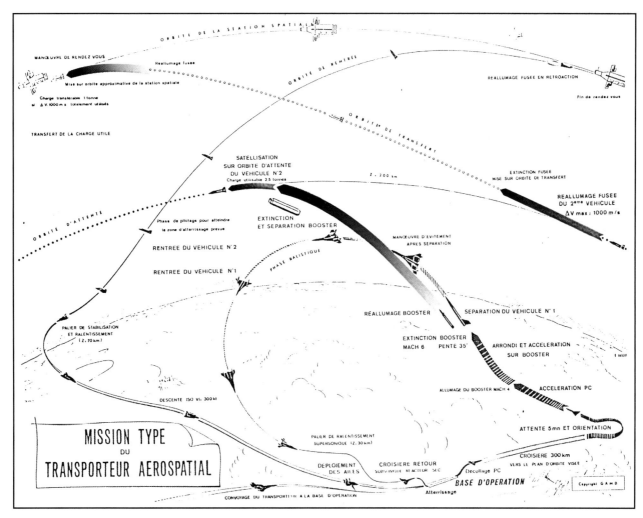

ABOVE Typical mission of the Dassault Aerospace Transporter. It takes off from its operating base, accelerates and climbs. The Vehicle 2 (orbiter) separates from Vehicle 1 (launcher), the booster thrusts the orbiter into a 'waiting' orbit, the booster then separates from the orbiter. On this lower orbit, a payload of 5,512lb (2,500kg) could be carried. The orbiter can then activate its rocket engine to reach a higher orbit to rendezvous with a space station. When the orbiter detaches from the space station it follows a re-entry orbit, slows down until at is back to an altitude of 8,850ft (2,700m). At this point, it converts to aerodynamic flight, deploying its wings only when nearing its landing base. The take-off and landing base for both vehicles is shown to be the same. *Dassault Aviation*

RIGHT Illustration of the orbiter's range. The launch base is in Europe—but not necessarily in France. The map is unclear on this point. *Dassault Aviation*

BELOW Orbits and trajectories of the Aerospace Transporter. The ferry trajectory appears to arrive at the Kourou base (which was not yet built at the time). Again, the lack of precise national reference is noticeable on this map. *Dassault Aviation*

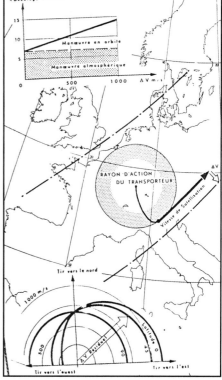

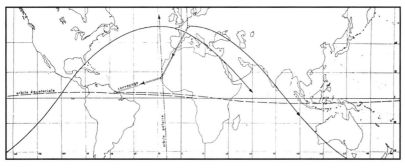

The 're-entry and landing system'

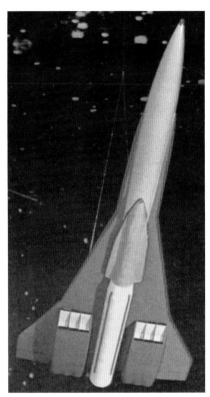

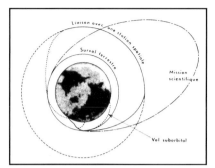

ABOVE Schematic of the types of missions which could be covered with the Transporteur Aérospatial, with the associated orbit. Legend: Vol suborbital = suborbital flight; Survol terrestre = round-the-globe overflight; Liaison avec une station spatiale = Shuttling with a space station; Mission scientifique = Science mission. *Dassault Aviation*

LEFT A model, illustrating the smaller, '150-tonne' composite. In grey the first stage (aircraft) launcher; in yellow the consumable booster; and in orange the recoverable orbiter. *Author's collection*

Dassault orbiter	
Status	Project
Span	–
Length	–
Gross Wing area	–
Gross weight	18,739lb (8 500kg)
Engine	1 × ? lift engines?
Maximum speed/height	–
Load	2,205lb (1,000kg) payload to a 186mi (300km) orbit

Dassault partially recoverable (150t TAS) (launcher)	
Status	Project
Span	72ft 2in (22.00m)
Length	170ft 7in (52.00m)
Gross Wing area	–
Gross weight	235,895 lb (107,000kg)
Engine	6 × ?
Maximum speed/height	–
Load	1 × 24,251lb (11,000kg) fully recoverable spaceplane + 70,107lb (31,800kg) booster

BELOW The model representing the '150-tonne' composite, here depicted in realistic colours and markings. The model is 'flying' across the Earth's atmosphere, climbing to separation altitude. *Dassault Aviation*

Dassault Transporteur Aerospatial Weight Table

Weight (expressed in lb/kg)	Partially recoverable orbiter			Fully recoverable orbiter	
	Launcher	Orbiter	Booster	Launcher	Orbiter
Airframe: wings	29,321/13,300	1,764/800			
Airframe: fuselage	24,912/11,300	4,189/1,900			
Airframe: fins	3,307/1,500	-			
Airframe: engine pods	4,409/2,000	-			
Airframe: landing gear	6,173/2,800	220/100			
Airframe: thermal protection	-	220/100			
Airframe: total weight	68,123/30,900	6,393/2,900	5,291/2,400	109,129/49,500	17,857/8,100
Engines	30,424/13,800	661/300	882/400	46,297/21,000	992/450
Equipment: controls	2,425/1,100	441/200	-	19,842/9,000	3,197/1,450
Equipment: fuel (?)	5,071/2,300	-			
Equipment: support systems	3,527/1,600	882/400			
Equipment: guidance	3,307/1,500	661/300			
Equipment: crew/equipment	1,323/600	882/400			
Empty weight	114,200/51,800	9,921/4,500	6,173/2,800	175,268/79,500	22,046/10,000
Kerosene	66,138/30,000	-	-	-	-
LOX	8,598/3,900	12,125/5,500	54,234/24,600	181,881/82,500	128,970/58,500
LH$_2$	47,400/21,500	2,205/1,000	9,700/4,400	-	-
Total	236,335/107,200	24,251/11,000	70,107/31,800	357,149/162,000	151,016/68,500

was detailed for the full recovery version: wings, controls, landing skids and lift engines with their fuel. One gets the impression that the difference between the two variants is that only the second, fully recoverable version, gets to receive lift engines. In other material, and in discussion with Michel Rigault, it appears that the lift engines were not actually intended for vertical landing but to brake the final flight sequence just before landing, kill speed and thus shorten the landing run.

For Vehicle 1, H$_2$/O$_2$ + Kerosene were used as fuel.

Like the other manufacturers, Dassault expected some trouble with the take-off gear and thought of replacing the undercarriage by a 'Hovercraft' system: *'a launcher fitted with an air cushion could work using a pressure of 4.27lb/sq in (300gr/cm²) on any flat ground (no running track necessary). The energy of heated gas generated by small turbojets added to fresh air in trompes [trunks] could supply the pressure to the air cushion'.* Anyway, conventional gear would be provided for landing.

Separation was studied, based on three possible configurations:

i Vehicle 2 attached to the nose of Vehicle 1

ii Vehicle 2 in a semi-recessed position on the back of Vehicle 1

iii Vehicle 2 in a semi-recessed position in the belly of Vehicle 1

The conclusions were that:

■ The front position generated *'inextricable problems'*. The centre of gravity of the two vehicles connected together does not coincide with the position of the landing gear of Vehicle 1, and resonance vibrations are likely to appear during the take-off run.

■ The dorsal position is not as good as the ventral position, which allows for a separation with positive G accelerations of which Dassault gained experience from launching the French nuclear bomb from Mirage IVAs. The separation of the X-15 from the B-52 mothership was also given in example.

Dassault fully recoverable (230t TAS) (launcher)

Status	Project
Span	-
Length	-
Gross Wing area	-
Gross weight	357,150lb (162,000kg)
Engine	6 × ?
Maximum speed/height	-
Load	1 × 149,915lb (68,000kg) fully recoverable spaceplane

■ In both dorsal and ventral positions, the displacement of the centre of gravity caused by the separation is minimal; the frontal area of the composite is again not much different in the composite and in the Vehicle 1 solo; and finally, accessibility for the maintenance crew is good.

Michel Rigault added, *'One should not forget the general context of the TAS study: it was done by a company which had previously developed and flown a great number of supersonic machines, either as prototypes or production aircraft. They gave Dassault an expertise in various architectures: swept wings, delta wings, variable geometry, vertical take-off and landing. Dassault had also acquired experience in hypersonic flight, re-entry trajectories and guidance with the MD-620 missile. High temperatures had to be dealt with in both the MD-620 and the Mirage IV supersonic cruise flight sequence. Dropping a load while at supersonic speed had been validated with the Mirage IV. Other features of the TAS came from technologies which were studied at the time and which, it was hoped, would soon be mastered, like variable-geometry and vertical thrusters to assist the landing.'*[30]

Dassault elected to have the separation occur at Mach 4.5 at an altitude of 34 miles (55km), an estimated 8 minutes after take-off.

For fuelling the engines of Vehicle 2, Dassault designers selected the combination H_2/O_2 on account of its excellent specific impulse (the specific impulse is a measure of how effectively an engine, usually a rocket, uses fuel and carburant; basically, it is the amount of thrust generated divided by the propellant mass flow rate or weight flow rate) even though it took up more volume than other propellants. They thought that it was worth working with SEPR to improve the specific impulse of those liquids, noting that an increase of 5% in specific impulse would be enough to reduce the total take-off weight of the composite from 507,060 to 308,650lb (230,000 to 140,000kg).

On Vehicle 1, Dassault choose to use the 'turbo-rocket' because it had a low weight (which they considered a requirement for the balancing of the weights), a good specific thrust at supersonic speed while being acceptable from 0 to Mach 1.5, and a good fuel efficiency. These preoccupations had also been those of Nord and Sud-Aviation but the CPE had not retained the turbo-rocket on the basis of its high fuel consumption during the landing phase.

Like Sud-Aviation, Dassault noted that full-rocket propulsion would result in a considerable increase of the take-off weight, estimated at 165,350lb (75,000kg). For their turbo-rockets, Dassault expected to use liquid oxygen (LO_2) and kerosene, indicating liquid hydrogen (LH_2) coupled with liquid oxygen (LO_2) would be more economical but that they had no data relating to their use in this kind of turbo-machines.

Finally Dassault studied the question of re-entry. Regarding Vehicle 1, the heat was considered easily manageable with ablative material in a few hotspots: 3mm of a material called AVCOAT (AVCOAT 5026-39/HC-G was used in the heat shields of the Apollo modules, still very innovative at the time…and untested in 1964) protecting light alloys was considered sufficient for the leading edges of the wings. The question of the engine nacelles was glossed over quickly. That is surprising, as it was considered a significant roadblock by CPE—four years later. Two of the turbo-rockets were to be shut down after separation, the other two powering Vehicle 1 until landing which *'will be easy because of the low wing loading.'*

Regarding Vehicle 2, the favoured design was a *'long fuselage with short wings,'* lateral stability being achieved by large 'crocodile' flaps at the rear of the fuselage. Those flaps were to be used both as air brakes and as elevators, lateral control being achieved through elevons.

Final braking before landing would use lift engines, one at each extremity of the vehicle which should reduce the landing speed to under 100 knots (115mph [185km/h]) and position the flight axis nearly parallel to the ground. It seems the said engines would be cut when speed was reduced to under 100 knots while aerodynamic controls would still be active due to the speed. *'Another option would be to extend variable geometry wings when the speed has dropped under Mach 1.'* This appears to have been the option finally selected by Dassault.

These early conceptions were later refined in a brochure describing two different aircraft:

Dassault consumable interstage	
Status	Project
Span	–
Length	–
Gross Wing area	–
Gross weight	70,107lb (31,800kg)
Engine	1 × rocket engine
Maximum speed/height	–
Load	Spaceplane

BELOW Trick photography depicting separation of the aircraft mothership and the orbiter at very high altitude, viewed from above. *Dassault Aviation*

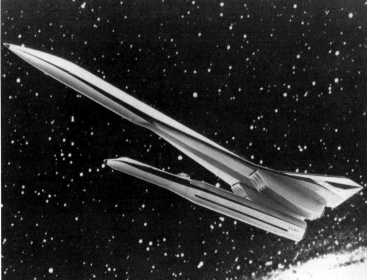

BELOW Side view of the separation of the upper aircraft and the lower orbiter. Dassault was confident separation would not be a problem, having accumulated experience with supersonic drops from the Mirage IV bomber. *Dassault Aviation*

ABOVE A beautiful piece of art depicting separation of the booster and orbiter from the launcher. This is a representation of the '150-tonne' composite. As in the model, the colours are symbolic. *Dassault Aviation*

- A composite, fully recoverable, with two separate vehicles, corresponding more-or-less to the machine described above in 1964.
- A second design, in which the smaller orbiter was boosted by a 70,107lb (31,800kg) rocket intermediate stage.

More specific details were given regarding the orbiter: '*The crew is of two men, who are positioned at the front, on seats which can be ejected through the launcher*' (no indication was given as how this feat could be accomplished).

'*The equipment case follows the crew compartment – which can be redefined as a rescue pod with a reduced weight penalty – and precede some of the LH$_2$ and LO$_2$ tanks, the cargo bay positioned at the centre of gravity, followed by more tanks and the engines. The airframe is a cylinder in the tank/cargo bay area. It is fitted with two extremely low aspect ratio wings to give the vehicle enough manoeuvrability during the re-entry phase. For landing, a variable-geometry, high-lift surface gives the orbiter a lift-to-drag ratio similar to a standard aeroplane (about 5) during the approach to the landing field. To facilitate this delicate manoeuvre, we provide for a small turbojet with enough fuel to enable the pilot to fly around the landing track. We estimate this incurs a weight penalty of between 220 and 441lb (100 and 200kg).*'

For propulsion, the Dassault engineers were obviously in a quandary: they rejected quickly the full-ramjet solution but stated that '*turbojets in association with ramjets or deviated flow reheat were probably the best solution.*' But they also considered that '*the turbo-rocket [...] is an interesting solution. This solution is*

BELOW Now the model orbiter separates from its booster against a starry background. *Dassault Aviation*

BELOW Three-view drawing of the '150-tonne' composite. *Dassault Aviation*

FRENCH SECRET PROJECTS: FRENCH AND EUROPEAN SPACEPLANE DESIGNS 1964-1994

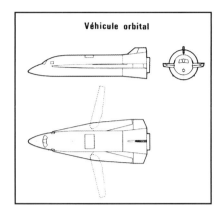

ABOVE Three-view drawing of the orbiter with wings extended (in doted lines). *Dassault Aviation*

ABOVE The nose of the orbiter was extrapolated from the nosecap of the MD-620 missile, seen here in a Dassault assembly hall. The MD-620 two-stage missile was first flown in March 1966. *Dassault Aviation*

BELOW Another influence on the Dassault TAS was the Mirage IV (here the -01). Experience on launching the A-bomb container at supersonic speed was used to select the belly attachment of the second stage. *Dassault Aviation*

BELOW While the orbiter goes into orbit, the mothership returns to its base. Dassault's promotional offices have always produced very realistic artist's impressions, sometimes evoking the best Gerry Anderson shows. *Dassault Aviation*

BELOW The orbiter floats into space. The 'galactic' background looks very sci-fi-ish but was probably required to separate easily the 'space' settings (for the orbiter) from the 'atmospheric' settings (for the mothership plane). *Dassault Aviation*

ABOVE A curious drawing published by *Aviation Magazine* in its issue No 423 of 15 July 1965. The cargo bay appears opened and the wings are extended—a rather unlikely occurrence. Besides, the wings appear to have reached an inverted sweep angle, which was not a configuration planned by Dassault. Aviation Magazine, *author's collection*

RIGHT With wings fully extended, the orbiter banks in the approach to its landing track. *Dassault Aviation*

RIGHT Model of the fully-recoverable orbiter separating from the '230-tonne' launcher. Actually four (gold-plated) models of the orbiter are used to depict the separation trajectory. *Author's collection*

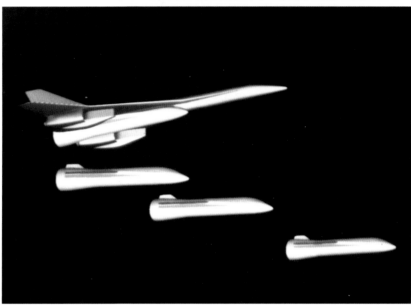

BELOW Three-view drawing of the '230-tonne' composite. The annotations to the right were made by Henri Lacaze while working at the Centre de Prospective et d'Evaluation. *H Lacaze collection*

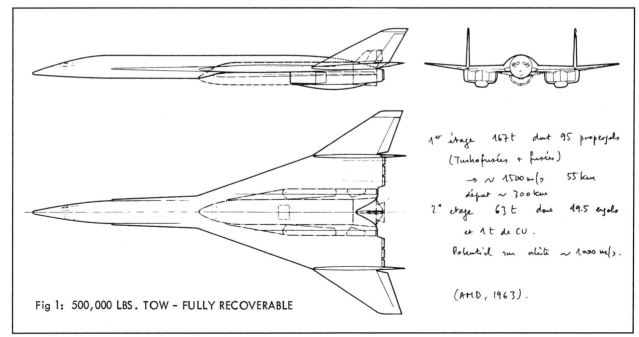

ABOVE For the 1987 Paris Air Show, the model of the '150-tonne' variant was dusted off and exhibited, stressing Dassault's long experience of space projects. Dassault had been allocated the design of the Hermes spaceplane late in 1985.
Author's collection

particularly interesting because it uses very lightweight engines, which facilitates the weight repartition in the first stage.'

They focused on a 147.6-tons (150-tonne) launcher and a partially-recoverable orbiter, but a larger (246-ton [250-tonne]) vehicle – in which the orbiter was fully recoverable – was also illustrated.

Presentation of the Dassault project in the CPE 1968 report does not completely agree with the description given by Dassault itself four years before and does not acknowledge the two variants proposed in the brochure. The engines of choice were again the turbo-ramjet and in the opinion of the CPE, *'it much resembles Mistral. The rocket is not stored in a bay but semi-recessed in the belly of the aircraft.'*

This configuration, with the rocket second stage transported outside the aircraft body, enabled an interesting flight plan in which the engines of the second stage were fired at Mach 4 and boosted the combination up to Mach 7, using fuel stored in the aircraft's body. During the 'fifties, SNECMA had designed Coléoptère-type missiles which engines were fed from tanks in the carrier aircraft and thus could act as booster during combat flight. The separation occurred at Mach 7.

The difference, which apparently did not strike CPE at the time, was that the third stage of the Dassault proposal was actually a manned, recoverable space glider and not a mere, generic 5.9-ton (6-tonne) satellite. Because of the similarity with Mistral, CPE had not much faith in the weight table given by Dassault and re-assessed it at 485,025lb (220,000kg) and not the 330,700lb (150,000kg) claimed by Dassault.

In summary, the Dassault projects appear somewhat visionary in their depiction of a manned orbiter with variable-geometry wings. Certainly the payload appeared much reduced (just 2,205lb [1,000kg]) where the other groups announced 13,228lb (6,000kg), but those 6 tonnes included the whole weight of the satellite including a would-be orbit-correcting engine, batteries and other support systems. To have fair comparison one should take into account, not the 2,205lb (1,000kg) announced by Dassault, but the weight of the complete in-orbit mass; that is, the 24,251lb (11,000kg) spaceplane. Then the Dassault projects were very much lacking in details. The type and number of engines to be used were still in debate; the thermal protection certainly received only cursory consideration (it was obviously a problem for the two other companies and they did not have to deal with the question of the re-entry); and dimensional data was not provided.

---

[24] *France and the Peenemünde Legacy* by Jacques Villain 43rd Congress of the International Astronautical Federation 1992 Washington DC.

[25] See *Du Gerfaut au Griffon: Conquêtes du Statoréacteur et de l'aile Delta en France* by André Turcat and Armand Jacquet testimony of Serge Kaplan; Philippe Ricco, *Avia*, Edition 2006.

[26] Interview with the author, 8 February 2019.

[27] Coralie was the name for the French-designed second stage of the Europa launcher, for which a association called Nord-Vernon was created. It regrouped Nord-Aviation and the Laboratoire de Recherches Ballistiques et Aérondynamiques de Vernon.

[28] See the author's *French Secret Projects Vol 2*, Chapter 6.

[29] Mail exchanges with the author, October 2020.

[30] Mail exchanges with the author, October 2020.

Chapter Five
French Hypersonics

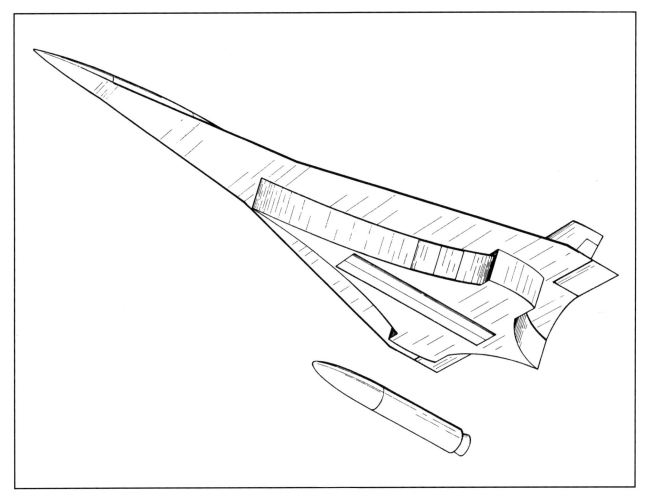

ABOVE Artist's impression of the LSM2 labelled as a 'classical layout' for a satellite launcher with air-breathing first stage. *Sud-Aviation via H Lacaze*

In France, the study of hypersonics began at the end of the 'fifties under different influences.

- Sud-Aviation, having broken (if unofficially for some) many records with its Trident rocket interceptor, hoped to push it into higher Mach numbers.
- Nord-Aviation, having developed the turbo-ramjet for its Griffon research aircraft, wanted to experiment with this new type of propulsion for all types of aircraft (see Chapter 3).
- From 1951 ONERA began researching very high speeds with experimental missiles and sounding rockets.

And in the background, there was the need to 'feed' the research required to develop missiles for the *Force de Dissuasion* (nuclear deterrent) and other military projects, a fact that the Centre de Prospective et d'Evaluation did not hide in the Transporteur Aerospatial report as seen in the previous chapter. Yes, studying hypersonics was not just for science and the beauty of man in space.

Sud-Aviation's Bureau des Projets à Long Terme (Long-Term Projects Research Office) agreed with this military orientation, pointing out in a 1968 report (unsigned) on propulsion systems for high-speed aircraft that, *'The turbo-ramjet could be used for an aircraft cruising at Mach 5, accelerating up to this speed having a low fuel cost. If the selected criterion is not the mass used up, but the take-off mass, then it is the*

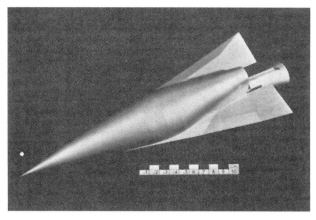
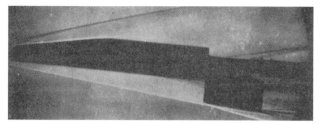

ABOVE View of a similar (possibly the same) model being flown at Mach 7 in the R2 wind tunnel at Meudon. *ONERA*

LEFT Besides its work on ramjet engines, ONERA also integrated ramjets and airframes. Here is a picture of a wind tunnel model intended to represent a ramjet-powered vehicle flying at Mach 7. *ONERA*

rocket-plus-ramjet which is the better option. However, while cruising at Mach 5 is again possible, it is the return leg which is an issue. The fuel consumption during this phase may curtail the cruise time. For this reason it appears particularly appropriate for an aircraft requiring speed bursts like an interceptor. Besides, this combination appears easier to design and build than others.'

The Centre de Prospective et d'Evaluation was worried that the development of the SST would make the Mach 2 interceptor ineffective. So it asked Dassault to study the feasibility of a Mach 3 *Police de l'Air*

LEFT Another model 'flown' by ONERA in wind tunnels. This design, which was tested in the early 'sixties cannot be associated with any research programme. *Author's collection*

BELOW This large model dominated the ONERA booth at the 1991 Paris Air Show. Although it probably refers to the various SSTOs being designed by Aerospatiale at that time, the author has not been able to associate it firmly with a specific project. *JC Carbonel*

(Air Police) system. When Concorde and the Tu-144 turned out to be the only SSTs to fly, the venture was not further pursued. This programme had a target speed of Mach 3 and a specific military objective, contrary to Nord- and Sud-Aviation research which had more open objectives. It has been covered in the first volume of *French Secret Projects,* so its details will not be repeated here.

1 Hypersonic Research at ONERA: Re-entry Studies

Very little has been released regarding studies carried out by ONERA about orbital and re-entry bodies, possibly because they were tied-in with the design of nuclear warheads for the *Force de Dissuasion*. Photographs of models of hypersonic vehicles were released, but they were not associated with written documentation.

A model, not unlike a classic 'Flash Gordon' winged rocket, attached to a bulky engine fairing, was tested at Mach 7 in a wind tunnel as revealed in a 1970 report by René Cérésuela.

2 Hypersonic Research at Sud-Aviation: Beyond the Trident

The **SO 9050 Trident III** had been a successful (if inconclusive) design which narrowly missed being put into production. The SO 9050-07, which was to be the prototype, was built (as were the -08, -09 and -10) but never flown. The Trident III would have been the operational variant of the SO 9050-SE Trident II. The Trident III programme was cancelled on 20 June 1958, as part of budgetary cuts. The **Trident IV** had been in development since late 1957 and Lucien Servanty, head of the SNCASO Bureau d'Etudes, was very eager to go on with this family of aircraft which had gained him the status of expert in high-speed aircraft at a time when this was the ultimate of aviation design.

In his book *Le Dictionnaire Fanatique du Trident,* Paul Gauge, himself a former Trident engineer, quoted a late 1957 meeting. *'M Vinsonneau asks that detailed planning be done for [the Trident IV], including operational research (notably the complete weapon system) associated with this project in our four-year programme. M Servanty presents the studies being presently carried out to solve the problems related to heat, using new solutions (sandwich-type materials with an outer skin in thin steel sheet covered by thin protective layer). Studies using wind tunnel tests are being initiated to explore, heat transfer. M Servanty reminds us that the Trident IV is associated with the SALP [Sol Air Longue Portée: Long Range Surface-to-Air missile; under this acronym Lucien Servanty probably referred to a remotely-controlled variant of the Trident]. Six thousand work-hours per month are planned for those two projects until mid-1958. The idea is to get a contract from the State for these projects.'*

This project culminated during the summer of 1958, with a design called 'IM 3' or **Interceptor Mach 3** which appears to have been designed in two versions. Both used a large, unspecified turbojet for its main propulsion and one had rocket engines to boost the combat speed to Mach 3, while the other (seemingly favoured by Servanty) used ramjets for the same result. Only a three-view drawing of the machine, with one missile under each wing, survives today. More material has been preserved regarding the 'EM 3,'

BELOW The SO 9050 with its SEPR rocket engine running. Although fitted with a missile mockup, this aircraft was still a prototype. *Author's collection*

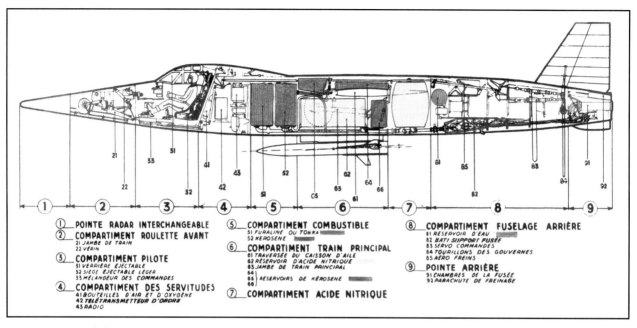

ABOVE Interior of the SO 9050 Trident II SE. Legend: 1 Pointe radar interchangeable = Radar nose; 2 Compartiment roulette avant = Nose gear bay; 3 Compartiment pilote = Pilot's compartment; 4 Compartiment des servitudes = Support equipment; 5 Compartiment combustible = Fuel tank; 6 Compartiment train principal = Main landing gear bay; 7 Compartiement acide nitrique = Nitric acid tank; 8 Compartiment fuselage arrière = Rear fuselage; 9 Pointe arrière = Rear cone.
Sud Advanced Hypersonics via Paul Gauge

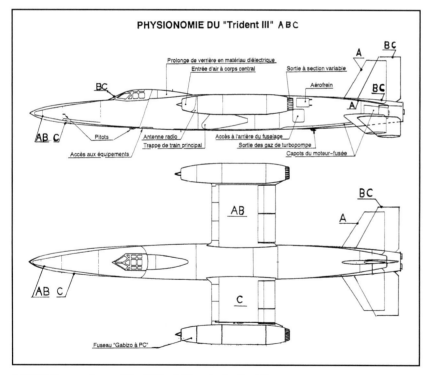

ABOVE Comparison of the Trident III variants. *Paul Gauge*

Sud SO 9050 Trident III	
Status	Built
Span	21ft 10in (6.95m)
Length	43ft 6in (13.26m)
Gross Wing area	155.9sq ft (14.50 sq m)
Gross weight	13,007lb (5,900kg)
Engine	2 × Turbomeca Gabizo 3,375lb (15kN) and two SEPR 63 3,375lb (15kN)
Maximum speed/height	Mach 2 at 68,900ft (21,000m)
Load	1 × missile MATRA R511

3 Hypersonic Research at Sud-Aviation: Beyond the Transporteur Aérospatial

Hypersonic studies resumed at Sud-aviation in 1965 when its Bureau des Plans à Long-Terme (Long-Term Projects office) received a contract for air-breathing launchers powered by ramjets with supersonic combustion as part of the research on the Transporteur Aérospatial.

Writing in 1968, the CPE, which had initiated the research, concluded, '*At that time, it was a quite futuristic study, as the actual ability of a supersonic combustion ramjet (as described by Professor Ferri[31]) to function was not*

the **Experimental Mach 3**, which used the same airframe as the IM 3 for flight-testing the formula. One can see here the Servanty touch—the same method had been used on the Trident, with three successive iterations (SO 9000, SO 9050, SO 9050 SE) to go from the technology demonstrator to operational testing.

At this juncture, Servanty was appointed to lead the Supersonic Super-Caravelle project and further development of the Trident faltered.

Sud EM 3	
Status	Project
Span	29ft 6in (9.00m)
Length	59ft 1in (18.00m)
Gross Wing area	312sq ft (29.00sq m)
Gross weight	24,059lb (10,913kg)
Engine	Super ATAR and two ramjets
Maximum speed/height	Mach 3
Load	?

proved.' However CPE wanted to know if such launchers, flying up to Mach 12 or even to orbiting velocity, would have better performance than conventional air-breathing, recoverable launchers.

As Mach 12 appeared at first the upper speed at which ramjet could work, this was set as the separation speed. The objective was, as in the earlier studies, to put a satellite weighing 13,228lb (6,000kg) into a 186-mile (300-km) orbit.

Sud-Aviation answered with brochures PLT 192, 204 and 222. The first designs were two-stage composites but to simplify the study, single-stage designs were also researched.

The study pointed out that above Mach 8, it was difficult for a subsonic ramjet to work efficiently (the 'subsonic' and 'supersonic' qualifiers refer to the speed of the air inside the combustion chamber of the ramjet). Another conclusion was that at Mach 12, the airflow inside the ramjet could be increased by 1.75 if the intake were to be located under the wing because of the 'awning' effect…but the counterpart of this was an increase in drag and the very high temperature reached by the leading edge of the 'awning.'

With help from ONERA, the length of the combustion chamber was computed to be around one metre but the report also added that more research was required. A difficult question raised by this study was the ability to build a combustion chamber which could resist the intense heat expected (between 2,000 and 3,000°K [1,726 and 2,726°C] while being subjected to an airflow of Mach 1.5 for approximately 10 minutes. The usual cooling agents such as kerosene were expected to be insufficient. And

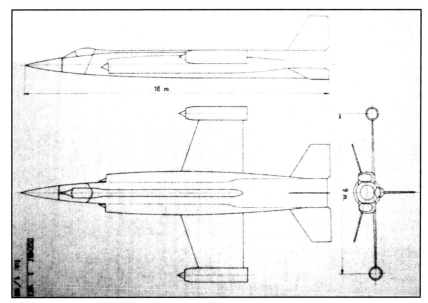

ABOVE The 'Trident IV'—actually the 'EM3' Experimental, Mach 3. *Paul Gauge*
BELOW Illustration of the 'LSB' 'Two-stage Launcher with Side-Opening Intakes.' The upper view represents the two stages docked together, seen from the side. The middle drawing is of the two stages docked together viewed from above; the lower drawing is the first stage alone. *Sud-Aviation via H Lacaze*

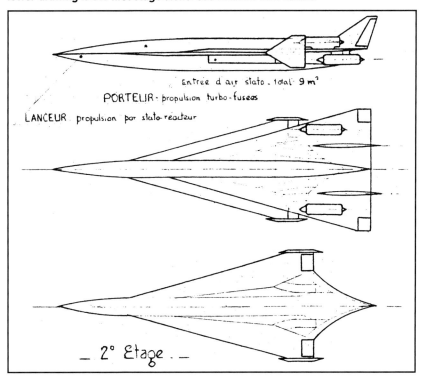

naturally there were the difficulties inherent in the design of any ramjet-like fuel injectors and the lighting device.

The question of the bi-dimensional exhaust nozzle was also identified as requiring further studies.

ONERA examined the possibility of designing a ramjet able to work in subsonic and supersonic modes. In such a ramjet, the combustion chamber would begin with a divergent section, followed by a constant section opening into the nozzle through a part with a slightly reduced section. Combustion was to begin in the constant-section at low Mach and then move forward

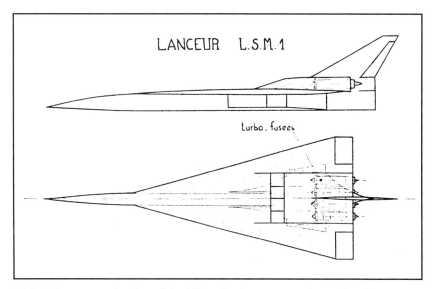

ABOVE Two-aspect drawing of the LSM1 'Single-stage Launcher with Side-Opening Intakes' from report PLT192. *Sud-Aviation via H Lacaze*

BELOW And now, LSM2 from report PLT204. The arrow indicates the position of the turbo-rocket engines. *Sud-Aviation via H Lacaze*

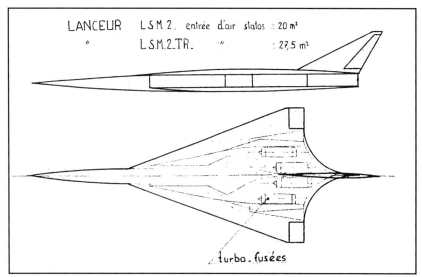

Sud LSB (PLT122)

Status	Project
Span	75ft (23m) (approx)
Length	213ft (65m) (approx)
Gross Wing area	-
Gross weight	-
Engine	4 × turbo-rockets (1st stage), supersonic ramjet (2nd stage)
Maximum speed/height	Mach 5 (1st stage), Mach 12 (2nd stage)
Load	Rocket 2nd stage. 5.9 tons (6 tonnes) on orbit?

Sud LSM1 (PLT192)

Status	Project
Span	88ft (27m) (approx)
Length	265ft (81m) (approx)
Gross Wing area	-
Gross weight	-
Engine	4 × turbo-rockets + supersonic ramjet
Maximum speed/height	Mach 9.7
Load	Rocket 2nd stage 5.9 tons (6 on orbit?)

towards the divergent section as speed increased. But that was just simulated aerodynamics and a lot of work was required to build the real thing.

Therefore, some of the PLT scientists were questioning the opportunity to study a Mach 8 ramjet when the same performance could be obtained through an easier-to-build rocket engine.

And that was not the only difficult question. The thermal protection would be very difficult to build: PLT recommended forming the skin of the launcher using a base structure in René 41 alloy, covered by a 0.47 in (12 mm) thick coat of low density insulating material – a product called 'Dynaquartz' was suggested – with an outer skin of niobium alloy protected from oxidation by a coat of Silicure. But this was admittedly more conceptual than real, as the lifespan of the niobium alloy in a hot, highly-oxidising atmosphere was uncertain, while it was expect that keeping those materials together would be difficult, considering their different dilatation ratio. And that was for the main body of the machine which was not expected to go above 1,200°K (926°C). The leading edges were expected to reach even higher temperatures, the greatest by the ramjet walls, where active cooling was expected to be required to limit the temperature to 1,800°K (1,526°C) and still use quite exotic material (porous zircon ceramics reinforced by tantalum nida[32]).

Propellant was considered for the role of active cooler: kerosene was removed from the options because not only was its ability to support restricted to heating to up to 700°K (426°C) but 191,800lb (87,000kg) of it would be required as coolant while only 143,300lb (65,000kg) was needed to feed the engines. Hydrogen was also out of question because of its bulk. In the end, liquid methane appeared to be the favoured product as it was not too bulky, and its cooling effect was good and expected to be over 2.6×10^6 kilojoules/tonnes. PLT estimated that 35,274lb (16,000kg) would be required for cooling, which was less than needed to feed the supersonic ramjets.

CPE was not satisfied by the study. There was too much imprecision; too many technological issues; many dubious options (like the choice of a straight leading edge for the 'awnings' above the intakes); and some probable underestimation of the overall weight of the designs.

RIGHT LSM3 from report PLT222 was not indicated as such but labelled 'Mach 12 Launcher.' Using four turbo-ramjets for propulsion it appears to have been a favoured design. If the image looks familiar, that is because the very same drawing was used for the proposal PLT 219 (which see), with a speed intended to be only Mach 7.
Sud-Aviation via H Lacaze

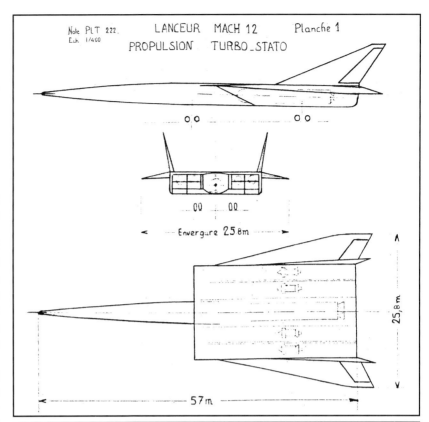

'To sum it up, using an air-breathing propulsion system on the acceleration vehicle, instead of a liquid H_2/O_2 rocket, above Mach 7 generates very important structural issues which are not compensated by an obvious performance improvement.' And the conclusion was definitive, '*Launchers with supersonic ramjets (for speeds above Mach 7) have poor performance; are very complex to design; and are of dubious feasibility. Above Mach 5, it appears a better option to use rockets, as suggested by Société des Avions Marcel Dassault, than air-breathing engines. Naturally this is applicable only to boosters (launchers); if a hypersonic cruise at high Mach was required, conclusions may be different.*'

PLT had proposed five different architectures:

- **Two-stage launcher with side-opening intakes** (called 'LSB' and presented in the PLT 185, 191 and 222 brochures). The first stage was propelled by a turbo-rocket up to Mach 5 at which point the second stage separated. This second stage, propelled by a supersonic ramjet was deemed to have a sufficiently low drag to be accelerated up to Mach 12. Two different variants of this second stage were studied, differing by the type of fuel used: kerosene (launcher LSBK) and hydrogen (launcher LSBH). Calculation showed that LSBH could reach a speed of Mach 16.

LSB was illustrated in the **PLT 222** report. No dimensional data were

RIGHT A compilation of other concepts illustrated at the end of PLT222. Only the lower two designs received a more detailed study. *Sud-Aviation via H Lacaze*

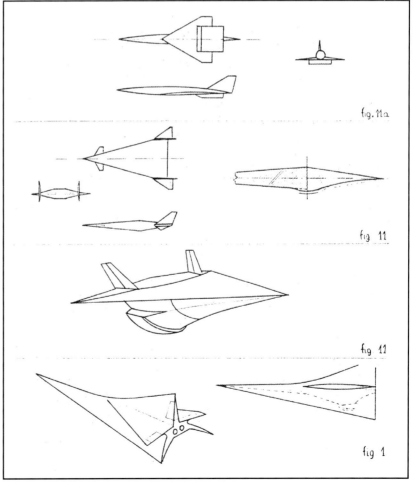

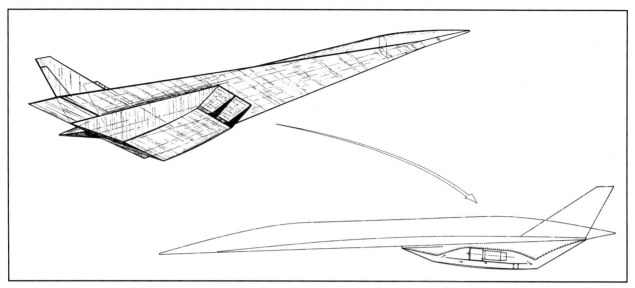

ABOVE Undated drawing from the same archive as the PLT reports. This appears to have been a 'concept' quickly drawn to illustrate a hypersonic transport (read 'Super Concorde') using the same technology as the space launchers.
Sud-Aviation via H Lacaze

indicated. Measurements taken on the report illustration in 1/500 scale give an approximate span of 75ft (23m), for a length of 213ft (65m).

■ **Single-stage launcher with side-opening intakes** (LSM1 illustrated in **PLT192**). As originally designed, there were two separate intakes, each closed flush by a movable panel when not in use. The primary propulsion was turbo-rockets in a pack of four above the rear fuselage while the secondary propulsion was a supersonic ramjet inside the fuselage body. During the course of the study, the surface of the side intakes was increased from 215 to 248sq ft (20 to 23sq m); the top speed increasing at the same time from Mach 9 to Mach 9.7.

Again, no dimensional data were indicated. Measurements taken on the report illustration in 1/500 scale give an approximate span of 88ft (27m), for a length of 266ft (81m).

BELOW Captioned upside-down, this is an 'artist's conception' of a Mach 10-cruise aircraft. The original document from which this drawing is extracted is unfortunately not available. Note, however, that a somewhat similar design with a 'ring'-type belly ramjet appears among the 'other concepts' illustrated in PLT222. *Sud-aviation via H Lacaze*

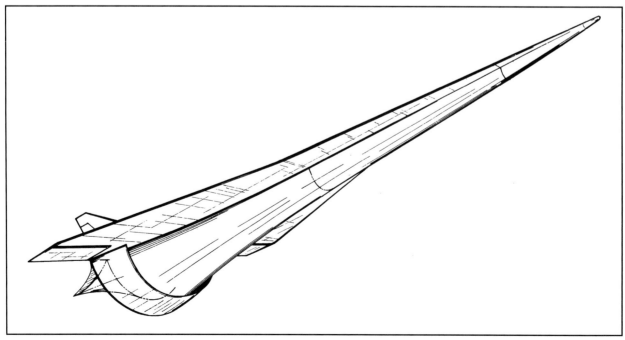

RIGHT **Artist's impression of the last of the 'other concepts' presented in PLT222. This was among the designs flatly rejected by Centre de Prospective et d'Evaluations because it did not allow for separation by dropping down the second stage.** *Sud-Aviation via H Lacaze*

Sud LSM2	
Status	Project
Span	–
Length	–
Gross Wing area	–
Gross weight	–
Engine	4 × turbo-rockets + supersonic ramjet
Maximum speed/height	Mach 12
Load	Rocket 2nd stage 5.9 tons (6 on orbit?)

Sud LSM3	
Status	Project
Span	82ft 0in (25.00m)
Length	187ft 0in (57.00 m)
Gross Wing area	–
Gross weight	374,775lb (170,000kg)
Engine	4 × turbo-ramjets
Maximum speed/height	Mach 12
Load	Rocket 2nd stage 5.9 tons (6 on orbit?)

A second iteration of the same concept (**LSM2**) was designed. A larger, 269sq ft (25sq m) intake, was used for both propulsion systems, resulting in less drag being generated by the intake. The turbo-rockets were relocated inside the fuselage. PLT hoped to reach Mach 12 with this design. No scale was indicated on the drawing.

A third iteration – the single stage launcher with belly intake (**LSM3**) – was offered. It was similar to the preceding except that turbo-ramjets instead of turbo-rockets were used up to Mach 7. Span was to be 84ft 8in (25.80m) for a length of 187ft 0in (57.00m)

Other shapes were proposed: designs with spindle-shaped engine pods and designs with '*propulsion d'intrados*' (the belly of the fuselage is the engine; the combustion being provoked outside the airframe). These last concepts were rejected by CPE because they did not allow a downward separation of the satellite from the launcher. Henri Lacaze, who was at CPE at that time explained to the author in 2019, '*The question of the separation was never studied in detail. We did not expect problems here, as long as the separation was from the underside: the second stage would be air-dropped like a bomb…and here we had a lot of knowledge about this question following the in depth work done for the separation of the nuclear weapon from the Mirage IVA.*'[33]

CPE flatly rejected the 'Mach 12 launchers' on account of the take-off weight reached by those machines, estimated to be above 771,625lb (350,000kg), which questioned their feasibility. Here CPE was probably too pessimistic as the Lockheed C-5 Galaxy, flown in 1968, could weigh up to 992,075lb (450,000kg).

Fig. 5(b) Lanceur de satellite à premier étage atmospherique, configuration évoluée
Satellite launcher with airbreathing first stage – advanced layout

---

[31] Antonio Ferri was an Italian aerodynamicist who, after the war, went to work at NACA Langley where he studied the questions of atmospheric re-entry and hypersonic propulsion. Questions which are germane to our subject, it is therefore quite logical CPE referred to him as part of the reflection on the Transporteur Aérospatial.
[32] 'Nida' is short for *nid d'abeille:* honeycomb structures.
[33] Interview with the author 8 February 2019.

Chapter Six
Getting into Hardware: Nord Véras

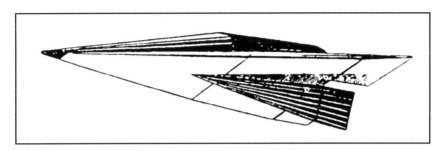

ABOVE Sketch of Nord VERAS project.
NASA

Véras: The First Full-Size Spaceplane Hardware Produced in France

Around 1961, the French Ministry of Defence initiated research into bodies for atmospheric re-entry. Obviously, the project had its origin in studies for the missiles that would equip the naval element of the *Force de Dissuasion* but shapes other than warhead cones were also investigated. Notably ONERA, under Philippe Poisson-Quinton (head of its department of aerodynamics) tested models of space gliders in the Cannes 10-foot (3-metre) subsonic wind tunnel. In a much later interview, Mr Poisson-Quinton reminisced that, '*one of the strangest configurations [we tested] was a magnificent model of a space glider with which we wanted to study the landing glide of such a vehicle. One of the modellers asked me why I had requested such a flat-iron shape. Actually it had about the same wing loading [of a real flat iron] but had to be able to support much higher temperatures. The whole modelling team thought I was crazy.*' Unfortunately the result of these tests – which might have been associated with design work done by aircraft manufacturers – has not been located by the author.

As pointed out by Mr Poisson-Quinton, heating was to be a major problem: re-entry was planned at around Mach 10, which generated surface heating, so excluding use of common aeronautical materials.

In 1964 (other sources give the date as 1965), Nord-Aviation was tasked by the Ministry with building an experimental re-entry vehicle which was called '*engin hypersonique susteňté expérimental Véras*' (Véras itself stood for '*Véhicule Experimental de Recherches Aérothermodynamiques et Structurales*': Experimental Vehicle for Structural and Aerothermodynamic Studies). The very name suggests the importance given to investigation of the heat problem.

Besides Nord-Aviation were Sud-Aviation (probably for the question of how to launch Véras), ONERA, Carbonne-Lorraine, INSA de Lyon (Institut National de Sciences Appliquées: National Institute of Applied Sciences, most probably for the materials engineering department), CEAT (Centre d'Essai Aérospatial de Toulouse), University of Poitiers (CEAT was attached to this university) and Pechniney.

Véras was to be only a demonstrator, unmanned and not reusable, but recoverable. The initial project was to build a static-test model of the design, which was not intended to fly. The shape selected was a '*highly-swept low delta wing and cylindro-conical fuselage. The wing was fitted with ventral fin and prolonged by two elevons.*' A ventral fin was added because during a hypersonic flight, it was expected that there would be a drop in pressure and airflow on the upper surfaces. The total weight was to be 3,307lb (1,500kg) of which 2,205lb (1,000kg) was payload (instrumentation). The flight trajectory would have been: launch by a rocket booster; horizontal flight at Mach 10 at 150,000ft (45,700m) for 500 seconds; then a gliding descent with an angle of attack corresponding to the maximum lift-to-drag ratio, during there would be a 100-second turn under a 1.65 load factor at a velocity equal to 0.85 times the steady level flight speed. The total flight duration would have been 1,640 seconds.

A major concern of the whole programme was the extreme temperatures which the vehicle had to resist during its flight. To minimise the thermal stresses and ensure sufficient stiffness, the structure had been designed on three-levels. The skin panels, which are divided into 'shingles' (in the roofing-tile sense) were attached by hinges and cover strips to an orthogonally shaped sub-structure which rested freely on the load-carrying structure by means of sliding attachments. Each member of the primary structure had a corrugated web directly welded to the flanges at the corrugation apices.

The model was built in the *atelier* (workshop) A14 at Châtillon and consisted of several parts which were to be tested separately. From the beginning, investigation of new materials and construction methods was a key objective of the project. A first element, comprising the wing and the main (rear) fuselage section, was designed and built.

The whole project turned out to be mostly an exercise in investigating the resistance of (exotic) materials. The structural frame was built of the Nickel alloy named René 41, while the coating used molybdenum alloy TZM (ultimately not retained) and Niobium alloy P333. Chemical giant, Pechiney was in charge of producing the alloy. Alloy P333 was based on nobium with additions of titanium, vanadium and zirconium. It was selected because it presented a good resistance to intense heat while able to be laminated into thin sheets and being acceptably receptive to soldering, thus making it easy to work with. Alloy was produced in ingots of 88lb (40kg). However it was found that the evenness of the sheets produced was unsatisfactory.

A heat testing facility implemented for Concorde structural ground testing was modified to deliver:

- Much higher heat flux at temperature levels (9 BTU sq ft/ sec, instead of .05 BTU/sq ft/ sec).
- Higher heating rates (2,000°F in 140 seconds instead of 150°C in 20 minutes).
- Smaller test article sizes (100sq ft [9.3sq m] instead of 12,000sq ft [1,115sq m].[34]

Two Infrared heaters were developed to simulate the in-flight temperature variations equivalent to 3,500kW. Prior to the mounting of the model on the rig, 74 thermocouples were connected to the wing and fuselage structures. The payload container wall was simulated

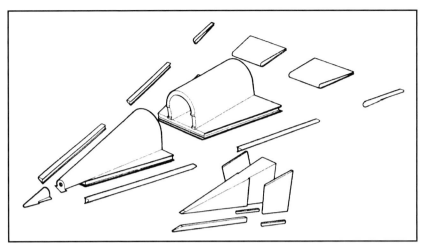

ABOVE Drawing illustrating components of the complete VERAS glider. ONERA

by an inorganic, heat-resistant felt-matting. The test specimen then received its load-introduction and heat-shield elements. With its equipment installed, the mockup was placed upside down in the test rig, under the lower surface heater, and attached to a dynamometric table. The upper surface heater with its supporting frame was positioned underneath the Véras. Twenty-three displacement pick-ups were required.

ONERA was tasked with testing the resistance of this alloy to oxidation. It was found that at the heat reached during re-entry (over 1,000°C) there was a possibility that the alloy would

interact with the little oxygen present in those outer layers of atmosphere and create a brittle layer of oxide over the metal sheets. This could lead to cracks developing in the metal sheets The P333 alloy was found wanting in this regard, even if it fared better than nobium alone. Therefore a protection had to be designed for the tiles. Two solutions were investigated: improving the alloy by adding titanium or zirconium. This answer appeared satisfying for a flight article, but could not be tested on the ground because atmospheric pressure could not be replicated, and there was a need for multiple testing of the same sample.

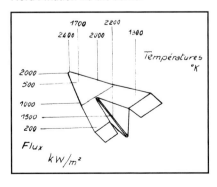

BELOW This rather rough drawing illustrates the high temperatures that were expected during re-entry. VERAS was mostly an exercise in identifying the materials which could be used in such a harsh environment. *Nord-Aviation via the author*

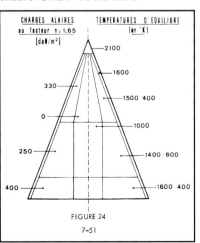

BELOW Another illustration of the temperatures which Veras had to endure. *ONERA via the author*

ABOVE VERAS, tuned upside down and with the wing frame still uncovered. The fuselage section has most probably received a few sensors, hence the various power cables hanging from it.

ABOVE Close-up of a structural longeron from which the VERAS structure was built.

Therefore a second solution was investigated: covering the alloy sheets with an oxidation-resistant coating. Tests of silicide of niobium and aluminide of niobium were deemed successful, although some of the parts were damaged by repetitive heating at 500°C. One possible cause of this damage was the removing and re-attaching the test tiles, which may have damaged the coating. At re-entry temperatures, another risk was 're-crystallisation' where, at the expected re-entry temperature, the metal – whose crystalline structure had been broken and reformed during the laminating and machining process – could crystallise again and adopt unwanted properties. Tests proved that no undesirable attributes appeared.

The undersurface of the wings was covered in tiles measuring $10^5/_8 \times 9^7/_{16}$in (270 × 240mm) made of P333. Fabrication of these tiles required the creation of a special machine by Nord-Aviation. On the whole, there were very few rejects and Nord was most satisfied with these tiles, which were heat-tested successfully up to 1,200°C. Although some damage was observed on the tiles, that could have been caused by the conditions of the test (tiles being regularly manipulated, detached, reattached, etc) or by an accident occurring during a test (projection of liquid water on heated tiles).

In January 1967 the early results of the feasibility study were presented to the Ministry. In the later part of the year two different tests were carried out:

i ONERA conducted vibration testing.
ii Centre des Essais Aérospatiaux de Toulouse designed a test bench on which a special heating installation applied thermal stress to the structure; this simulated the intense heat to be expected during re-entry.

Before being sent to the Centre d'Essai Aérospatial de Toulouse, a sample test article was examined by ONERA. It represented a sub-scale version of the half-wing rear section torque box consisting of two main longerons with corrugated webs and four spars also with corrugated webs. The skins were elementary two-face corrugated sandwich shingles of René 41, hinged to both sides of the torque box, and therefore carrying little or no load.

BELOW Complete structure of the VERAS wing being assembled at Nord-Aviation.

BELOW Close-up of one of the shingles covering the wing of VERAS.

Tests at ONERA were run to define the behaviour of such an unusual structure under mechanical vibration and to be able to set the installation for the CEAT test site experimentation with a minimum of tuning time. ONERA tests were performed at the ambient temperature; at a medium temperature of 930°F (500°C); and up to the design operating temperature of 1,830°F (1,000°C). The evolution of the modes and damping coefficients were observed and were found not to be too dependent upon the average temperature—indeed, level as a general rule. However, local plastic over-strains might have occurred, changing the stress distribution and preventing detailed analysis.

Finally, the full-scale model was heat-tested at the Centre d'Essai Aérospatial de Toulouse with 'very good results.' Other tests were carried out in the Fontenay wind tunnel, where another model was subjected to a gas jet reaching Mach 18. These tests required a lot of energy in in comparison to the power input of the whole Paris Area, but only for $1/_{50}$th of a second.

In January 1968 it was planned to launch it over the Centre d'Essais des Landes (Landes Test Centre) circa 1971–72. The flying model was about 10ft (3m) long and weighed 2,200lb (1,000kg). It was intended to reach Mach 5 above 9,840ft (3,000m) during a 500-second flight. The Véras would be launched using an Emeraude sounding rocket. Unfortunately these ambitions still remained unfulfilled when the whole project was cancelled in 1971.

ABOVE The VERAS airframe being tested at CEAT.

Véras was revealed to the World in February 1968. At that point more than 1 million Francs had been devoted to the programme. A model of Véras was exhibited at the 1969 Paris Air Show.

According to Henri Lacaze, who at that time observed this from the Ministry, '*I doubt there was ever any will to launch Véras. The main goal of Véras, under the leadership of Naval Engineer Bellot, was to learn how to work with the required exotic alloys. It was also necessary to check that these materials were actually available on the market; it is something knowing some material exists, it is something else entirely to order square meters of metal sheets made of it. The project was stopped because the requirement for a hypersonic glider did not exist.*'[35] The requirement to which Mr Lacaze refers was most probably a military one for warhead vehicles, not for a space vehicle.

BELOW A picture of the heat rig at CEAT. The shape of the VERAS vehicle has disappeared behind the many heat sources applied to it.

BELOW Less romantic than pictures of a fully-fledged spaceplane, this is a photograph of the electronic banks behind the CEAT tests.

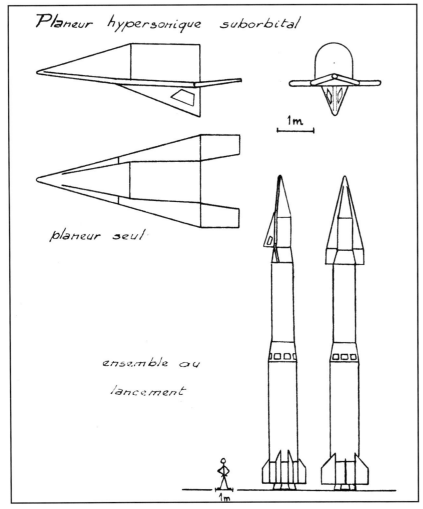

ABOVE Drawing by Henri Lacaze of the VERAS vehicle on top of a rocket booster. The exact identification of the rocket drawn here is unknown. *Henri Lacaze*

ABOVE Two photographs of the VERAS vehicle exhibited at the Paris Air Show, 1969. By the time those pictures were taken, the project was dying.

Reviewers among the various news magazines of the time *(Flight, Aviation Magazine, Air & Cosmos)* noted the similarities between Véras and the American programme ASSET (Aerothermodynamic Elastic Structural Systems Environmental Tests).

Re-entry Rotary-Wing Style?

Gyravions Dorand was also investigating re-entry methods but, in keeping with its technological know-how, used rotary wings.

In his 13 December 1965 patent, Ingénieur Marcel Kretz first pointed out that the usual recovery method, using an ablative shield for braking the fall in the higher atmosphere coupled with parachutes for the landing, had a few disadvantages: high G sustained by the passengers or load and the difficulty in controlling precisely the landing spot. As for hypersonic gliders, the wings added a lot of weight to the machine. The solution proposed by Gyravions Dorand was a rotary-wing device which could oscillate in comparison with the spacecraft. To resist the heating, the rotor blades were to be constructed of metal *'in a group comprising molybdenum, tantalum, tungsten, niobium, refractory steels and super alloy.'* It was proposed to build the upper and lower parts of the blades in different materials, able to bear a temperature difference of up to 200°C. A successful experiment was made using refractory steel for the upper surface (heating up to 900°C) while the lower surface was of a nickel-base super alloy (heating up to 1,000°C). It showed that the two metals used in this combination had similar expansions.

The second patent, applied for in 1968, expanded upon the first, with the spacecraft being encased in a rotating heatshield to which were attached blades similar to those described above. The new design even included landing gear.

However neither patents clarified how the rotary blades were stored during launch.

RIGHT This drawing from patent US3412807A illustrates René Dorand's re-entry solution for recovering a rocket—a rotary wing!
Gyravions Dorand, author's collection

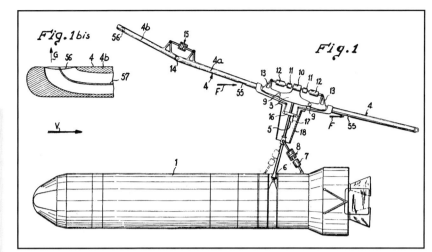

Nord Véras (initial design)	
Status	Static test model
Span	–
Length	–
Gross Wing area	–
Gross weight	3,000lb (1,360kg)
Engine	none
Maximum speed/height	Mach 10 at 150,000ft (45,720m)
Load	2,000lb (907kg)
Nord Véras (later design)	
Status	Flying model (untested)
Span	9ft 10in (3.00m)
Length	–
Gross Wing area	–
Gross weight	2 205lb (1,000kg)
Engine	none
Maximum speed/height	Mach 5
Load	?

RIGHT Another drawing from US3412807A shows how the rotary wing system would be protected from hot re-entry airflow (plasma) by its location out of the flow of air. Temperature gradients from 3,273° to 6,273°K (3,000° to 6,000°C) are shown.
Gyravions Dorand, author's collection

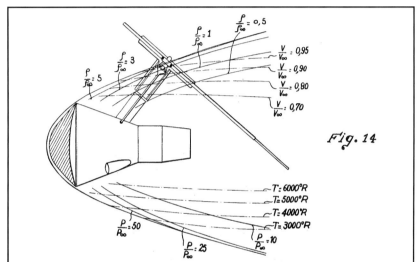

RIGHT Here we see a drawing from US3558080A in which the spacecraft is enclosed in an outer shell with rotor blades are attached. The true payload is enclosed in part 9. This conical hub is articulated so that the outer shell could oscillate around it, so moving the centre of gravity. The vehicle is shown fitted with a three-legged undercarriage, one of the legs supporting a rudder. The vehicle would descend in autorotation. It is depicted during re-entry and no explanation is given regarding storage of the rotor during the ascent.
Gyravions Dorand, author's collection

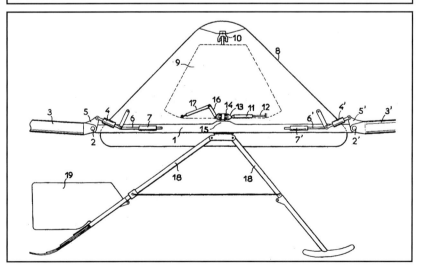

---

[34] This and other details come from *Thermal Ground Testing of Concorde and Véras or Improvement In French Test Methods and Facilities* by G L Leroy (SNIAS, France), N'Guyen (SNIAS, France), M Perrais (CEAT, France), H Loiseau (ONERA, France).

[35] Interview with the author 8 February 2019.

Chapter Seven
Competing for Hermes

ABOVE The recoverable first stage is shown here returning to its base—most probably Kourou, in view of the landscape. *Aerospatiale via H Lacaze*

A – CNES Early Thinking About Man in Space

While the European aircraft manufacturers were busy designing winged launchers that never took off, rocket technology had matured. On the organisational scene, things had been restructured: the European Launcher Development Organisation (ELDO, which had been unsuccessful at putting into service the Europa launcher), was merged with the European Space Research Organisation (ESRO, which had put seven satellites into orbit, albeit using US launchers) on 31 May 1975 to become the European Space Agency. With the Europa launcher in trouble and the Americans unwilling to launch satellites which could adversely affect their own commercial interests (they had refused to launch Symphonie, a telecommunications satellite developed jointly by France and Germany), the French had proposed in 1973 to develop a new launcher: the L3S, later known as Ariane. With Ariane on the way to success, including a commercial success as it began to launch satellites for non-European countries, CNES, the French space agency, began thinking about ways to put *spationauts* (as the French elected to call their equivalents to the American astronauts and Russian cosmonauts) into orbit.

Therefore, as early as **1975**, during a workshop initiated by CNES to discuss the possible evolution of the Ariane family, proposals were made regarding a manned version. A capsule *à la* Apollo was briefly considered but immediately rejected on the grounds of recovery problems. Recovery would have to be made at sea and would require a full fleet of ships, including probably, an aircraft carrier—something that France could not spare. And prestige considerations could also be taken in account: Americans and Russians had been using capsules for fifteen years, so Europe would have just been a follower. Thus, the spaceplane – which could use 'cross-range manoeuvrability' to reach a European airfield from any orbit – was back in consideration after having been studied with ups and downs during the 'sixties.

And by developing such a machine, France and Europe would 'boldly go where no one has gone before,' which was an additional benefit in terms of international prestige. The will of German companies to get involved with American shuttlecraft most probably played a role in the choice of going with the spaceplane. Henri Lacaze, who had followed the TAS programme from his post at the Ministry and would later become Technical Director on Hermes concurred, saying, '*After the abandonment of the Aerospace Transporter (which was never much developed), the hypersonic glider was forgotten. It only came back when CNES launched an initiative which would materialise into Hermes.*'[36]

Early studies were spearheaded by the CNES *Direction des Lanceurs* (Launchers Directorate) led by Frédéric d'Allest. The initial CNES design came up with a 7-tonne spaceplane to be launched by an Ariane 5 (At that time the Ariane 5 was still a beefed up Ariane 4).

In April **1977** CNES approached Aerospatiale for help and advice regarding the spaceplane project. The first question to be settled was the debate between reusable space glider and an expendable capsule. Michel Rigault, who led the early work at Dassault on Hermes, noted in this connection, '*Re-entry trajectories of both types of vehicles are similar. The difference lays in the aerodynamic performance of each. The cross-range manoeuvrability defines the maximum distance between the orbit projected on the Earth's surface and the geographical position of the landing site. Increasing the aerodynamic performance increases the cross-range ability and reduces the maximum load factor. A high cross-range ability (it was 1,491 miles [2,400km] in the initial Hermes specification) allows definition of the return site in advance—even at short notice, like in an emergency. For a capsule, there are less opportunities, which implies the management of a variable recovery area, at sea for the USA; in Siberia for Russia. It also implies a high load factor which limits crew membership to highly trained personnel.*'[37]

Once the space glider had been selected, the second task was to stabilise its parameters which were defined by the weight-to-orbit capability of the launcher and by the aerodynamic

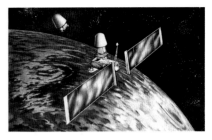

ABOVE **The Solaris project followed the Minos project to investigate the feasibility of 'space factories'. Re-entry vehicles were mostly imagined as capsules *à la* Apollo, as seen here.** *CNES via Capcom*

constraints of the re-entry. The first point was settled at 22,046lb (10,000kg) in 'low earth orbit' (between 186 and 1,243 miles [300 and 2,000km]) and resulted in a glider carrying a crew of two plus one passenger and 882lb (400kg) of payload for a seven-day mission. That translated into a 9.8-ton (10- tonne) spacecraft, while the target weight had been 6.9 tons (7 tonnes). This was a problem because structurally, the Ariane 4 launcher was not able to support such a weight. In particular, the third stage was too frail for the job.

The spaceplane concept was still

BELOW **...however as part of Minos some space glider concepts were investigated.** *CNES*

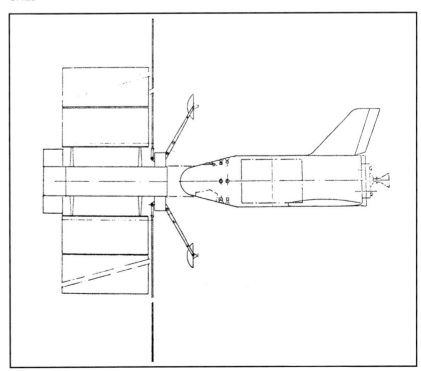

undecided at the end of the 'seventies. During the second half of **1978**, Aerospatiale, under contract from CNES developed the project MINOS (in Greek mythology, King Minos was the father of Ariane) *Modules Industriels pour Operations Spatiales*: Industrial Modules for Space Operations, which it was hoped, would be used to produce special materials for metallurgic and electronics industries. MATRA and Aerospatiale were tasked with studying the concept and came up with a 7,055lb (3,200kg) space factory positioned in a 497-mile (800-km) equatorial orbit. The basic frame included a power unit delivering 25 kW using solar panels. Those modules were expected to have a lifespan of seven years.

Besides the material production aspect, it was intended to use the base frame to create observation and communication satellites. By 1980, MINOS had evolved into SOLARIS (*Station Orbitale Laboratoire Automatique de Rendez-vous et d'Intervention Spatiales*: Automated Orbital Laboratory Station for Rendezvous and Space Interventions). This larger design was expected to last fifteen years. Although initially the return of material produced in orbit was to be via modules recovered by parachutes, a timid attempt to introduce return by spaceplanes appeared as early as 1978 in the MINOS documentation. However, for MINOS, the basic re-entry vehicle was a large module '*in the shape of a car headlight, like the Soyuz*' although an '*automatic variant of Hermes*'[38] was also considered.

The name Hermes was coined by Frédéric d'Allest's son who suggested naming the spaceplane after the ancient Greek messenger of the gods, bringing 'messages' (payloads) to the 'modern-day-gods' (astronauts in the space station). Later, d'Allest was told by a German colleague that Hermes was also the god of thieves[39].

Hermes was revealed at the **1979** Paris Air Show. At that time, it was described as a 9.8-ton (10-tonne) vehicle, much inspired in shape by the Rockwell Space Shuttle orbiter, to be launched by Ariane 5. It would have a dry weight of 5,953lb (2,700kg), a span of 24ft 3in (7.40m)[40], a length of 39ft 4in (12.00m) and a height of 14ft 9in (4.50m). It would carry either five *spationauts* or two crew members and 3,307lb (1,500kg) of payload. Typical missions would involve either observation at an altitude of 124 miles (200km), transferring a module[41] from 124 to 248 miles (200 to 400km) orbit (there had been, during the 'seventies, a study regarding a 'space tug' to be designed by Europeans to participate in the assembly of the US-led International Space Station), or to bring astronauts to and from the US Space Station. The maximum mission duration was to be seven days. A notable side-effect of the vehicle being designed by people not used to dealing with humans was that the Hermes mission would be fully automated; the astronauts could pilot it, but only 'if necessary.'

It is interesting to note that the same approach had been taken by Soviet designers at the time of the Vostok and by the US designers of the Mercury spacecraft (resulting in a 'revolt' by the astronauts who reminded NASA they were pilots). Thermal protection was to be assured by reusable silica-based tiles directly inspired by the solution adopted for the Rockwell orbiter (therefore not following directly the work done on Véras). The internal structure would be based on aluminium alloys. A 4,496lb st (20kN) rocket engine was provided for orbital flight.

An additional 179,850lb st (800kN) podded rocket engine would serve as an escape device, in case of failure of the launcher. Fuel cells would provide power in orbit. Three Freon radiators would dissipate excess heat.

In 1979 it was estimated that ten years would be required to build three Hermes, comprising two prototypes and one operational machine. The cost of the development was not estimated. At this point, Hermes lacked a direct operational role; it was actually a machine designed to support the satellites which were the real operational machines. At least, that was d'Allest's vision and he tried to push it through CNES, but only as a French program. This 'repairman' job was key to the vision of Hermes' mission by CNES to the point that the drawing of parallels to the US Orbiter (by calling it a 'mini-shuttle' as the press was keen to label Hermes) was strongly resisted by CNES, which insisted on naming it *avion spatial*. Basically, CNES wanted Hermes to be a people-mover, not a satellite-launcher like the US Shuttle, so the cargo bay was underplayed, as just an area where repair tools for servicing satellites or space stations were stored. In CNES's view, the

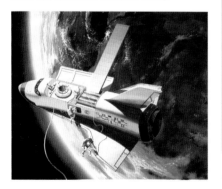

BELOW An early artist's impression of the spaceplane envisioned by CNES, with consulting by Aerospatiale. Note the great outward similarity of shapes with the Rockwell orbiter. *CNES*

LEFT The same early design is shown in this artwork, docked with an ESA space station—itself an early appearance of the future Man-Tended Free-Flyer. *Aerospatiale*

tasks were well separated. Moving people and equipment to Space Stations, repairing satellites in orbit was Hermes' job; launching satellites was Ariane's role.

It should be noted that, even at this early point, a 'rescue' role was considered for Hermes with up to six persons being carried 'in case of emergency.'

Later, in its January 1985 brochure, CNES specified the role it intended for Hermes. After identifying two large classes of mission – '*missions with a commercial feature*' (that is launching commercial satellites, which CNES intended to allocate to Ariane); and '*missions with a technological feature*' (servicing space stations, repairs in orbit) – it justified Hermes thus:

■ *An integrated system in a single vehicle which can fill all the needs. That's the case of the American Shuttle which orbiter is fitted with a powerful engine. Its drawback is that it impacts badly on the cost of commercial flights because a crew should be aboard whatever the mission—which requires a very costly environment, both on-board and on the ground. Of the 88 tons (90 tonnes) put into orbit by the Shuttle, one third at most, is payload.*

■ *A system of the Soviet-style in which the manned spacecraft is a load like every other that can be positioned atop the launcher. This system separates the question of manned spaceflight from automated flights. However, the Soviet system does not*

BELOW A cutaway of this early variant. Note the engine nozzles are more realistically presented here than the earlier large 'bell'. *Aerospatiale*

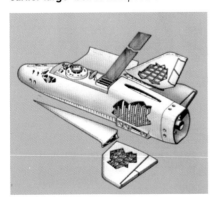

ABOVE A 1979 announcement in the French specialist press was this black-and-white drawing in *Air & Cosmos*. *Capcomespace.net*

easily allow for daily re-entry windows and is non-reusable.

■ *The Hermes spaceplane is designed to avoid the inconveniencies of both previously mentioned systems, being positioned atop the launcher only when a man's presence is required. During automated commercial flights it is replaced by payloads [...] Its wings give it a wide cross-range ability (+ 1,553 miles [2,500km]) and longitudinal (+ 5,592 miles [9,000km]), the spaceplane Hermes may, at any time, come back to a landing strip, land softly onto it and therefore be reusable. Fitted with powerful electronic equipment for orbital operations, it controls the launcher during its powered flight and therefore can dispense with an equipment case, which is required for automated flights.*

While the technical project was advancing, politics were also evolving. In **1981**, Président François Mitterrand replaced Valéry Giscard d'Estaing. Giscard d'Estaing initially was not really pro-space and while his mind evolved during his seven-year mandate, Mitterrand was definitively enthused.

In **1982** Jean-Loup Chrétien became the first Frenchman to fly into orbit aboard Salyut 7.

The incoming president of CNES, Hubert Curien, presented the new five-year plan of the Centre on 30 June **1983**, announcing, '*we are now practically certain that France and Europe will involve itself in manned flights.*'⁴² From this observation Curien wanted to

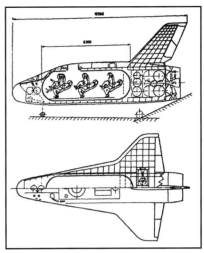

ABOVE Also from 1979, this two-view drawing reveals the similarity of shape with the Rockwell orbiter, together with internal details including pressure chamber for the crew and adjusting seats. *NASA*

BELOW Cutaway of the early CNES Hermes from a 1981 CNES document. Note the evolution of the shape of the crew pressure chamber and that the engine nozzles are no longer enclosed in the fuselage. *CNES*

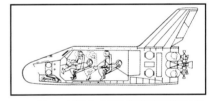

Hermes (1979)	
Status	Pre-project
Span	24ft 3in (7.40m)
Length	41ft 2in (12.55m)*
Gross Wing area	–
Gross weight	22,046lb (10,000kg)
Engine	4,496lb st (20kN) rocket engine of unspecified type
Maximum speed/height	249 miles (400km)
Load	5 × spationauts; or 2 × spationauts + 3,307lb (1,500kg)

boost the Hermes programme. Preliminary studies were to be finalised in 1987 so that the project could be presented to ESA for 'Europeanisation' that year. Development should then proceed up to the first flight, planned for 1996, with operational missions beginning the following year.

After MINOS and SOLARIS had

* In 1981, *FlugRevue* indicated 12.00m.

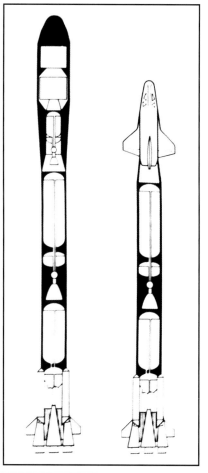

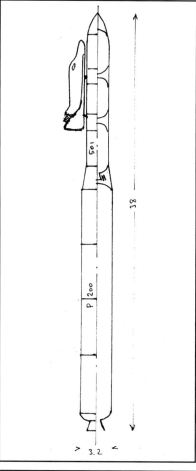

FAR LEFT In those days, the leading European launcher was Ariane. Here we see a comparison of an Ariane with a large shroud for launching satellites side by side, and an Ariane carrying a Hermes. *CNES*

LEFT Some consideration was given to attaching the spaceplane to the side of the launcher. The author has not located a drawing of Hermes/Ariane in this configuration, but here is an undated (most probably going back to the beginning of the 'seventies) drawing by Henri Lacaze depicting a launcher (most probably Europa) carrying on its side a spaceplane with the profile (reduced to around half-size) of the X-20. *Henri Lacaze*

BELOW In 1983, Aerospatiale presented the complete Ariane line-up. At this point the 'Ariane 5' AR54LH was actually based on the Ariane 4 stages L220 (first stage) and boosters L34. The variant at the right depicts an Ariane 5 with the twin-fin Hermes on top. *Aerospatiale*

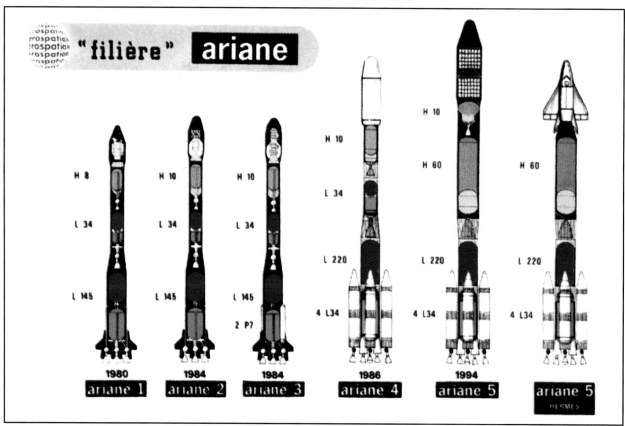

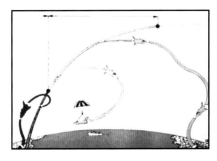

ABOVE Hermes launch (red trail), return to Earth (to a Spanish airfield) (yellow) and abort launch (blue). *Aerospatiale*

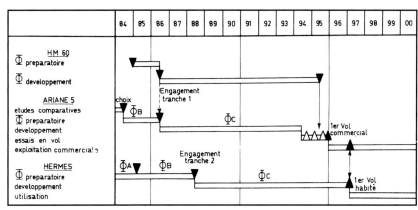

ABOVE In 1984, CNES issued this intended development plan which optimistically announced a first flight of Hermes in 1997. CNES

BELOW Around 1985, the CNES design evolved, most notably sprouting twin fins and presenting a much revised profile and planform. CNES

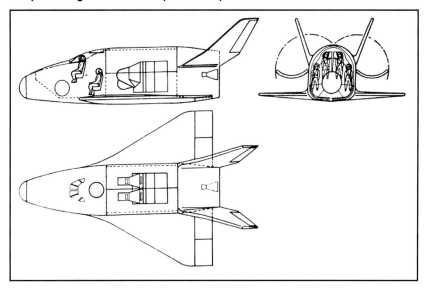

shown that space factories were a possibility, there was some fear that the older space powers would be reticent to launch French astronauts if commercial interests were at stake. As recounted above, this had already happened to the Franco-German communications satellite Symphonie which had to be limited to experimental role before it could be launched by an American rocket. *Spationaut* Patrick Baudry (and Alain Souchier) wrote in *Ariane*: 'We need to guarantee the access of European astronauts to the lower orbit and to the Station and soon to other stations. If the economic interests of producing material in orbit are confirmed (there are talk of hundreds of billions [of French Francs] of annual revenue by the year 2000), it is possible that the present operators of manned spacecraft, USSR and USA, will strongly regulate access to the stations'.

The first step was to select an industrial contractor to design and build the spacecraft. A specification was issued in February **1984** which completely upturned the simple, people-mover concept of earlier days. Now, emphasis was on the cargo bay which diameter was specified as 9ft 10in (3.00 m) and its volume at 1,236cu ft (35cu m). The payload was to be up to 9,921lb (4,500kg) in mass. Fuel cells should provide 240 kWh to power internal systems and experiments, providing an extended flight duration of ten days. Aircraft life was to be one hundred missions (ten years at 10 missions per annum).

Other requirements were the all-up weight (14.7 tons [15 tonnes] was the maximum Ariane 5 could lift, although by that time an Ariane 'Super 5' able to lift 19.7 tons [20 tonnes] was already being considered); surface (the wing area was limited to 915sq ft [85sq m] for fear a larger surface could be caught by the wind during take-off and destabilise the rocket); and protection of the windscreen during re-entry (this particular area could not resist a heat stress of over 500°C, the maximum clear material could sustain).

Hermes (1984)	
Status	Call for proposals
Span	–
Length	49ft 3in to 59ft 1in (15.00 to 18.00m)
Gross Wing area	–
Gross weight	36,817lb (16,700kg)
Engine	2,248 to 3,372lb st (10 to 15kN) rocket engine
Maximum speed/height	–
Load	Up to 9921lb (4,500kg)

RIGHT Sectional drawing of the same design dating from the early 'eighties. *Aerospatiale*

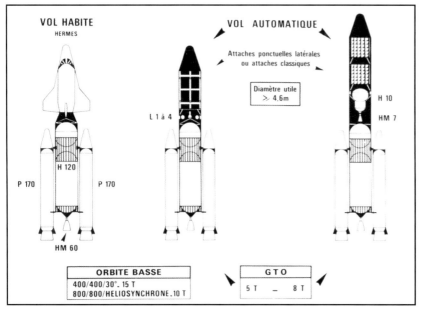

ABOVE Although appearing in 1985, this document describes a somewhat earlier configuration in which Ariane 5 had evolved but the spaceplane was still the 'mini-Rockwell-Orbiter' shape. *Aerospatiale*

BELOW In this brochure, distributed in June at the Paris Air Show but dated January 1985, the 'twin-fin' Hermes is now shown atop the Ariane 5. *CNES*

B – The Question of the Launcher

Ever since the early 'eighties, as soon as the future of Ariane appeared assured, ESA began thinking about what could be done after Ariane. Notably the question of a recoverable launcher became the subject of intense debate.

From 1980 to 1984, Aerospatiale, under an ESA contract studied first in 1980 to 1983 the concepts for Future Launch Systems (FLS), then in 1983-84 examined the feasibility of these concepts. '*ESA would have felt inferior to its own task, if it had not looked at advanced "exotic" solutions,*' recalled Henri Lacaze who himself worked on the FLS.

In 1984 Hermes was coalescing into a real spaceplane project. It was still to be launched vertically by a non-recoverable rocket (Ariane 5) following a pattern established twenty years previously with the American X-20 Dyna-Soar. In that same year, BAe proposed HOTOL, which – being fully recoverable; single-stage-to-orbit (SSTO); and with horizontal take-off and landing – appeared a considerably more advanced concept (see Chapter 9). Up to that point ESA had rejected SSTO concepts – as exemplified by the BETA spacecraft (developed into ITUSTRA and ISTRA) – proposed by Dietrich Koelle from the University of Stuttgart since 1971.

Early in the development of Hermes, Dietrich Koelle, now at MBB, put forward the horizontal take-off and landing Sänger II which he presented as the next step after Hermes (see Chapter 9). This new advancement encouraged ESA to follow-up FLS with the Future European Space Transportation Investigations Programme (FESTIP) but this was only established in 1994, although contribution to this programme appeared much earlier—for example, with Dassault STAR-H (which used Hermes as its second stage), or MBB Sänger which, until cancelled in 1995, hoped to be integrated into FESTIP. At the same time, the research on hypersonic ramjets which had dominated the field during the 'sixties and early 'seventies (see Chapters 3 and 4) saw a resurgence with the work done by François Falempin for the PREPHA programme, beginning in 1992 (see Chapter 12).

B1 – ESA and the Future Launch System

In 1986, Aerospatiale issued a report on *Lanceurs Futurs* (Future Launchers). This analysed all the material produced in France since the beginning of the 'sixties, including the Transporteur Aérospatial studies (see Chapter 3). The conclusions regarding the TAS, were not totally negative, although they conceded *'feasibility and attractiveness of an airbreathing supersonic ramjet-powered launcher, remains to be demonstrated.'*[43]

Studies carried out for ESA between 1980 and 1984 were presented. Those studies were intended to investigate launcher concepts with a target date for its operational roll out between 1995 and 2000. Since these studies were parallel to the early phases of Hermes design, one of the requirements was the *'capacity to carry manned payloads.'* As an opening to these studies ESA had asked two questions:

- *'In practical and technological terms, which reusable launcher concepts can be constructed in Europe?*

- *For these concepts, what are the recurring launch costs to be expected and how do they compare with those of non-reusable launchers? If those costs are more advantageous, then which development costs are to be expected or technological research necessary?'*

These studies were actually carried out in four phases: three 'prospective' phases in 1980-81, 1982-83, 1983-84 and a feasibility study afterward.

The 1980-81 study compared an airbreathing first stage with a rocket-powered first stage, both winged. Vertical and horizontal launches were also evaluated.

RIGHT A beautiful model of the 1985 CNES configuration: twin-fin Hermes atop Ariane 5. *Author's collection*

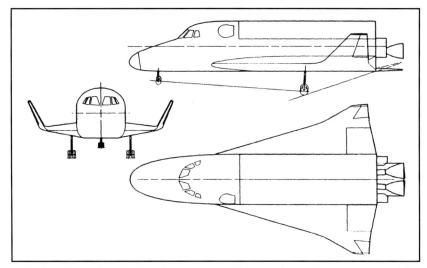

ABOVE In November 1985, CNES's concept of Hermes had taken a new step with disappearance of the twin fins from the fuselage, those now attached to the wingtips. *CNES*

BELOW From the same brochure, an exploded drawing illustrating the structural elements of Hermes as it was in late 1985. *CNES*

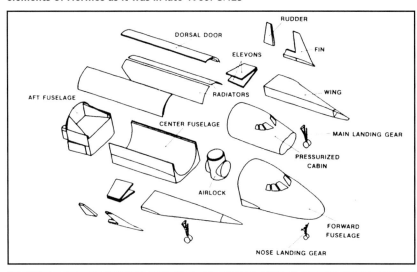

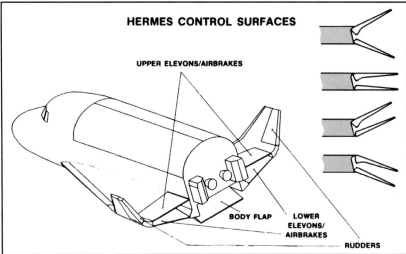

ABOVE Illustration of the controls on the wing trailing edges. Note the 'body flap' which was directly inspired by the Rockwell design. *CNES*

The conclusion followed the analysis made earlier in this report about the TAS program. '*An air-breathing first stage with a horizontal take-off, because of the very high specific impulse, would have a reduced take-off mass, but to use this technology would require considerable development work on high-thrust turbo-ramjets.* Impractical then—especially when rocket propulsion appears better adapted to vertical launch. A cryogenic fuel engine would be an attractive proposal for the first stage, '*but its high cost would make it impractical for a single-use first stage.*'

Horizontal take-off with air-breathing engines having been deemed unfeasible, the later studies focused on vertically-launched rockets. The second phase, carried out in 1981-82, dealt with two-stage rocket launchers in which both stages used cryogenic fuel and were recoverable. The target was 9.8 tons (10 tonnes) in a 124-mile (200-km) orbit. The first stage was winged; returned to its base on small turbojets; and landed there on a standard landing strip. The second stage was intended to return 'ballistically' 'near its launch base.'

The conclusion indicated that the first stage with a gliding re-entry '*did not raise any unreasonable technical problems.*' It was the second stage, with its ballistic re-entry that generated new questions because the aerodynamic stability during the re-entry required it to have a large diameter. '*The re-use of the second stage requirement has an important impact on the size and therefore the cost of the first stage for a given payload.*'

BELOW The heat protection of Hermes is described here with the different materials used. *CNES*

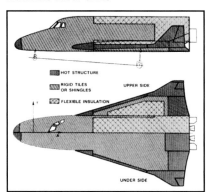

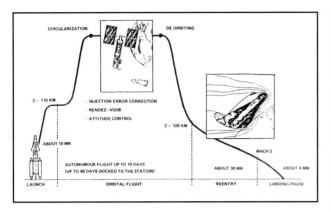

ABOVE Schematic of a standard Hermes flight. *CNES*

ABOVE RIGHT Abort trajectories for Hermes. Note that the silhouette of the spaceplane still sports twin fins. *CNES*

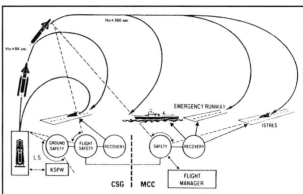

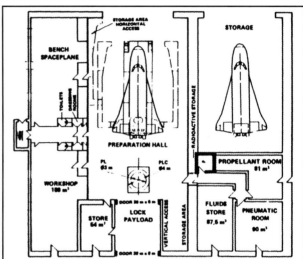

RIGHT CNES even designed the hangar in which Hermes would be maintained and stored.

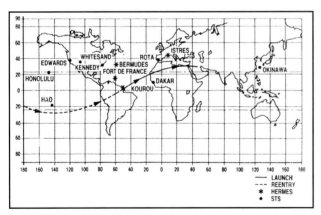

ABOVE A map showing a standard orbit with landing fields for Hermes and for the US Shuttle (annotated STS). *CNESCNES*

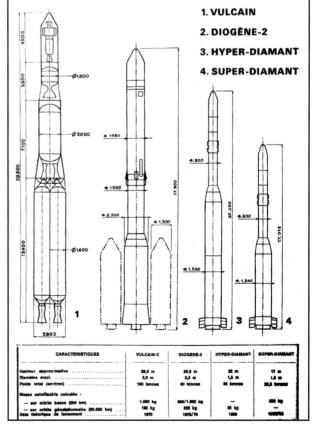

RIGHT SEP launcher projects from the mid-'seventies. These were all abandoned in favour of Ariane.
Aerospatiale via H Lacaze

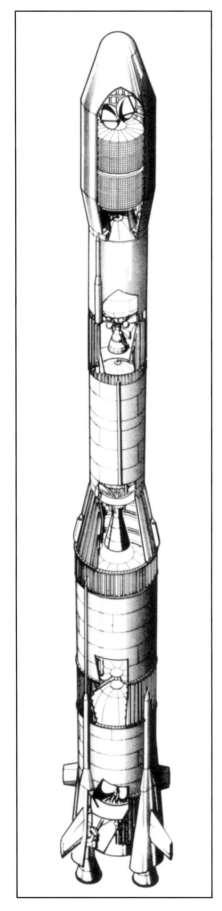

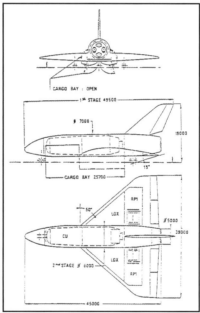

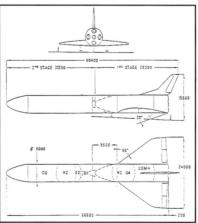

ABOVE Cover of the 1984 Aerospatiale report brochure about recoverable launchers developed from the Ariane family (Future Launch System – FLS). *Aerospatiale via H Lacaze*

LEFT As part of the Future Launch System (FLS) many configurations were studied. Here the Horizontal Take-off (HTO) proposal. *Aerospatiale via H Lacaze*

For the third phase (1982-83), Aerospatiale was teamed with other companies MBB-ERNO, Dornier Systems and Marconi Avionics. '*ESA was eager to have industrial companies working in pools,*' remembered Henri Lacaze.

This time, configurations of the launcher were studied: how to best mix cryogenic fuels, standard liquid fuel and solid fuel (powder). The question of whether it was worth recovering the second stage was also examined. Four configurations were therefore investigated with a target of putting 14.7 tons (15 tonnes) in a 124-mile (200-km) orbit:

i CR = both stages cryogenic fuel, both recoverable.

ii C/SR = both stages cryogenic fuel, but only the first stage recoverable.

LEFT And here is the Vertical Take-off (VTO) proposal. *Aerospatiale via H Lacaze*

LEFT Ariane 1 was the selected French/European launcher from the early 'eighties. Ariane 1 first flew in 1979. *Aerospatiale*

iii S/SR = first stage with liquid fuel and Viking engines, second stage with cryogenic engine, only the first stage recoverable.
iv P/SR = first stage using solid fuel and the second stage a cryogenic engine, only the first stage recoverable.

The conclusions of this phase were:

- Cryogenic fuel engines are well adapted to recoverable launchers.
- Fully-recoverable launchers are no better, economically speaking, than a partially recoverable launcher because of the weight penalty induced by the recoverable second stage. The technology involved is also more complex.

For the feasibility studies, the team was enlarged with the addition of Fokker, Aeritalia and CASA.

This time the study observed that the following points needed to be detailed:

- Re-entry controls.
- Aerodynamic coefficients to be tested.
- Resistance of carbon/polyimides composite materials to heat.
- Internal protection of fuel tanks.
- Carbon-carbon composite materials' resistance to oxidation and the capacity to produce thin sheets of such material.
- Aeroelastic floating.
- Atmospheric turbulence.

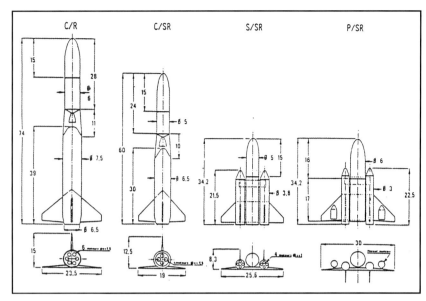

ABOVE For the Vertical Take-Off configuration various options were examined during 1983–84. *Aerospatiale via H Lacaze*

A specific trajectory called the 'dead leaf procedure' in which the launcher was pointed toward its return point before de-orbiting was investigated. A bonus of this trajectory was elimination of the requirement for auxiliary turbojets. However, this was compensated by the need to enlarge the de-orbiting retro-rocket and the accompanying fuel tank. So, in the end the 'dead leaf procedure' was not retained.

The final conclusion was that the partially reusable launchers could be built in ten years, employing available technology and with a cost only 25% higher than the development of an all-new, non-recoverable launcher. It was expected that the extra cost could be recovered after a dozen flights.

BELOW This schematic illustrates the development case for the 'twin-body first-stage' variant. *Aerospatiale via H Lacaze*

BELOW Another schematic for the 'parallel configuration' variant. *Aerospatiale via H Lacaze*

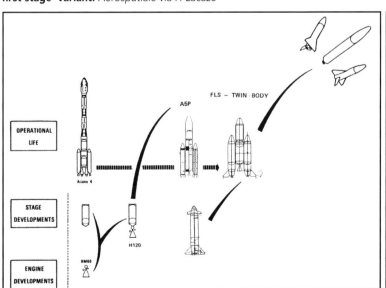
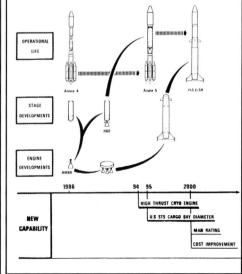

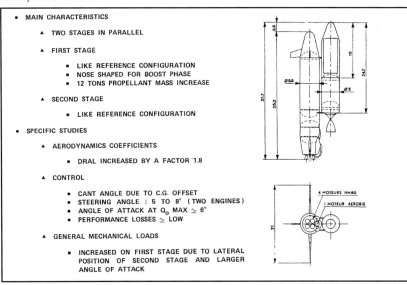

ABOVE More detailed drawing of the twin-body first-stage variant.
Aerospatiale via H Lacaze

BELOW ...and here, the same drawing of the 'parallel configuration' variant.
Aerospatiale via H Lacaze

B2 – The Case for the SSTO

From 1970, Dietrich Koelle of MBB promoted a single-stage-to-orbit vehicle, extrapolated from a Douglas design from the preceding decade. The Ballistiches Einstufiges Träger Aggregat (BETA: Ballistic Single-stage Carrier[44] Vehicle) used the 'plug nozzle' concept to serve as heat shield for re-entry while the dozen (or more) rocket engines were arranged around it. Six retractable landing pads were provided at the ends of spindly legs. It was expected that only four engines would be required for landing. Trajectory control would be by throttling a single engine among the dozen in action at take-off (or the four used during descent). There was no specific payload fairing, the fairing being adapted to the payload and part of it. According to the designer '*this arrangement means much less geometrical restrictions than a cargo compartment within a winged shuttle.*'[45]

Main advantages of BETA were '*the ability to launch it from Europe, since no stage or part falls away during ascent. Vertical take-off and vertical landing in combination with the landing leg system allows continuous abort capability in the critical launch phase. It also allows an outstanding means of test operations, which continuously lead from ground testing to flight tests.*' The use of such a vehicle was expected to reduce the cost to less than $200/kg orbited.

While pointing out all those advantages, Koelle could not hide the fact that designing a 33-ft (10-m) diameter plug nozzle with 12 to 20 rocket engines represented '*a relatively large development effort.*' Koelle claimed that, '*The BETA concept seems to be the final solution to the space transport problem since it combines operational simplicity with lowest cost, both for development and specific payload cost.*'

Payload of BETA was to be 8,818lb (4,000kg), but Koelle proposed a BETA II with a payload of 22,046lb (10,000kg), a BETA III to deliver 44,092lb (20,000kg) and even a BETA IV taking 220,460lb (100,000kg) into orbit.

BETA was later developed into ITUSTRA and ISTRA under the sponsorship of the University of Stuttgart. ITUSTRA took the BETA concept and added a ring of air-breathing engines to be dropped at a height of 28 miles (45km) and a speed of Mach 6. This ring of engines would be recovered by parachutes. ISTRA replaced the turbojets by a simpler ramjet.

These studies were investigated by Aerospatiale as part of the Future Launch System studies for ESA. The results were unconvincing. The examiners doubted the performance table for ITUSTRA/ISTRA, remarking, '*Those performance [estimates] lack credibility; the weight of the air-breathing engines is under-estimated. The structures have not been seriously studied and uncertainties plague the trajectories and injection to orbit curves.*' Alain de Leffe concluded: '*The studies on the SSTO did have a use: they pointed out the difficult questions and they paved the way to modern two-stage launchers.*'[46]

An early version of HOTOL (see Chapter 9) was also examined but there was just not enough data available to give an assessment of the design: '*The HOTOL system is very attractive but it is difficult to give an appreciation of it without a more detailed knowledge of its propulsion system.*'

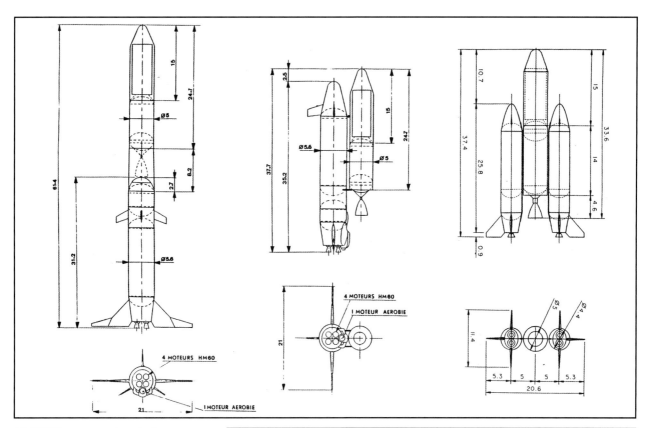

ABOVE These were the winning configurations at the end of the feasibility stage. *Aerospatiale via H Lacaze*

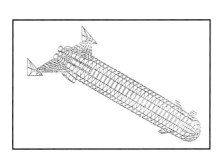

ABOVE The launcher as a line drawing. *Aerospatiale via H Lacaze*

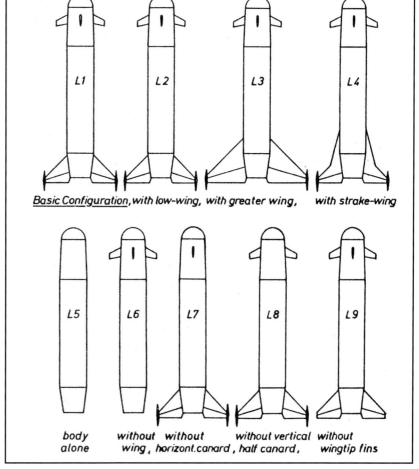

RIGHT Various architectures with, and without wings were evaluated by Dornier. *Aerospatiale via H Lacaze*

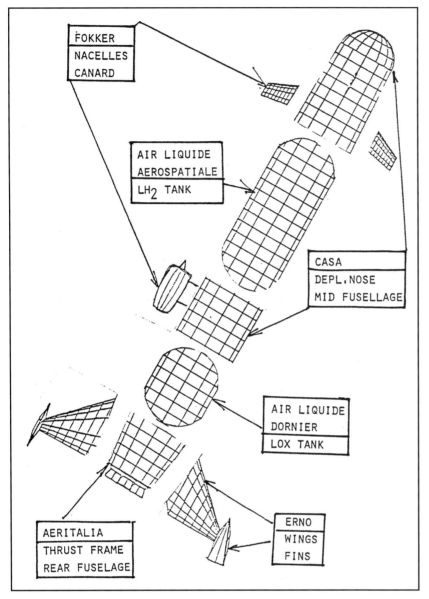

LEFT This stylised sketch indicates which company was in charge of designing which section of the vehicle. *Aerospatiale via H Lacaze*

B3 - The Ariane Launcher

The launcher selected for Hermes was the Ariane 5.

Ariane 5 was studied for seven years by CNES before being presented to the industry on 4 and 5 October 1984. It was a completely new vehicle whose commonality with Ariane 4 was limited to its design team: Aerospatiale, SEP, Matra, SNPE. From the beginning, Ariane 5 was intended not only to be a launcher of large geostationary satellites, but also of manned vehicles 'like the Hermes mini-shuttle.' After studying about twenty possible architectures, CNES finally selected three:

i   Ariane 5 *de référence* (AR5R: baseline Ariane 5): a direct derivative of Ariane 4 with the first stage being also used for the second stage.

ii Ariane 5 *cryogénique* (AR5C: cryogenic Ariane 5): in which all stages used LH_2/LO_2 propellants.

iii Ariane 5 *à poudre* (AR5P): in which the cryogenic HM-60-powered first stage was assisted by two solid-propellant boosters. This was the configuration selected, as it presented the '*best compromise on the technical, economical, calendar, logistical and industrial levels.*'[47]

The cryogenic engine used in the first stage, named HM-60 Vulcain (Vulcan) was developed by SEP with contributions by FIAT, MBB-ERNO and Volvo. This engine weighs 2,425lb (1,100kg) with a height of 10ft 2in (3.10m). Its nozzle is 5ft 11in (1.80 m) wide and can gimbal 4° in every direction. It gives a thrust of 100 tons (102 tonnes) for 519 seconds.

LEFT An Aerospatiale painting illustrating the winged FLS injecting a load into orbit as seen from a future European space station. *Aerospatiale*

The two *étages d'accélération à poudre* (EAP: solid-propellant boosters) P170 develop a thrust of 443 tons (450 tonnes) each for 118 seconds. They are 82ft 0in (25.00 m) tall with a diameter of 10ft 2in (3.10 m). The current version is the P230, giving 531 tons (540 tonnes) of thrust.

The third stage is available in both 'low energy' (L4) or 'high energy' (H10) versions, depending on the mission. It is not used for Hermes launches. In its three-stage configuration (satellite launcher) Ariane 5 is 159ft 5in (48.60m) tall with a diameter of 17ft 9in (5.40m). Its mass is up to 536 tons (545 tonnes) while the Vulcain engine and the two solid boosters together generate a thrust of 984 tons (1,000 tonnes).

First flight occurred on 4 June 1996, well after the abandonment of Hermes, and ended on a failure due to a mistake in a computer programme. The second Ariane 5 was launched on 30 October 1997, but could not achieve orbital insertion of the load. The third launch happened on 21 October 1998. It was a total success. The third launch had been originally planned as the maiden flight of Hermes. Instead it carried the Atmospheric Re-entry Demonstrator, an Apollo-style re-entry vehicle which performed perfectly as well.

But as Philippe Couillard, who would in 1986 become the *Directeur Hermes et vols habités* (Director Hermes and Manned Space Flight) at Aerospatiale wrote, 'Preceding studies had shown that a spaceplane is an oddity – neither satellite, nor launcher, nor aircraft – it is a little bit of all that, all rolled out into a single vehicle. Notably it associates space

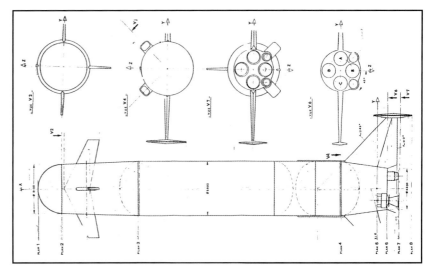

ABOVE A detailed drawing of the recoverable first stage with sections of the rear fuselage, showing the intakes and air ducts for the turbojets. *Aerospatiale via H Lacaze*

BELOW On the left, close-up of the nose with landing gear and bay; on the right, the main landing gear bays and turbojet pods. *Aerospatiale via H Lacaze*

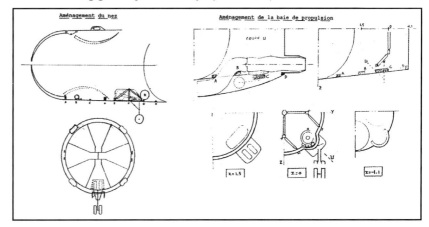

and aeronautical aspects.'[48]

CNES originally wanted to deal with Aerospatiale which was its traditional partner as the 'integrator' of space systems and which had an aircraft division which could work on the aircraft aspects ('glider') of Hermes. As Philippe Couillard put it, 'Aerospatiale is a big company where divisions are very independent and the Space Division, not convinced of the aeronautical aspect of Hermes, displayed little enthusiasm for involving another division. And the Aircraft Division was not much attracted by this project as it was busy developing, successfully, the Airbus family.'[49]

BELOW These two drawings illustrate how CASA of Spain investigated the practicalities of retractable turbojets. *Aerospatiale via H Lacaze*

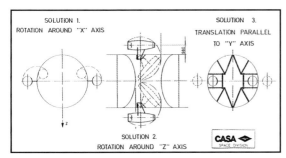
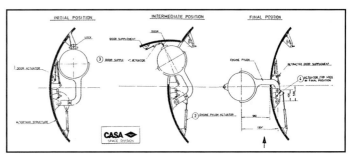

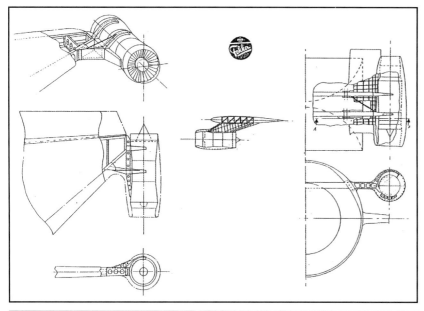

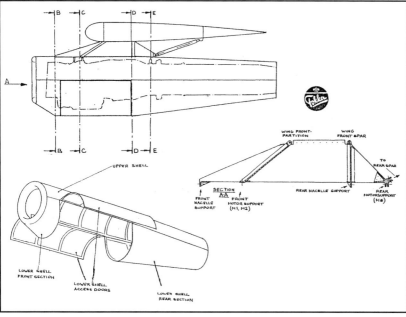

ABOVE Fokker, for its part, studied the positioning of turbojets in pods.
Aerospatiale via H Lacaze

Indeed, the Aircraft Division, although it had designed Concorde and was still keeping an eye on SSTs, had lost all interest in fighter planes[50], to which it likened the size and performance of Hermes. So CNES went to see Avions Marcel Dassault, which was regarded as the expert in high-performance military aircraft in France—and indeed, has been the sole supplier of fighters and bombers to the French Air Force[51] since the late 'fifties. This move was seen with disquiet by Aerospatiale which feared it would be left out of a major programme.

Conversely, Dassault was keen to enter the space market. Bruno Revellin-Falcoz, Directeur Général Technique (Technical Director) at Dassault said, '*All the major aircraft manufacturers in the World have a space division. Our ambitions in this sector should not come as a surprise, we just waited for the opportune hour*'[52]... but was less keen to co-operate on equal terms with any other manufacturer.

Therefore, CNES elected in March 1984 to issue a call for proposals, to which both manufacturers were invited, with the papers to be submitted in June 1986. Henri Lacaze, who had joined SNIAS in 1975 from ELDO entered the game at that time, recalling, '*CNES wanted to give the whole project to Dassault which was a dynamic company, while CNES perceived Aerospatiale as 'sleepy'—which was, in part, true. Dassault pushed forward its work on the Transporteur Aérospatial in the early sixties, but in truth, they had done nothing significant in the space sector. But they were good at selling their products and produced nice brochures.*'[53]

C – Early Hermes Concepts: Aerospatiale

Design of Hermes was the responsibility of the 'launcher division' based at Les Mureaux. Although in the press, the aircraft division's contribution was put forward, emphasizing work previously carried out on Concorde and on the then-projected *Avion de Transport Supersonique Futur* (ATSF = Future Supersonic Transport Aircraft), in truth, very little input came from it. The factory at Nantes was mentioned as contributing its experience with composite materials. The 'missile division' at Cannes is also said to have contributed to these early studies. At this point, Hermes was, in the words of Henri Martre, CEO of Aerospatiale '*un satellite habité*' (a manned satellite).

Originally, Aerospatiale did not invest much in the project, feeling it would not get into anything and would hide behind the necessary Europanisation. '*It is a big project around which could be built a European motivation and a European spirit. It is up to the French authorities to take the initiative of making contact with the various European partners. It is only after that we or Dassault, will be invited to take contact with our European partners,*'[54] said Henri Martre at that year's Paris Air Show, playing a waiting game in so far as this project was concerned. Eventually, Aerospatiale was forced to step up its efforts after the call for tender was issued and Dassault became involved in the affair.

After studying various shapes, Aerospatiale submitted 'Hermes 122'

RIGHT AND MIDDLE Fokker also proposed using a single, large turbojet in a bulge underneath the fuselage. Note the moving fins on the second drawing. *Aerospatiale via H Lacaze*

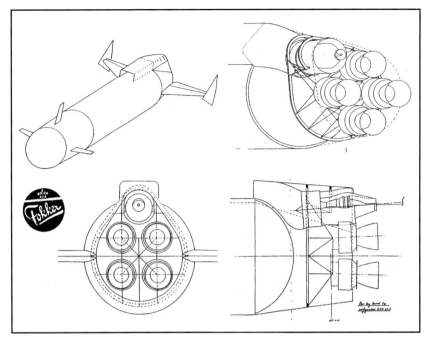

which could be recognised by its single fin, similar in profile to Rockwell's Orbiter. '*The vertical fin contributed to the equilibrium of the vehicle and also to a nose-up attitude thanks to the "crocodile" control flaps (a Payen patent). CNES's original project had V-fins but in the end, industrial designers favoured wingtip fins.*'[55]

The upturned nose was also a feature of this design. '*Work from Aerospatiale's Projets à Long Terme (Long Term Projects) and ONERA had shown that an upturned nose was better adapted to a hypersonic re-entry vehicle. So, we integrated this shape to the Aerospatiale proposal. Dassault had initially a more drooped nose but they agreed with us and the final project had a slightly upturned nose.*'[56]

Michel Rigault of Dassault adds, '*The curve of the nose directly influences local thermal flow. The material selected for the nose (carbon-carbon) implies a minimum local curvature. But this mostly concerns the lower surfaces and the curvature of the upper nose can be higher. (This is very noticeable on IXV[57] which was much smaller than Hermes). Then the curvature also impacts the aerodynamic efficiency, and therefore the cross-range performance. So, there is a need to reduce the nose section. The shape of the undersurface is also defined to satisfy the longitudinal aerodynamic stability constraint. The Shuttle orbiter was so much bigger, it is much easier to comply with the flow requirement; this may have helped to give it a simpler shape.*'[58]

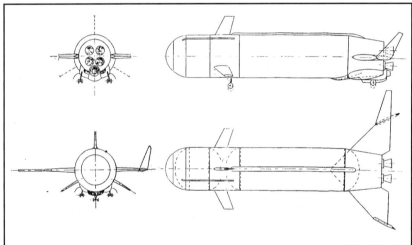

The vehicle had a double-delta wing with a span of 39ft 4in (12.00m) (without the winglets), a length of 50ft 10in (15.5m) and a height of 23ft 0in (7.02m). The cockpit was based on the one used in the Airbus A320, with CRT monitors and control stick on the right console.

RIGHT Aeritalia studied how to fit turbojets inside the fuselage in the same bay as the main rocket engines. *Aerospatiale via H Lacaze*

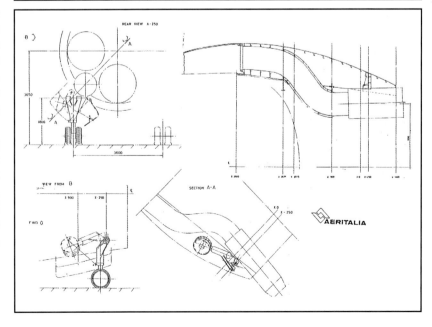

ABOVE This painting from 1984 illustrates separation of the two recoverable boosters and the consumable stage. *Aerospatiale via H Lacaze*

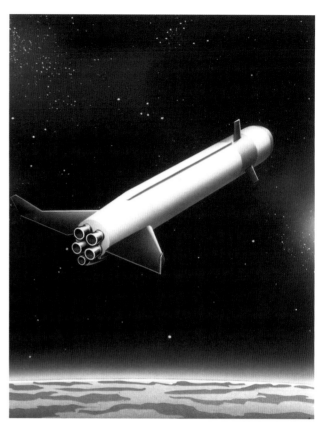

ABOVE The recoverable first stage begins its return flight. *Aerospatiale via H Lacaze*

Seats were adjustable in height to give the best outside view to the pilots during the approach and landing phase. The seats could be reclined for optimum resistance to the stresses of take-off and could be flattened for rest periods. Voice control was even considered.

While Hermes operations would be greatly automated, Aerospatiale insisted that '*the pilots will remain 100% in control, automation being there to assist them at will.*'[59] The 16 × 10ft (5 × 3m) cargo bay would offer a total volume of 1518cu ft (43cu m). A remote manipulator was positioned inside the bay as well as radiators as part of heat control system while in orbit. Thermal protection would rely entirely on passive materials: carbon-carbon for the areas (notably the nose) bearing the brunt of the heat stress during re-entry and Norsial panels for the underside.

Orbit insertion, manoeuvring and de-orbiting would be assured by two 1.96-ton (2 tonnes)-thrust rocket engines.

As could be seen, most of the design relied on solutions similar to those used on the US Shuttle orbiter, only the undersurface heat protection derived from earlier work by Nord-Aviation on Véras (see Chapter 6). Alain de Leffe explained to the author, '*We did not use the older researches like Véras because we had access to much more up-to-date information from the USA and the USSR.*'[60]

D – Early Hermes Concepts: Dassault

To reply to the call for proposals, Avions Marcel Dassault set up a new 'space department' under Jean-Maurice Roubertie. Aerodynamic studies were carried out by Pierre Perrier under the direction of Pierre Bohn.

Dassault claimed to have based its work on its TAS Orbiter proposal but never considered variable geometry and

BELOW A lot of research was conducted into cross-feeding the two stages of launcher. This technology was also intended for use in the later Taranis/STS-2000 studies. *Aerospatiale via H Lacaze*

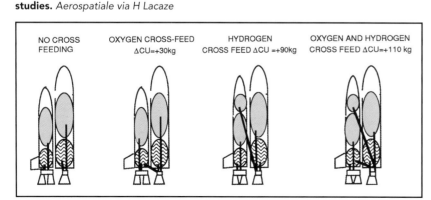

RIGHT Studies of recoverable boosters for the basic Ariane 5.
Aerospatiale via H Lacaze

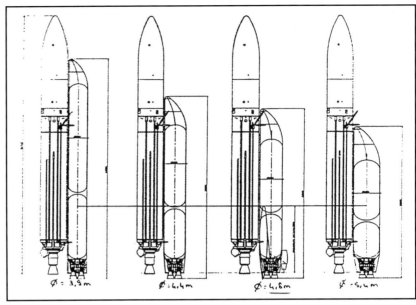

soon abandoned the TAS geometry for a more-or-less delta wing plan. Michel Rigault, who led the early Dassault studies, remembers, '*Aerospatiale's reply to the 1985 call for proposals was based on the American orbiter, reduced in size (double-delta wing and single fin/rudder assembly). Dassault selected a different approach, thinking that the larger size of the fuselage, when proportioned to the wing, could not lead to a controllable configuration. Dassault indeed based its initial work on the TAS studies. The earlier shape Hermes 1 was a scale-up of the TAS orbiter, itself based on the re-entry head of the MD-620 missile. All the computer programming related to hypersonic aerodynamics designed during the 'sixties, was re-used, as were the trajectory computation computer programmes in addition to all the computer tools Dassault had developed since.*'[61]

Hermes Aerospatiale 1985	
Status	Reply to cfp
Span	39ft 4in (12.00m) (without winglet)
Length	50ft 10in (15.50m)
Gross Wing area	–
Gross weight	–
Engine	1.9 tons (2 tonnes) thrust rocket engines
Maximum speed/height	–
Load	–

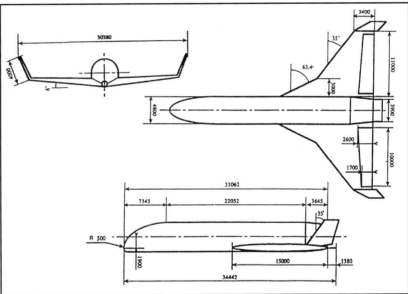

ABOVE Reference aerodynamic shape for a large winged booster.
Aerospatiale via H Lacaze

BELOW Cutaway drawing of that vehicle.
Aerospatiale via H Lacaze

BELOW Three-view drawing of the same vehicle, with twin turbojet pods. *Aerospatiale via H Lacaze*

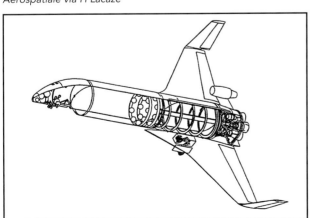

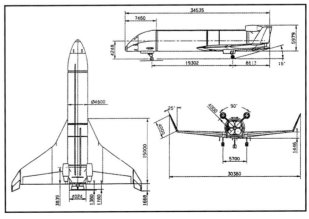

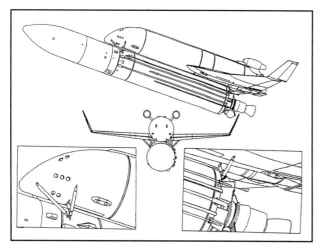

ABOVE Line drawing depicting the winged launcher attached to Ariane 5. *Aerospatiale via H Lacaze*

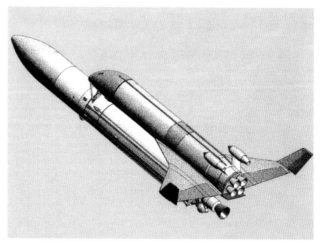

ABOVE Ariane attached to its winged booster. *ESA*

Hermes Dassault 1985
Status	Reply to cfp
Span	33ft 6in (10.20m)
Length	58ft 9in (17.90m)
Gross Wing area	-
Gross weight	36,927lb (16,750kg)
Engine	-
Maximum speed/height	-
Load	-

The original TAS shape was said to be too heavy while having insufficient 'cross-range' ability, insufficient lift and insufficient rudder (in fact, it had none). So, the first thing was the addition of a large fin and rudder assembly at the rear, but this (Hermes shapes 21 and 25), too, proved inadequate during re-entry. Therefore, instead of a large fin, the Dassault Hermes grew marked upturned winglets (although twin fins were considered for a time in Hermes 16 to 23). Michel Rigault explained, 'At the beginning of the re-entry trajectory, with an incidence of 40°, the fin/rudder assembly of the orbiter was inefficient. During a large part of the re-entry, flight control was obtained through vernier jets. Winglets are never masked by the fuselage and are therefore efficient from the beginning of the re-entry trajectory. They can stabilise the vehicle. The rudders mounted on them are also quite efficient for lateral control.'

LEFT Line drawing of MBB BETA II, the 10,000-kg payload SSTO design postulated by Dietrich Koelle. *MBB*
BELOW Cut-away of BETA II. *MBB*

BELOW ISTRA was Koelle's development of BETA which was entered in the Future Launch System study with unconvincing results. *Aerospatiale*

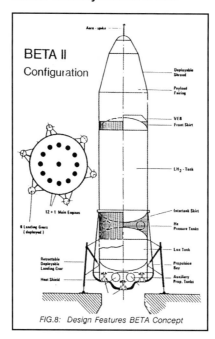

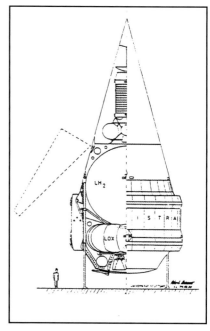

RIGHT Stylised art presenting Aerospatiale's entry in the CNES 1985 call for proposals. The specific feature of this design was the high tailfin associated with shorter winglets. *Aerospatiale*

This configuration, originally Hermes 35, then renamed Hermes 5D, studied by Michel Rigault and François Lemainque, was selected as the reply to CNES in late 1984. *Air & Cosmos* reported that among the references given by Dassault to support its project were '*simulation programmes for theoretical aerodynamics based on the profiles of the US Shuttle which compared up to Mach 26 to real measurements recorded by NASA during the second flight of the Shuttle*' and '*thermal protections which resisted 100 simulated re-entries at 1,200°C for a US call for proposals for Aerospace Transporter.*'[62] Indeed, shingles were taken from studies Dassault had done for Grumman on this company's Space Shuttle project.

According to Henri Lacaze, Dassault had considered attaching Hermes to the side of the launcher. '*Dassault had designed his Hermes like an aeroplane: the rear structure was built light because it was not required to bear any important stress. For an orbiter positioned on top of a rocket, this is different: this area must be reinforced*

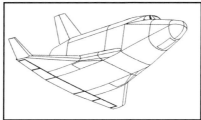

ABOVE Line drawing of the Aerospatiale design seen from below. From this angle, landing gear bay doors and trailing edge flaps are noteworthy. *Aerospatiale*

BELOW A large model photographed by the author at the 1985 Paris Air Show. Of note are the solar reflectors inside the payload bay doors; the upper airlock; and the manipulator arm. *JC Carbonel*

RIGHT Also seen on the Aerospatiale display at the 1985 Paris Air Show, the Aerospatiale spaceplane atop Ariane 5. *JC Carbonel*

ABOVE As usual, Dassault's promotional team came up with superb SFX photographs to depict its proposal. © Dassault Aviation

because it had to absorb the efforts of launch. Therefore, Dassault proposed to attach the orbiter sideways on the launcher. But it was not that easy; the rocket is mostly tanks and is not structured for attachment of big objects. It had already been quite difficult to attach the boosters to Ariane 4.'[63]

It should be noted that this question has been overlooked by CNES itself: in a January 1985 brochure, which described Hermes positioned atop Ariane, it was still described as having '*the same architecture as an aircraft airframe*' while acknowledging that '*use of composite materials will be generalised, except for a few parts using titanium or aluminium alloys*.'

The vehicle had a span of 33ft 6in (10.20m), a length of 58ft 9in (17.9m) and a height of 16ft 9in (5.10m). The cockpit offered six astronauts a space of 636cu ft (18cu m). Control was by way of two joysticks: the left one for translation, the right one for rotation. The control panel featured six CRT screens and two touch screens. Like Aerospatiale, Dassault offered to use voice control for some operations. Controls, including this voice control and fly-by-wire technology, came from systems already in use or tested on Dassault's jet fighters—Mirage 2000 and Mirage 4000 being expressly referred to. Dassault put forward at this occasion that it was '*the only aircraft manufacturer in the World to design and build the flight controls of its aircraft*.'[64]

BELOW Numerous wind tunnel tests were carried out to finesse the shape of the spaceplane. © Dassault Aviation

BELOW Another wind tunnel test, this time representing the complete launch configuration with Hermes atop Ariane 5. © Dassault Aviation

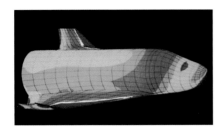

ABOVE Thermal studies of an early Hermes shape at Dassault.
© Dassault Aviation

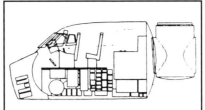

ABOVE Cutaway of the crew cabin inside Dassault's Hermes. On the right, the large cylinder is the upper airlock.
© Dassault Aviation

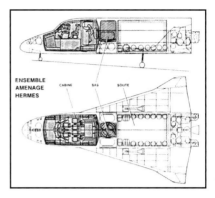

ABOVE A cutaway of the complete machine issued in 1986. Legend: Cabine = Crew cabin; Sas = Airlock; Soute = Cargo bay. © Dassault Aviation

The 9ft 10in (3 m)-wide cargo bay would offer a total volume of 1,236cu ft (35cu m). A docking unit was positioned inside the bay as well as radiators as part of heat control system while in orbit. This solution allowed for longer radiators.

The thermal protection would rely entirely on passive materials such as a composite structure for the areas supporting the brunt of the heat stress during re-entry; nose and leading edges and ceramic shingles, filled with different types of insulating material for the undersurfaces and the sides of the nose. Orbit injection, manoeuvring and de-orbiting would be facilitated by two rocket engines.

E – A Military Hermes?

In November 1984, Aerospatiale Division Systèmes Balistiques et Spatiaux (Space and Ballistic Systems Division) issued a few reports regarding a Véhicule d'Intervention Rapide en Orbite (Quick in-Orbit Intervention Vehicle). The documents indicated that they were based on a Hermes 'glider' feasibility studies. Not much is known about the VIRO project; it is presumed to have been a military programme, oriented toward reconnaissance from space but the 'intervention' in the acronym suggests that offensive action against enemy satellites could have been contemplated.

At the same time, the French military were concerned about the possible militarisation of space because they questioned the strategic value of something on such a pre-determined path as a satellite. 'Satellites have such kinetic energy given to them by their launcher that it is nearly impossible for them to move from their orbit. Their trajectory is entirely predictable, which is not a quality for a military vehicle. Against any power mastering space technologies, satellites will be vulnerable,' said Ingénieur Général Bousquet in a conference.[65] Although the rest of the article indicates that France had military observation and communication satellite programmes, there is at least an implication in General Bousquet's conference that any space power, as France aimed to be, would need offensive anti-satellite weapons. Could VIRO have been such a weapon?

According to the surviving documents, VIRO was designed for space reconnaissance…but nothing more is known about what kind of equipment could have been fitted in it. It would have been launched at high altitude from a mother aircraft (an Airbus A310 was quoted as having adequate technical features), lighting its rocket engine to reach a high altitude (VIRO was designed as a 'sub-orbital plane') and return by gliding from a point 2,485 miles (4,000km) away from its landing field. The whole mission was to take less than 24 hours.

One of the reports gave a short description of the vehicle. 'The suborbital vehicle would have a steel or special materials structure covered in ablative thermal protection. On each side are fitted large, droppable fuel tanks (manufactured in 15CDV6 steel) with a length of 18ft 1in (5.50m) and a diameter of 4ft 3in (1.30m). It features:

- Variable geometry wings
- Piriform fuselage
- Three-wheel landing gear
- Two symmetrical fins with controlled rudders
- Fuel for the Larzac turbojet stored in a bay behind the pilot and in the wings […]
- Propulsion systems:
  Rocket engine (½ Wiking): 1,323lb (600kg)
  Larzac turbojets: 2 × 662lb (300kg)
  Fuel and fixtures: 2,205lb (1,000kg)
  […]

The total weight dropped from the Airbus would be 55,116lb (25,000kg) out of which 11,464lb (5,200kg) would be the vehicle and 43,651 lb (19,800kg) the fuel tanks.'[66]

Of note is the fact that this report does not give any hint about the equipment to be fitted inside VIRO. The whole content of the airframe is bundled under 'equipment' weighing-in at 1,543lb (700kg). Because 'pilot' is singular, one could deduce that crew is limited to one person.

Viro (VG)	
Status	Pre-project
Span	45ft 3in (13.80m)
Length	24ft 7in (7.50m); 28ft 3in (8.60m) over the side tanks
Gross Wing area	–
Gross weight	11,464lb (5,200kg); to 55,116lb (25,000kg) with tanks
Engine	–
Maximum speed/height	2 × Larzac turbojets 3,327lb st (14.4kN) each + ½ Wiking rocket engine 75,310lb st (335kN)
Load	?

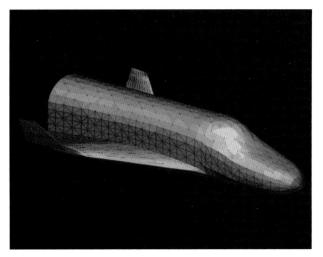

ABOVE This computer simulation comes from a 1987 brochure, but illustrates an earlier shape from 1985. © *Dassault Aviation*

ABOVE As the manufacturer of Ariane, Aerospatiale promoted it as the launcher of Hermes. However this undated drawing must have been created after the award of contracts. *Aerospatiale*

It should be further noted that another report dated 9 November 1984 illustrates a completely different machine, far more similar to Hermes, but without any precise dimensional data.

On 19 November the launch options were considered, including a fully-recoverable, two-stage launcher (but not air-launch as detailed in the December report) and single-stage launchers. '*A single-stage craft placing in orbit a load of this size, appears beyond the reach of French (or European) industry for many years. The use of air-breathing engines during the atmospheric flight, would considerably reduce the value [expressed in the report]. However, here too, it seems engines powerful enough would not be available for years.*'⁶⁷

The following day, another report was published aiming to evaluate the development costs of such a vehicle. It pointed out that, '*some of the technologies required [for VIRO] are not mature or available.*' In this report three prototypes were contemplated and detailed development was expected to begin in 1988.

The report's conclusion was that, using the PRICE methodology to compute the costs of such a project does not give reliable results because too many parameters are based on 'intuition'.

Nothing further was heard from VIRO.

F- Hermes for Europe?

While the two manufacturers worked on their respective proposals, CNES, having understood that France alone could not finance such a large project as Hermes, initiated the process of having it approved by ESA. So, in June 1984 it submitted a programme to be reviewed at the next (January 1985) ESA Council Meeting. France would then come to this meeting with the new Ariane 5 launcher, including Hermes as an optional load; while Germany would propose Columbus, a European module to be part of the International Space Station; and the UK submitted for European adoption a scientific Earth Observation Program.

But getting Hermes approved would not be easy; it was costly and, in the view of a few countries, it was 'too French.' Notably, the Germans, while

BELOW Another Aerospatiale illustration from the same set depicting separation of the boosters. *Aerospatiale*

shifting slightly from their earlier position of putting everything in the collaboration with the USA, had doubts that Europe could afford all these project—even if Hermes was interesting 'in the long term.' And various Franco-German meetings, up to Presidential level (Mitterrand and Kohl) would not change the situation. D'Allest, who had become CNES Director General in 1982, announced a few days before the Rome ESA meeting, *'If we can get agreements from our partners and friends, we will go ahead with the [Hermes] project in the framework of ESA. If not, France will go for it alone.'*[68]

Alain de Leffe said about the Rome meeting, *'D'Allest required me to accelerate the design of Hermes so it could be presented to ESA at the Rome meeting, at the same time as Ariane 5.'*[69]

On 4/5 October, at Arianespace headquarters in Evry (south of Paris), CNES presented to the European industrial partners and to ESA the development of Ariane, with the different variants of Ariane 5, including as launcher to a spaceplane named Hermes.

At the Rome meeting on 30/31 January **1985** various decisions were taken, initiating development of both the Columbus and Ariane 5 programme. But for Hermes the meeting was less fruitful: ESA only took *'note with interest of the French decision to undertake the Hermes manned spaceplane programme and the proposal by France to associate her European partners interested in the programme in detailed studies and invite France and associated partners to keep the Agency informed of progress of these studies with a view to include this programme, as soon as feasible, in the optional programmes of the Agency'.*

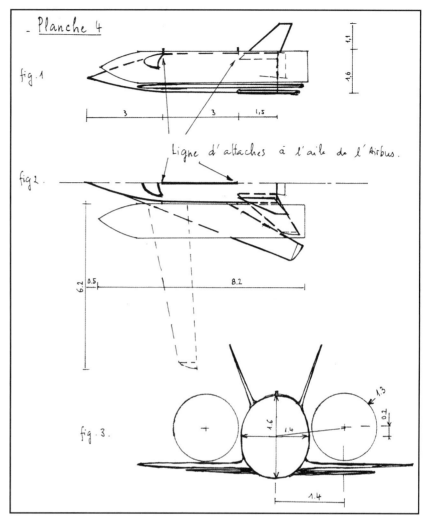

ABOVE Rough sketch of VIRO, the Rapid Intervention in Orbit Vehicle showing the variable-geometry wings, drop-tanks and the twin fins inherited from Hermes studies. *Aerospatiale via H Lacaze*

BELOW Illustration of a VIRO attached under the wing of an Airbus A310. *Aerospatiale via H Lacaze*

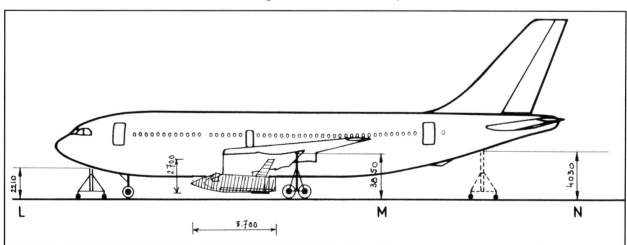

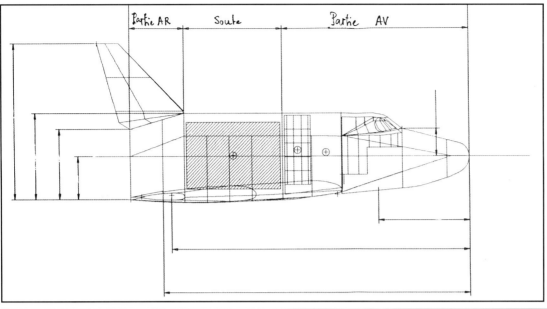

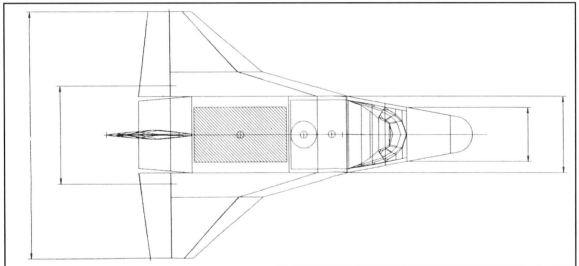

ABOVE Two drawings from the VIRO file…which appear to depict the Aerospatiale Hermes shape minus winglets and the domed rear deck. *Aerospatiale via H Lacaze*

A near similar wording was used for a new project, revealed at this meeting: BAe's HOTOL. ESA *'take[s] note of the studies under way in the United Kingdom of the future generation HOTOL project and invites the United Kingdom to keep the Agency informed.'*[70] (See Chapter 9.)

Philippe Couillard was keen to underline, *'The tastiest bit of the affair was the phrase "the French government has decided to launch the Hermes spaceplane programme." As far as I know, there was never a formal official decision to launch Hermes before the Rome conference. Maybe Prime Minister Laurent Fabius or President François Mitterrand had given a favourable impression during private conversation – of which I know nothing – but there was no government-level decision. Frederic d'Allest had obtained from his Minister the most he could get.'*[71]

On a more positive note, Belgium, Italy, Sweden and Switzerland announced at the Rome meeting their will to participate in Hermes. CNES had hoped to involve NASA in the project, but the Administration considered the Shuttle fulfilled all its needs, so a smaller personal mover was not required.

At this point CNES wondered about the competition it had launched. Should it continue? There was a risk that it spread inside Europe and that some ESA members would be dismissed because their national companies would have bet on the wrong horse. Which was clearly not possible because of the 'geographical return' rule of ESA[72]. But then the schedule had suddenly accelerated, so CNES could not afford to wait until the next year to select the industrial programme leader. It now needed to be identified quickly and presented to the European authorities. Therefore, in March CNES asked both industrial companies to present their case; now. Dassault presented its project to CNES on 18 April 1985; Aerospatiale, its own the following day. CNES said that the winner would be announced in September of that year.

In the same month a working group was initiated with CNES and ESA under Michel Bignier, ESA Director of Space Transportation Systems and Jean-Claude Husson, Directeur du Centre Spatial de Toulouse (Director of CNES Space Centre at Toulouse).

On the money side, Aerospatiale expected its Hermes to cost 17,120 million Francs to develop and fly, while Dassault announced 'only' 14,000 million Francs, both for three vehicles and all the accompanying support and launch equipment.

At the Paris Air Show it was announced by Frederic d'Allest that the first flight of Hermes would occur on 1 April 1995. Interestingly, this was an advancement of the schedule when compared with the January brochure which announced a first flight in 1997.

In June, Philippe Couillard was appointed Hermes Project Director at CNES while Bernard Deloffre became Hermes Programme Director at Aerospatiale.

[36] Interview by the author 8 February 2019.

[37] Mail exchanges with the author, October 2020.

[38] That's how Alain Souchier and French *Spationaut* Patrick Baudry call it in *Ariane* (Flammarion 1986) but it is obviously an abuse of language as the spaceplane was not yet named at that time.

[39] The anecdote was told by d'Allest himself to Luc van der Abeelen who reported it in his book *Hermes Europe's Dream of Independent Manned Spaceflight* (Springer 2017).

[40] Some of these data come from *Flug Revue* 3/1981 but do fit with other elements announced in 1979.

[41] As part of the MINOS programme.

[42] Quoted in Alain Souchier and Patrick Baudry *Ariane* (Flammarion 1986).

[43] *Lanceurs Futurs* R&D 85/86 Aerospatiale 18 March 1986.

[44] Koelle's own translation. Today we would tend to write 'launcher.'

[45] *BETA a Single-Stage, Reusable Ballistic Space Shuttle Concept*; Dietrich E. Koelle, MBB available from www.spacefuture.com.

[46] Phone interview with the author, 25 November 2020.

[47] *Air & Cosmos,* 13 October 1984.

[48] *Rêve d'Hermes,* by Philippe Couillard, April 1993.

[49] Ibid 5.

[50] In 1970 when the Dassault-Breguet-Dornier Alpha Jet had been selected over the Aerospatiale (then SNIAS) proposal, Marcel Chassagny, the CEO had objected strongly to seeing his company being pushed out of the military combat aircraft market. It is likely the pain was still felt inside Aerospatiale and may explain this rejection of anything approaching a fighter jet.

[51] But not the French Naval Air Force, which had bought Vought Crusaders during the 'sixties.

[52] Reported in *Air & Cosmos,* 30 March 1985.

[53] Interview by the author 8 February 2019.

[54] *Air & Cosmos,* 1 July 1985.

[55] Henri Lacaze interview by the author 8 February 2019.

[56] Ibid.

[57] IXV (Intermediate eXperimental Vehicle) is a re-entry vehicle tested by ESA in 2015. IXV is detailed on page 193, 194 and 195.

[58] Mail exchange with the author, October 2020.

[59] *Air & Cosmos,* 4 May 1985.

[60] Telephone interview with the author, 25 November 2020.

[61] Mail exchange with the author, October 2020.

[62] *Air & Cosmos*, 30 March 1985.

[63] Interview by the author, 8 February 2019.

[64] *Air & Cosmos,* 18 May 1985.

[65] Quoted in *Aviation Magazine,* 1 June 1989.

[66] *VIRO Avion Suborbital*, Tran-Thuan, Aerospatiale, 14 December 1984.

[67] *VIRO Premier Dimensionnement des Moyens de Lancement*, 19November 1984.

[68] Quoted in Luc van der Abeelen's book *Hermes Europe's Dream of Independent Manned Spaceflight,* Springer 2017.

[69] Phone interview by the author 25 November 2020.

[70] Both from Luc van der Abeelen's book *Hermes Europe's Dream of Independent Manned Spaceflight,* Springer 2017.

[71] In *Rêve d'Hermes*.

[72] According to the French Senate report on *Les enjeux et perspectives de la politique spatiale européenne* (Challenges and perspectives of the European space policy 'ESA uses for its contracts a rule of geographical return which is not compatible with the EEC. This rule of geographical return aims at an equitable participation by each State member to the execution of the space programmes, on the basis of its financial participation. Differently phrased: The more a state participates in the ESA budget, the more ESA contracts its national industry receives.' That rule would, indeed, cause many difficulties in the program.

Chapter Eight
Designing Hermes

ABOVE This very evocative model of the 5M2 Hermes docked with an MTFF demonstrates that the opening panels on the back are for solar reflectors, not payload bay doors. *Aerospatiale*

To assess the two manufacturers' offerings, CNES set up a Groupe d'Evaluation (Evaluation Group) under Philippe Couillard and a Commission du Choix (Selection Commission; actually a special meeting of the Board of CNES). As is often the case when two committees have to study the same question, they diverged: the Groupe d'Evaluation favoured Aerospatiale, based on this manufacturer's long experience with the Ariane launcher (Philippe Couillard was to take the position of Director of Manned Flight at Aerospatiale, so it was only logical it would propose Aerospatiale); while the Commission du Choix saw fit to favour the new entrant, Dassault. In the end, the managers won over the technicians and Dassault was selected.

That was not the end of the story because CNES – being a state entity – required validation of its choice at government level. And French President François Mitterrand and his Prime Minister Laurent Fabius were wary of Dassault, which was perceived as an arms dealer versus the (nearly) purely civilian and partly state-controlled Aerospatiale. This put CNES into a quandary, only resolved by the announcement on 18 October 1985 that, *'the responsibility for project management (the Maitre d'Oeuvre Industrielle: MOI) of the spaceplane is entrusted to Aerospatiale. This responsibility encompasses all works required to build the aircraft.*

'The position of delegate project manager for aeronautics (Maître d'Oeuvre Délégué à l'Aéronautique: MODA) is entrusted to Avions Marcel Dassault-Breguet Aviation. This responsibility encompasses all works required for the success of the atmospheric flight (aerodynamics, aerothermal, flight qualities, general design of structure and aerodynamic shapes, design and provision of thermal protection). The responsibilities are given to these two companies through a contract co-signed by the two companies and CNES, which is the manager of the complete Hermes system encompassing the spaceplane, control and mission centre, payload preparation centre, crew training, launch & landing installations and logistical support.'[73]

This organisation became known as *la direction bicéphale*: the two-headed directorate. Basically, Aerospatiale was in charge of the project, which satisfied everyone in term of politics, and Dassault was in charge of the spaceplane which was also to its own satisfaction.

The press was quick to point out that Aerospatiale and Dassault had co-operated in the past: Concorde, Mercure, Mirage 2000, business jets and composite materials. Interestingly, the Mirage VTOL in which Sud-Aviation supplied the whole fuselage was not mentioned; while the Concorde co-operation was limited to Super-Caravelle studies.

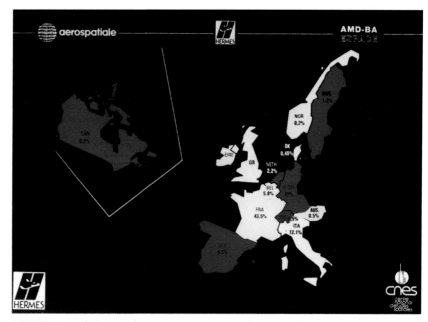

ABOVE **Map of Europe showing countries contributing to the Hermes programme. The significance of the colours is unknown.** *CNES*

BELOW **Model of the 1985-86 design model '5M1'.** © *Dassault Aviation*

BELOW **Rendering of a 1985 'Hermes 5D'. This was a shape very closely derived from the Transporteur Aérospatial re-entry glider of 1964.** © *Dassault Aviation*

BELOW **The rendezvous with an MTFF was a constant of Hermes illustrations.** *CNES*

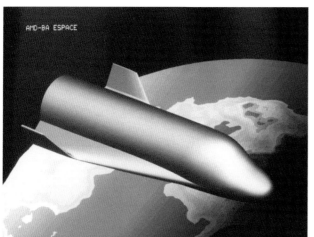

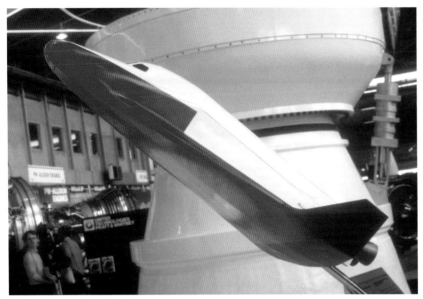

ABOVE At the 1987 Paris Air Show, Dassault had a good way of implying that re-entry would be a 'hot' affair: painting the whole underside of a model in bright orange-red! *Author's collection*

BELOW As the designer of Ariane rocket engines, SEP (Société Européenne de Propulsion) was keen to jump onto the Hermes bandwagon. This artwork was used to promote its wares in the late 'eighties. *SEP via author*

This new organisation was agreed on by the three participants on 5 October 1985 and accepted by the French Government and officially announced by CNES on 18 October 1985. Technically, and rather logically within the organisational set-up, it was Dassault's proposal which was retained as the basis for Hermes development. So the design accepted by CNES had the following features:

- Dimensions and weights: 33ft 6in (10.20m) span, 58ft 9in (17.90m) length, 36,927lb (16,750kg) take-off weight with an empty weight of 19,842lb (9,000kg).

BELOW This model of Hermes, signed by Spationauts Patrick Baudry and Jean-François Clervoy now resides in the salons of the Aero Club de France. *JC Carbonel*

- Capacity: six astronauts in a pressurised cabin and an unpressurised cargo bay with a diameter of 9ft 10in (3.00m) with a length of 16ft 5in (5.00m) carrying up to 9,921lb (4,500kg) of payload (for a mission to a space station; for a polar orbit this would be reduced to 2,205lb (1,000kg).

- Propulsion: two 4,496lb st (20kN)-thrust rocket engines identical to those used on Ariane 5 L-4 stage. For fuel, monomethyl hydrazine and nitrogen peroxide were to be used.

- Power management: 240kW would be available for vehicle operation and on-board experiments. Power would come from three fuel cells (H_2/O_2) which CNES intended to buy in the USA as no European supplier for this technology was available. Adding lithium batteries were also mooted.

- Heat management: only passive heat control was considered, using ceramic shingles[74] (silica shingles were also investigated) for the nose, the leading edges and the under surfaces. For upper surfaces (regarded as 'cold'), advanced flexible materials were used. Michel Rigault explained, 'Shingles (Dassault used the word 'tuiles') are rigid panels, made of ceramic-matrix composites (CMC), mechanically attached to the structure and protecting it through a multi-layered reflecting insulation, much lighter than silicates. Dimensions are 11¾ × 11¾in (0.30 × 0.30m). This system re-used the principles of Dassault's proposal for the thermal protection of the NASA orbiter (Dassault had worked with Grumman on NASA's call for proposals but Rockwell was selected in 1972). At that time, panels were made of metal and the system had been proved in laboratory tests. Ultimately, this solution was tested on the IXV that effected a piloted re-entry in 2015.'[75] For the windows (windscreen, observation ports) the technology would have been based on the materials used on the Rockwell orbiter.

ABOVE A model Hermes joined a model futuristic SST in Aerospatiale's display at the 1987 Paris Air Show. But by that time, the spaceplane was a pure Dassault design. JC Carbonel

- Environment management: the astronaut cabin was to be pressurised, and air-conditioned at 19°C. Water for both drinking and hygiene would be provided by the fuel cells: each crewmember would be allocated 4½ pints (2.5 litres) per day.

At that point, the question of the sub-contractor in charge of the heat protection was unresolved, although the Société Européenne de Propulsion (SEP) had announced its interest in that aspect.

Hermes was presented to ESA on 25 October and soon many countries were climbing aboard this bandwagon, not wanting to be left out of what was perceived as the major European space programme for the end of the century. Philippe Couillard recalled, '*I remember mostly the favourable welcome of our presentation by German industries when we met them in the Ministry in Bonn.*[76] *I also remember the enthusiastic and spontaneous adhesion of the Belgian industry delegates, like they had given to Ariane and SPOT. Actually, all the delegations including the British and Irish ones were very receptive to the French proposal. The appeal of manned spaceflight is unquestionable and no one can refuse to be in at first. Everyone wants to at least participate in the Programme Préparatoire Européen (European Preparatory Programme), soon to be set up.*'[77]

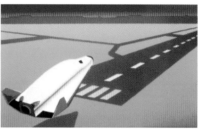

ABOVE Artwork showing Hermes about to land on an airfield. This illustration originates from a 1987 brochure. ESA

BELOW Sectioned views comparing the 5M1 (upper airlock + cargo bay opening to space) and the 5M2 (airlock moved to the rear, fully enclosed cargo bay). Author's collection

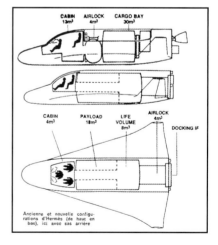

However, the German government delegates remained dubious; they thought France just could not afford the large investment required, especially at the beginning of the project, and doubted the rest of the ESA countries involved in the programme could step over. Germany however, proposed to have 30% participation if France agreed to postpone the first flight from 1995 to 1997.

30% was larger than its participation in Ariane. But the German position was ambiguous because, while the German government was cautious, German industry, universities and even the DFVLR (Deutsche Forschungs- und Versuchsanstalt für Luft- und Raumfahrt – which became the Deutsche Forschungsanstalt für Luft- und Raumfahrt in 1989 – German Research Establishment for Air- and Space Travel) were very keen to participate in Hermes. To try to smooth the relations between France and Germany, French Research and Technology Minister Hubert Curien suggested a German national could be part of the crew for the first flight of Hermes.

Finally, the programme budget was overfinanced by 11.4%—a rare occurrence. But then it was probably greatly under-estimated.

At that point three main missions were planned for Hermes:

i Between 7 and 30 days on a 249 to 497-mile (400 to 800-km) orbit for science experiments.

ii Up to 15 days on a 311-mile (500-km) sun-synchronous orbit for satellite maintenance.

iii Shuttling between Earth and space station to transport astronauts, supplies and samples. It could be used as a rescue vehicle for those stations, carrying up to 12 people in this role (with a special pressurised module in the cargo bay).

Although Hermes had been initially envisaged with a flight life of 100 missions (ten missions per year over ten years), CNES now only planned two missions per year for fifteen years. With two vehicles, that meant 60 flights.

In November the industrial programme team was established at Les Mureaux to design Hermes. As Philippe Couillard recalled in *Rêve d'Hermes* (Dream of Hermes), '*That was not easy. Cultures and approaches from Dassault and Aerospatiale clashed widely. And the wounds from the competition were still open.*' Basically, the design philosophies were opposed on two points:

■ The space people knew they needed to get it right from the outset, so they had to keep everything simple

ABOVE AND LEFT Three photographs of a model of the 5M2 configuration. During the life of the project many different desk models of Hermes were produced. *JC Carbonel*

Hermes 1985	
Status	CNES project
Span	33ft 6in (10.20m)
Length	58ft 9in (17.90m)
Gross Wing area	–
Gross weight	36,927lb (16,750kg)
Engine	2 × rocket engines
Maximum speed/height	–
Load	Up to 9,921lb (4,500kg)

CHAPTER EIGHT — DESIGNING HERMES

ABOVE Dassault, being foremost an aircraft manufacturer, studied in depth the purely aeronautical aspects of Hermes. Here a model is undergoing test in a subsonic wind tunnel in 1989. © Dassault Aviation

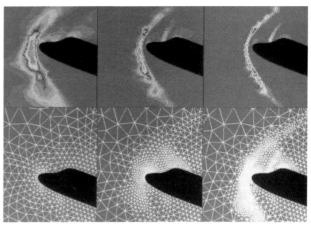

ABOVE To study the dispersion of shockwaves, Dassault used 'meshing' of the shockwaves on the nose of Hermes as seen in this picture. © Dassault Aviation

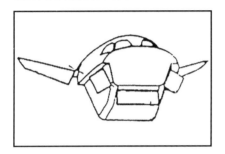

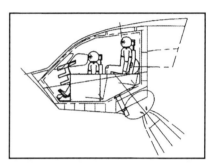

RIGHT Sectioned view showing an enclosed rescue module in Hermes. *Author's collection*

LEFT Line drawing of the jettisonable rescue cabin. *Author's collection*

and, if possible, incorporate many redundancies. A side effect of this approach was to automate things as much as possible and keep people 'off the loop.' '*Space engineers have a deep faith in the good working of automatic systems*.'

- On the other hand, aircraft people sought to approach their goal by increments, relying on a succession of prototypes which could themselves be improved after each flight. Aircraft are flown by pilots at the controls, so pilots should remain in charge as much as possible. '*The importance of the machine-human relationship came from aeronautical culture and was difficult for the space world to understand*.'

This opposition, notably in regard to the role of pilots in the machine, was not specific to the European project and had been observed in the past, notably during the American Mercury project, in which the astronauts had to fight the engineers to get control of the spacecraft.

In the end, Dassault brought its aerodynamic study, its concept of thermal protection and structure while Aerospatiale supplied the avionics and the idea of using solid rocket extractors, these having the dual role of third stage in a nominal flight or escape system if trouble occurred during the launch.

In **1986** something happened which had major repercussions on the Hermes programme: the accident of Space Shuttle *Challenger*. This was the first time NASA astronauts had been killed during a mission. The possibly of having people actually dying during a space mission directly impacted the minds of policy-makers and the public at large.

ESA took immediate notice and began developing various escape solutions for Hermes, these being beyond the four solid-fuel rockets attached to the Hermes-to-Ariane adapter that were considered up to that point. Ejection seats and even an escape capsule were now to be integrated into Hermes. Ariane 5 itself was reviewed: *Spationaut* Patrick Baudry had just been recruited by Aerospatiale as advisor on the Hermes program: he proposed to have Ariane HM-60 first stage redesigned with two H205 rocket engines instead of a single H155 Vulcain.

Alain de Leffe confirmed this approach, '*The requirement given to Ariane's development team was to keep the "lower composite" (first stage and boosters) common to both Hermes launcher and satellite launcher. The only concession to man-rating Ariane 5 was to improve the avionics with more computer power (the same computer planned for Hermes was used for Ariane 5 instrument case). Spationaut Patrick Baudry, who was our consultant for Hermes, asked that Ariane 5 receive a twin-engine first stage claiming that "one does not fly across the Atlantic Ocean on a single engine", but this was refused. The only thing that was granted to us was to fire the first stage before firing the boosters. It improves safety of the launch but loses 3,086lb (1,400kg) to geostationary orbit.*'[78]

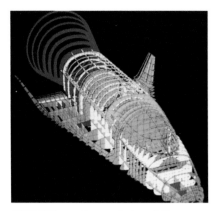
ABOVE Computer rendering of the spaceplane's structure. © *Dassault Aviation*

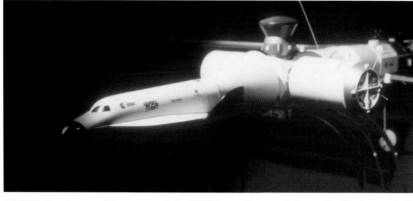
ABOVE The theme of Hermes and the MTFF was also carried over in display models at the 1991 Paris Air Show. *JC Carbonel*

In February, French President François Mitterrand met Chancellor Helmut Kohl and tried to convince him to get "on board" Hermes, with very moderate success. The official communiqué stated, '*The German Government agrees to examine with the French Government the conditions and the width of its participation in the Hermes project. The German Government's final decision will be taken during Autumn 1986 after supplementary bilateral talks, so that the realisation of the project can begin in mid-1987*.'[79]

In view of German indecision, French Research Minister Alain Devaquet did not hesitate to tell his German counterpart Heinz Reisenhuber '*France, if needed, will build Hermes alone.*'[80]

In March 1986 the official 'Europeanisation' request was submitted by CNES to ESA. It was approved on 23 October 1986, but with the caveat that CNES was not awarded complete delegation of authority from ESA. All orders had to be approved by ESA which, for this reason, established a special team in Toulouse in November. This way, CNES and ESA teams could work side by side. Thus, began the Hermes Preparatory Programme.

The Hermes schedule was at that time the following:

- Beginning of Hermes development: mid-1987
- Detailed design studies up to October 1990
- Ground qualification tests up to mid-1993
- Six subsonic flights in 1994
- First flight in April 1995
- Earliest operational flights in 1996

The early flights were already planned:

1995 two qualification flights
1996 two science and technology experimental flights

ABOVE But using Hermes to go to the Soviet (Russian) *Mir* space station was always a prime interest. As A De Leffe said to the author, while Hermes was being designed, the Russians had a space station in orbit and operating and a French spationaut went there; while the USA was still undecided on the question of European participation in their space station. *Author's collection*

RIGHT Two drawings of the 8M1 configuration.

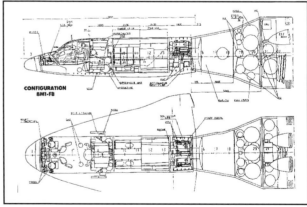

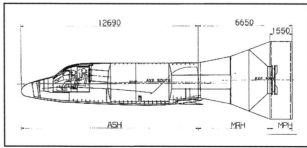

Hermes 5M1 1986	
Status	Project
Span	35ft 1in (10.70m)
Length	50ft 6in (15.40m)
Gross Wing area	915sq ft (85.00sq m)
Gross weight	-
Engine	2 × rocket engines
Maximum speed/height	-
Load	Up to 9,921lb (4,500kg)

Hermes 5L3 1986	
Status	Project
Span	-
Length	-
Gross Wing area	-
Gross weight	-
Engine	-
Maximum speed/height	-
Load	-

ABOVE The configuration 8R1, illustrated in model art by Dassault. © Dassault Aviation

1997 one science flight and one orbital intervention
1998 one science flight and one flight to a space station
1999 one science flight and one orbital intervention
2000 one science flight and one flight to a space station
2001 one science flight and one orbital intervention
2002 one orbital intervention and one flight to a space station
2003 one orbital intervention and one flight to a space station
2004 one orbital intervention and two flights to a space station
2005 one orbital intervention and three flights to a space station[81]

Meanwhile, ESA had asked the European Space Research and Technology Centre (ESTEC) to explore the feasibility of having Hermes as support vehicle to Eureca, Polar Platform and the MTFF (Man-Tended Free-Flyer platform; see Chapter 12). Eureca was an ESA programme for acquiring a better knowledge of the space environment. The Eureca spacecraft was designed and developed to be recovered from orbit after completion of the mission and returned to Earth. It actually flew between 1992 and 1993, being launched and recovered by the US Shuttle. The Polar platform was an unmanned spacecraft intended for Earth observation. MTFF was a 1987 project from ESA to have experimental platforms in orbit. MTFF and Polar Platform were part of the *Columbus* umbrella project. Those studies were highly critical of Hermes, pointing out a cargo bay of insufficient volume, a too short manipulator arm, and underdeveloped EVA and crew escape systems. ESTEC even went further; a working group led by its director Marius Le Fèvre concluded that Hermes was not feasible.

Missions to the Polar Platform were cancelled after the *Challenger* accident as indicated by Alain De Leffe. '*The Challenger accident caused a global reappraisal of the security of the astronauts. Besides the obvious question of the recovery of the astronauts in case of a catastrophic failure during the launch sequence, the question of the cosmic and solar radiation was re-examined. Polar orbits were much more risky, regarding radiation, than LEO orbits. This contributed to the abandonment of all Hermes missions with polar orbits, and therefore of interacting with a polar platform.*'[82]

That year saw the appearance of competition to Hermes in the form of the British HOTOL proposal and the German Sänger II (see Chapter 9).

Following wind-tunnel experiments, the outer shape of Hermes was modified becoming Hermes 5M1 in June with a fuselage that was both shorter and wider. However, this version was the extreme limit in term of acceptable wing surface at 915sq ft (85.00 sq m) (it had been earlier estimated that a larger wing surface could destabilise Ariane during the launch sequence: a sudden gust of wind impacting Hermes' wing could thus cause a total loss of control of the launcher).

Michel Rigault remembers, '*Indeed, it was a question of stability but also of flight controls (gimballing of the nozzles of the boosters). It was required that Hermes use the standard version of Ariane without any modifications. Modifying Ariane would have reduced its performance or increased the costs of commercial launches.*'[83]

That month, the CNES industrial design team was transferred from Les Mureaux to Toulouse.

In 1995 the mission programme was revised, the first orbital flight to be preceded by six manned, subsonic sorties. Operational flights would begin in 1996, at first with two a year, then after 2004, three or four flights per annum. The life of each Hermes spaceplane was still estimated as 30 flights.

In 1986 the list of sub-contractors participating in the Hermes programme was finalised:

Hermes sub-contractors	
Aeritalia (Italy)	thermal control
Alcatel-Thompson (France)	telecommunications
Dassault (France)	thermal protection
Dornier (Germany)	life-support and fuel cells with Elenco (Belgium)
ECTA (Belgium)	electrical systems
Fokker (Netherlands)	HERA manipulator
MATRA (France)	data management
MBB-ERNO (Germany)	propulsion systems
SEP (France)	hot structures

Hermes Mission Control would be located at Darmstadt in Germany.

Philippe Couillard thought this allocation of roles came too early. 'How could work be divided into thirteen or fifteen countries, while the product to be delivered is not frozen? A new project is to be worked on top-down. First must come detailed "system analysis." From the objectives and the needs of the mission, the concept must be defined in broad strokes. And this must be reiterated until the best compromise between mission and vehicle is obtained. Only then can one deal with sub-functions and equipment, together with their associated technologies. And even at this level, multiple iterations are required to fine-tune the balance between the mission objectives/requirements, the system and the required equipment.'[84]

Following the *Challenger* accident, ESA set up a Hermes Safety Advisory Committee (HESAC) in January **1987**. HESAC was chaired by Pierre Govaerts with vice-Chaimen André Turcat (French test pilot who gained notoriety with Concorde 001 maiden flight) and Ernst Messerschmid (the third West-German astronaut).

Henri Lacaze remembers, '*The question of the rescue of the crew in case of a mishap during the launch sequence became very important after the accident to the US Shuttle. Dassault had not planned anything in this respect, while Aerospatiale suggested ejection of the whole spaceplane.*'[85]

In February a new version of Hermes, the 5M2 was finalised. A major development was the suppression of the open, unpressurised cargo bay. The bay remained, but would not open directly into space for satellite delivery or repair.

Henri Lacaze felt it was a bad decision and elected to leave the project, moving to the Direction des Etudes Avancées (Advanced Studies Direction) of Aerospatiale, saying, '*CNES decided that Hermes would have an enclosed cargo bay which enabled a gain in mass, but also greatly reduced the possible uses of the orbiter. All our previous work had been carried out on the principle that the cargo bay would open into space—which "opened" a lot more possibilities for use.*'[86]

On the contrary, Philippe Couillard justified it. '*If an unpressurised bay is used, since the main mission is to serve a pressurised – by definition – space station, then the load should be carried in a pressurised container, which could be docked to the station. Instead of carrying this container up to the station and down from it at each flight, it appeared more interesting to use the mass allocated to the container and re-affect it to the structure of the plane, and contribute then to the stress resistance of the whole craft.*'[87]

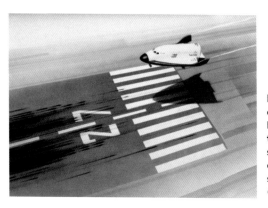

RIGHT Dassault was very concerned by the operation of Hermes as an aircraft. Here their house artist depicted the spaceplane about to touch down on a runway after a successful re-entry.
© Dassault Aviation

On the other hand, a docking ring, compatible with the ISS (at the time the Space Station was know as "US Space Station Freedom", today it is the ISS, the International Space Station. However, the actors interviewed by the author, all talked or wrote about the "ISS". Because of this the word "ISS" was used as short-hand to designate the object, whatever its real name at the time of the action) was installed in the rear fuselage. This would integrate an airlock with enough air for two successive cycles. Two EVA suits would be carried in Hermes. However, the foldable heat-exchangers, the manipulator arm and the high-gain antenna remained. The payload capacity was reduced to 6,614lb (3,000kg). This version introduced a major change: the deletion of the engine. Small bulges, like those on the Rockwell orbiter, were installed at the rear of the fuselage, containing small thrusters for attitude control. Added to the spaceplane was the Hermes Propulsion Module. This was the adapter section between Hermes and Ariane inside which were installed two gimballed 6,744lb st (30kN) rocket engines and the associated hydrazine and nitrogen tetroxide tanks.

For the 'hot' structures (leading edges, winglets and control surfaces) silicon carbide-silicon carbide (SiC-SiC) and carbon-silicon carbide (C-SiC) would be used, with carbon-carbon (C-C) for the nose. The rest of the spaceplane would be covered by Flexible External Insulation (for the cooler surfaces) and Rigid External Insulation for the undersurfaces. The REI would be constituted of shingles containing Internal Multilayer Insulation.

This new iteration introduced the escape module concept, in which the cockpit section could be ejected as a 'lifeboat' in case anything went wrong during the launch. This offered numerous advantages:

- All the crew in single system
- Ease of localisation of a module by comparison with dispersed astronauts
- Protection offered by the module against environment and possible explosion, debris, etc
- No action required by the crew
- Ejection takes place in a very short time

- But also many disadvantages:
- The module is aerodynamically unstable
- It may prove impossible to control it during and after ejection
- Difficulty in designing an effective shock-absorber to cushion the landing
- Impact on the spaceplane design: the centre of gravity will be displaced and the whole system will have a weight penalty

Balancing advantages and disadvantages of the system would prove a subject of discord between the pros and the cons for several months. Philippe Couillard, for one, was not averse to this new addition, saying '*It replaces the solid-rocket ejectors intended to get the plane away from the launcher in case of trouble. Those ejectors would have required an increase in Ariane 5 overall performance. Even used as a third stage, they did not "pay for their ticket" on board Ariane.*'[88]

This new version of Hermes showed a notable increase in weight, which climbed to 20.6 tons (21tonnes). At that time, it was planned that the first Hermes flight would be the third Ariane 5 launch.

The ministerial-level Council of ESA, originally planned for March, was pushed to November 1987, due in part to some of the 'preparatory phases' of Hermes, Ariane 5 and *Columbus* needing extension and also because of negotiations with NASA on participation in the ISS trailed off. During the summer of that year, the UK decided to participate in Hermes, investing £2 million in the process…while still promoting its own HOTOL (see Chapter 9). At the same time, the Germans were also completely undecided between Hermes and Sänger II (see Chapter 9).

During 9/10 November, the ESA Council met in The Hague. It approved three programmes: Ariane 5, *Columbus* (MTFF and the European Module to be attached to the US Space Station) and Hermes. Other programmes were just studies: European Manned Space Infrastructure (EMSI), an acronym covering an independent European Space Station and Future European Space Transportation Investigation Programme (FESTIP) to study possible post-Hermes vehicles (see Chapter 12). '*It is necessary to initiate studies about the future systems needed at the beginning of the 21st Century, so that options could be offered around 1996,*' said the ESA booklet summarising the Council.

Because the three programmes were quite interdependent a 'coherence board' was set up at ESA. Hermes was fiercely debated, with attacks from the British and German delegations but in the end it was passed. Ariane 5 was voted without demur. A major item of news was that Germany had accepted participation in the Hermes programme, financing 27% of the budget (against France 43.5%, Italy 12.10% and eight other countries the remaining 17%)

Hermes development was now separated into Phase 1, running from 1988 to 1990 intended to lead to freezing the definition of the spaceplane while Phase 2, running from 1991 to 1999 was to see the actual assembly of the spaceplane and its first launch. Philippe Couillard Indicated in *Rêve d'Hermes*, '*I personally realised that the test flights were to occur in 1997 and 1998: ten years to develop [Hermes] appeared to me to be enough. The longer the development, the costlier it is. Neither Apollo nor Ariane had a longer duration.*'

ABOVE The cover of this CNES-ESA brochure illustrates the whole flight of Hermes: taking off atop an Ariane 5; docking with an MTFF; and returning to Earth. CNES

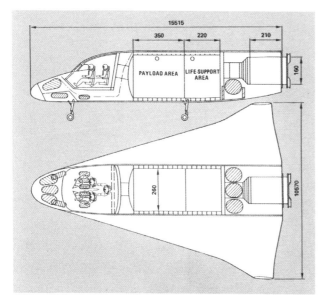

ABOVE A two-view drawing of the new Hermes 5M2 configuration with the upper airlock transferred to the rear of the hull and the payload bay fully enclosed. *Dornier*

ABOVE Artist's impression of the 5M2 configuration in orbit. The thin black lines on the back do not suggest opening bay doors but opening solar reflectors. *Dornier*

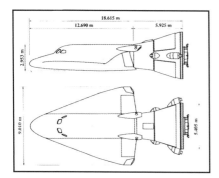

ABOVE Two-view drawing of the 8P8 configuration of 1990. Hermes was by that time approaching its final configuration. *CNES*

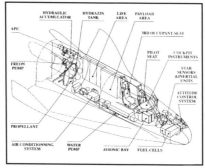

ABOVE 'Ghost' view of the interior equipment of the 8P8 configuration. *CNES*

Reversing its previous decision, the UK decided against participating in any of the three main programmes.

At that point the Hermes Programme was piloted by the Direction du Transport Spatial at ESA, led by Jean-Jacques Caspart. Responsibility for the spaceplane was delegated to CNES's Direction Hermes et Vols Habités (Directorate of Hermes and Manned Spaceflight) under Philippe Couillard. Industrial responsibility was shared between Aerospatiale and Dassault.

However as pointed out by Philippe Couillard, inside these broad delegations a few key elements were not assigned:

■ The structure for communication with the spacecraft—ie, ground stations, satellite relays.

■ CNES had a Centre de Contrôle en Vol (Flight Control Centre) but ESA set up a Mission Control at its historic Space Operations Centre at Darmstadt, Germany. The latter had its origins in 1967, to follow the satellites to be launched under the former ESRO organisation.

■ The *spationauts* 'hand luggage' (as Philippe Couillard put it), their interior suits (this was rescinded when the escape cabin was abandoned), the manipulator arm (HERA: Hermes Robotic Arm).

Project management was being questioned within that organisation. Methods and procedures used to guarantee the best viability and security of the various products to be designed and produced for the project were much discussed…which was not entirely unexpected, as everything in the Hermes programme was new and complex. But it still could be pointed out that sometimes it was more a question of being 'not invented there' than a practical affair. According to Philippe Couillard, '*MBB claimed that the product assurance rules used on Ariane could not be adopted for Hermes because they were French and therefore not European. What a sterile debate!*'

In early **1988** a new Hermes shape appeared: 5MX which departed from earlier design in pushing more and more equipment into the Ariane 5 adapter, now re-labelled Hermes Resource Module (HRM) while the spaceplane proper became the Hermes Spaceplane (HSP) (in French: Avion Spatial Hermes [ASH.]) The idea was to shed some of the weight gained in the previous iteration of Hermes but at the cost that everything placed into the HRM would be non-recoverable. Another reason for pushing as much equipment into the adaptor was to take into account the differential between the admissible launch mass (about 23.6 tons [24 tonnes]) and the admissible landing mass (about 16.7 tons [17 tonnes]) this being dependent upon the wing area.

There were three configurations possible for the 5MX:

i 5MX-A: baseline version with the airlock relocated in between the crew section and the payload bay. The HRM contained the rocket engines and a detachable payload module with a capacity of 1.47 tons (1.5 tonnes).

ii 5MX-B: the HRM no longer included any payload but contained heat-exchangers and various tanks.

iii 5MX-C: the payload bay was now fully pressurised and the airlock/docking ring was again pushed at the rear, inside the HRM.

After review by ESA, -A and -B were found lacking regarding the docking system and generally the Hermes/ISS interface. So the 5MX-C was divided into sub-variants, while the –D and –E configurations appeared.

iv Configuration -D reintegrated the docking ring/airlock in the spaceplane while putting more payload in the adapter (3,748lb [1 700kg]) than in the plane (2,866lb [1,300kg]).

v Configuration -E kept the same internal organisation while managing to reduce the overall span of the re-entry section.

A reason behind the indecision as to where to put the airlock may have been NASA insistence for standardising the ISS docking unit at 50 inches (127cm) which was quite large when trying to fit it inside a 99½in (253 cm)-wide fuselage (the outer section of the NASA-standard module was 80in [203cm] in width).

Philippe Couillard reported. '*I remember a meeting at CNES in September 1988. I had in front of me the Directors of Space Transport and Orbital Infrastructure with their teams. I fought to make them understand that these requirements*

RIGHT Cutaway of Hermes. *ESA*

made Hermes more difficult to design and they could be filled through other ways. But to no avail: Hermes was to be built to American standards or not serve the MTFF. CNES agreed to that; I still regret it.'[89]

Alain de Leffe noted, "*When Hermes was designed, the ISS was still being designed and the European participation in it was unclear. At the same time, the Soviets (Russians) had a space station that was real and running and which Spationaut Jean-Loup Chretien visited in 1988. So, it was more logical (and practical because it had a smaller diameter) to design Hermes with a Mir-compatible docking ring.*'[90]

In the end, configuration -C was preferred as its docking module/airlock in the HRM made it easier to comply with the NASA specification. But there remained the mass question: it was still too heavy for Ariane 5.

Hermes's main mission was then a fully European one: servicing the MTFF (although the MTFF could also fall back on the American Space Station for resupply).

Hermes 5M2 1987	
Status	Project
Span	33ft 0in (10.07m)
Length	50ft 7in (15.43m)
Gross Wing area	786sq ft (73.00sq m)
Gross weight	46,932lb (21,288kg)
Engine	None
Maximum speed/height	–
Load	Up to 6,614lb (3,000kg)

ABOVE **Model of the 8P8 configuration of Hermes.** *Author's collection*

From the last quarter of 1988, contracts were allocated to European industries for designing and testing Hermes components. This occupied the full year 1988 and continued into **1989**. The configuration of the HRM was still discussed: where to put the two rocket engines in relation to the airlock?

The heat-exchangers were also source of debates. Now that they had been pushed into, or more exactly onto the HRM, it appeared they needed to be protected from the heat stress incurred during the ascent phase. Finally, it was decided to use a heat-exchanger folded inside the HRM which would therefore not need thermal protection. But this was a source of problems regarding the docking procedure.

The payload continued to contract: it was now estimated at barely over 1.96 tons (2 tonnes): 3,527lb (1,600kg) in the spaceplane and 1,279lb (580kg) in the HTM. The difference between these and the earlier 6,614lb (3,000kg) laid in the manipulator arm, EVA suits and other equipment which was now considered as 'mission dependant' on the basis that they would not be needed on each mission. Originally, they had been considered as the fly-away kit of the spaceplane.

Indeed, as servicing the Man-Tended Free-Flyer platform (MTFF) was still Hermes's main role, tests were carried out to see how astronauts would interact with it. Under the SALOON (Study of Accommodation Logistics On-board) sub-programme, a full mockup of the Hermes cargo bay, interior of HRM and MTFF interior was built to investigate loading and unloading experiments module to and from the MTFF.

On 23 March, the design of the Crew Escape Module was discussed at an ESA meeting with various extraction devices being considered: side boosters or escape tower *à la* Apollo. The favoured solution was the one with side boosters; a smaller integrated rocket engine was also part of this design to cushion the landing. While approving this solution, the meeting recommended continued studies of other escape options like encapsulated ejection seats. Two variants were proposed: either ejecting the whole nose of the spaceplane (Type A) or only the cockpit section (Type B).

Between 27 June and 19 July the Revue de Définition Préliminaire-Avion (RDP-A: Preliminary Aircraft Definition Review) took place to evaluate and validate the design specifications of the Hermes spaceplane. Results were presented to the ESA Steering Board on 21-22 September. The configuration was basically approved but more than 1,000 questions were raised leading to 117 recommendations. A main focus of these recommendations was the Crew Escape Module.

Coincidentally, the Russians who had always been friendly towards the French in space co-operation, were now even more willing to collaborate under

RIGHT Ariane promoted its capacity to launch either satellites or Hermes. *ESA*

Mikhail Gorbachev's new 'glasnost' administration. During the Paris Air Show they forwarded information on the ejection seat designed for the Buran shuttle, which could be used up to Mach 2.5. Further co-operation on this subject appeared therefore very much *à propos*. Besides, TsAGI facilities were offered at a time when the use of US resources had proved difficult.

Co-operation with NASA was also investigated: in May, Hermes was offered as a Space Rescue System (SRS) for the ISS. A delegation of CNES/Aerospatiale (Dornier was also a contributor to these studies) representatives was sent to Lockheed in conjunction with a request by NASA to design an Assured Crew Return Capability (ACRC). It was proposed to sell to NASA three vehicles, one of which would remain permanently docked for two years to the ISS (two simultaneously docked vehicles would be possible but with 'high impacts on *Freedom*.'

This design, named Crew Escape Re-entry Vehicle (CERV) would have been modified by Lockheed to become a 'lifeboat' for the future ISS. Two scenarios for use were considered: either evacuation of a sick or hurt ISS crew member, in which case the rear section of Hermes-SRS could receive a stretcher; or all-out evacuation of the ISS. The payload section of Hermes would be completely redesigned to fit six seats to transport astronauts in case the evacuation of the ISS was necessary. This could even be expanded to eight astronauts by deleting galley, toilets and even…the pilot (*'pilot's role questionable,'* said the final review report). This variant of Hermes was 7,716lb (3,500kg) lighter

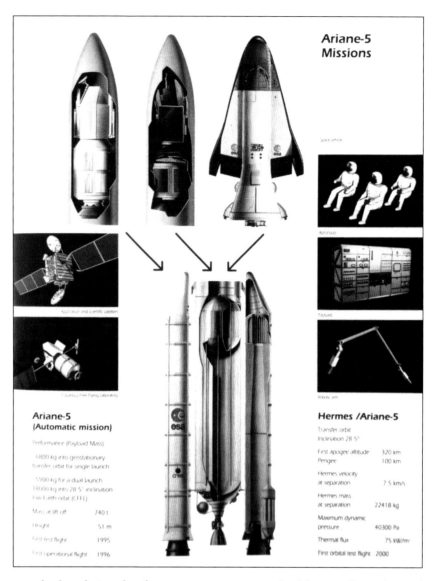

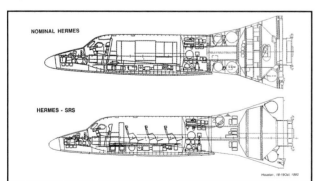

ABOVE Hermes was offered to NASA as a space rescue system for the Space Station. Here are two sectioned views comparing the nominal Hermes and the SRS. *Aerospatiale via H Lacaze*

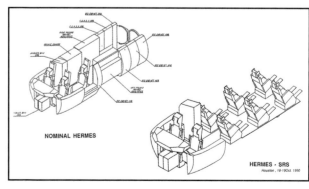

ABOVE Comparison between the interior fittings of the nominal Hermes and the SRS. *Aerospatiale via H Lacaze*

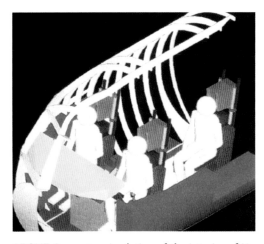 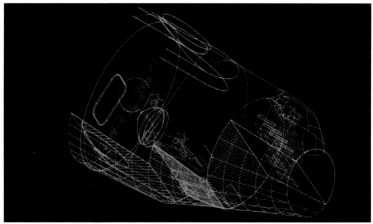

ABOVE Computer simulation of the interior of Hermes, as drawn by Aerospatiale circa 1985. The rectangular door on the left was the access to the cargo bay while the double blue circle on the top of the vehicle represent the airlock hatch/docking ring. *Aerospatiale*

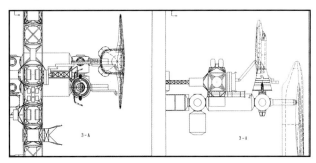 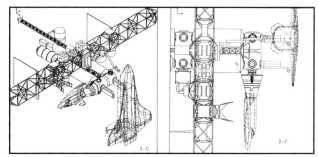 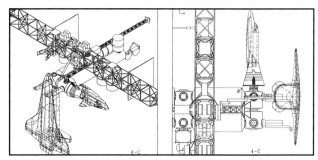

ABOVE AND RIGHT Three pictures extracted from an Aerospatiale SRS brochure showing the three ports which were deemed practical for docking Hermes-SRS to the Space Station. *Aerospatiale via H Lacaze*

than the baseline version. Other fittings such as life support equipment and fuel cells would need to be redesigned to extend the mission-span from 28 days (Hermes) to 730 days (CERV).

Although Philippe Couillard hoped that by benefiting from the European work at a good price, Lockheed and Aerospatiale had a great advantage when replying to NASA's call for proposals.

Talk about the CERV lasted one year and a final review meeting was held in Houston on 17/18 October 1990. Only two docking ports were found to be available for a Hermes-SRS: the 3C and 4C. By that time Hermes had evolved into the 8M1 variant.

One question raised by the Americans was the resistance of Hermes (and notably of its thermal protection) after two years among micro-meteorites and other orbital debris (a risk estimated at 1% during a two-year stay). A possible solution would have been to deploy protective blankets from the SRS. Other options examined included a double-skin pressurised module with a high thermal coupling between both aluminium skins; exchanging the radiating surfaces on the resource module for 'foam thermal protection' (FEI) 0.39in (10mm) thick; inducing an airflow from the station; and even…painting Hermes overall black.

Unfortunately, NASA refused to be dependent on foreign suppliers for major equipment.

In the end, the ACRV/CERV programme saw many different iterations, including prototypes like the X-38. ESA participated in several studies but NASA cancelled the whole effort in 2002 and has relied on Soyuz spacecraft for this role since the beginning of the ISS.

The new configuration defined by RDP-A, notwithstanding the many questions still attached, was labelled 8M1-E, itself an evolution of the 5MX-E of the past year. Its payload was still indicated at 6,614lb (3,000kg) but with this payload, its overall weight would be 54,322lb (24,640kg) while Ariane 5 was rated at 52,249lb (23,700kg)—therefore 2,072lb (940kg) must be shaved off the announced cargo mass which was then barely above 4,409lb (2,000kg). Moreover, the mass which could be returned to Earth was now limited to 1,279lb (580kg).

In this version, the docking ring/airlock returned to a dorsal position on the spaceplane which had already proved a difficult location in view of the NASA requirements. This airlock was only 31½ × 39½in (80 × 100cm)—presumably, therefore dedicated to MTFF operations. A second airlock, with a side door was provided in the HRM. Storage space was available in both the pressurised and unpressurised sections of the HRM. A Hermes Propulsion Module, supporting two rocket engines and associated fuel tanks, was integrated into the HRM.

Philippe Couillard, in his souvenir book *Rêve d'Hermes,* regretted that the mass requirement of Ariane 5 was enforced quite tightly. '*In 1989, a project review pointed out that the mass requirement imposed upon Hermes was so high that it led to an exaggerated use of new technologies. To reduce costs and development risks, the study suggested to relax the rule by 1.96 tons (2 tonnes) based on the following logic: experience shows that the performance of launchers increases during their operational life.*' However, he pointed out that it rendered Hermes dependent on Ariane 5 growth and the cost of this growth could then be charged to Hermes, making it even more costly. And there was also the question of forgoing research on advanced materials. This work was done mostly in Spain, Italy and Belgium, so reducing this research effort meant diminishing the appeal of Hermes for these countries

At the end of 1989, it was recommended that the decision to commit ESA to Phase 2 (the actual construction of Hermes craft and associated equipment) be delayed from the end of 1990 to June 1991.

In January **1990** two major questions were to be answered: What escape system? What propulsion system? The decision was made early to abandon the escape module solution. When compared with the simpler, leaner, ejection seats like those in Buran, there were just too many uncertainties concerning effectiveness of the module. It was more efficient to use something already developed and tested like the Russian seats. Naturally this would have impacts on the recovery procedure (the need to locate and pick-up three separate astronauts instead of one large module) the design of the Ariane launcher to incorporate alarm systems and 'degraded mode' operations.

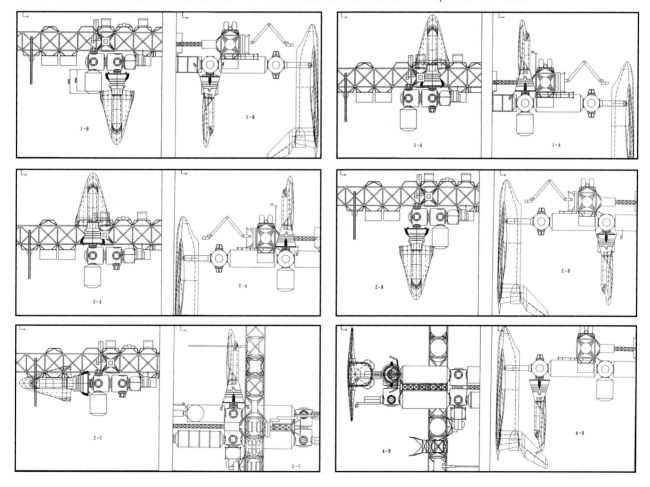

BELOW Six other locations were studied but were found unsuitable for the SRS. *Aerospatiale via H Lacaze*

Hermes 8M1 1989

Status	Project
Span	29ft 7in (9.01m)
Length	42ft 8in (13.00m); 60ft 0in (18.30m) with HRM
Gross Wing area	786sq ft (73.00sq m)
Gross weight	54,322lb (24,640kg)
Engine	2 × rocket engines (in HRM)
Maximum speed/height	–
Load	Up to 6,614lb (3,000kg)

LEFT Artwork of the Hermes cockpit, as designed by Aerospatiale circa 1985. Note the mini-stick on the piloy's right armrest. *Aerospatiale*

ABOVE Spationaut Patrick Baudry in the cockpit mockup, circa 1986. *Author's collection*

The psychological impact of this option was also to be incorporated in the astronaut selection process. And finally, this meant that the internal structure of Hermes should be hardened and made capable of surviving a heat shock like the one which caused the destruction of the shuttle *Challenger*. The deletion of the escape module enabled designers to shave a full ton (1.02 tonnes) from Hermes. Consequently, the contractor for the escape system (ejection seat and associated spacesuit) was selected outside Europe: the Russian manufacturer Zvezda, which had supplied Buran, was chosen.

Alain de Leffe told the author, 'Mach 3.5 ejection seats had been tested successfully aboard Soyuz. It could eject the astronaut over 1,312ft (400m) and could be activated even with the rocket still on the launch pad. Choosing the ejection seat solution prompted a lot of re-thinking: the original plan was that Hermes could bring anyone into orbit: scientists, technicians…Using ejection seats limited the population to fully trained crewmembers, in good physical condition…like military pilots. Ejection seats are a violent solution. So, in the end, using ejection seats was like going back to the old days of the conquest of space. Indeed, during the first flights of the American Space Shuttle, the test crews were on ejection seats.'[91]

Philippe Couillard confided in *Rêve d'Hermes*, 'Personally I regretted this choice. I don't minimise the complexity and therefore the costs of developing an escape cabin compared with the ejection seats, but I remained convinced that a rescue solution applicable during the whole atmospheric sequence of Ariane 5, that is until the solid-propellant boosters are ejected, was needed.' He added that the studies on such escape pods could find its use in hypersonic aircraft and hoped that the question would be assessed again when new studies are made on manned spaceflight in Europe. In retrospect this vision appears to have been quite optimistic, as hypersonic manned aircraft and European manned spacecraft are still things of the future.

Regarding the propulsion system, the Hermes Propulsion Module was deleted. It was estimated simpler to enlarge the fuel tanks of the Ariane 5 H150 stage to take 4.9 tons (5 tonnes) more fuel and thus assure the direct orbital insertion of Hermes. In this configuration Hermes could reach the MTFF or the ISS directly. A small additional engine package would, however, be needed to reach the Russian Space Station *Mir*.

During the second Industrial Day held in Munich on 1 February 1990, ESA announced that the Hermes and Columbus programmes would be delayed by six months. Actual industrialisation was postponed until 1 July 1991. MTFF was renamed as the Columbus Free-Flyer. This six-month delay forced Jörg Feustel-Büechl ESA Director of Space Transportation Systems to restate the reasons for Europe to build Hermes in an interview given to *Flight* magazine in March 1990: to master manned spaceflight; to become independent in space; and to be recognised as an equal partner in negotiations with the USA and Russia.

This year held various events which could influence the project: German elections in the winter (not forgetting that since the disintegration of East Germany in November 1989, Germany was engaged in a costly re-unification process) and the development of the ISS. Indeed the US Department of Defense wanted to use the ISS as part of the Space Defence Initiative (the so called 'Star Wars' programme), and the Congress wanted to trim the budget of the American Space Station—both actions which endangered the previously agreed European and Canadian participation.

The reference design for Hermes was 8R1 with a slight increase in size but a reduction in weight, and the abandonment of on-board rocket engines. Philippe Couillard explained in *Rêve d'Hermes* how this was possible, saying, '*It is necessary that the main cryogenic stage [of the launcher] falls back in the ocean, so that in case of any failure, its probability of hurting or killing people be significantly reduced. This consideration led to favouring a natural fall back in the Atlantic Ocean (from Kourou, launches aim toward the east). This led to reduction of the speed increase expected from this stage, and to compensation for that by adding an extra propulsion system in the adaptor between the plane and the launcher. That actually equals adding an extra stage to the rocket. This configuration remained in the Hermes design until 1990. Then more detailed studies showed that by selecting adequate orbital insertion parameters, the fall back of Ariane 5 central stage occurs in the Pacific Ocean and the probability of a fall back in Africa in case of failure was small enough to fulfil the criterion imposed for the safeguarding of populations*.'

One would have thought that at this point the overall shape would have been finalised but studies with a full delta wing, no winglet and a large fin still appeared.

In an effort to improve project management, a joint ESA-CNES Hermes Programme Directorate was set up under Michel Courtois who had been, up to that point, the head of Hermes Project Team at CNES. His ESA deputy was Jean-Jacques Caspart, who had formerly been the head of the ESA Hermes team. This configuration was questioned at ESA as some members objected to a CNES manager being at the same time the head of the ESA-CNES joint committee.

Philippe Couillard welcomed this new organisation, saying, '*Most delegations wished for a more European, less French programme, so the controls from the Agency required a lot more people and became just as big as the CNES team. Seen from the industry, the customer organisation had two stages and generated a lot of exchanges up and down between the two stages [...] Such a large and advanced programme required a demanding customer, but also one which could lead and be reactive, which it was not [...] so around the end of 1989 and into 1990, the idea emerged of merging both teams into one.*'

Yet this new organisation had an impact on the industrial side: ESA wanted to have one, and only one partner, and this meant Aerospatiale. According to Philippe Couillard this led to having Dassault on the side... physically. The Dassault engineers were regrouped in the newly opened Dassault space centre at Blagnac, not far but away from the main CNES team.

Regarding technical matters, it was decided to use aluminium for the 'cold structures' of Hermes. It was easier to work with, and the industry was familiar with it. But the real reason was that the European industry had been unable to come up with the composite material originally intended. This resulted in an increase in weight of 551lb (250kg) and a limitation in the acceptable temperature range. Another area in which European industry was found deficient was fuel cells.

In July ESA issued a request for proposals for Phase 2: the actual building of Hermes.

By the second half of 1990, Hermes had evolved into the 8P8. Recognisable features of this version were enlarged winglets and prominent, elongated thruster pods on the rear fuselage. Weighing 21.65 tons (22 tonnes) at launch, Hermes would be only 14.7 tons (15 tonnes) at landing. It was able to carry 6,613lb (3,000kg) into orbit and 3,307lb (1,500kg) from orbit. This shape was constantly refined through many iterations taking into consideration position of the centre of gravity, subsonic performance, and definition of the thruster pods.

Hermes 8R1 1990	
Status	Project
Span	32ft 2in (9.80m)
Length	45ft 7in (13.9m); 72ft 2in (22.00m) with HRM
Gross Wing area	904sq ft (84.00sq m)
Gross weight	48,502lb (22,000kg)
Engine	None
Maximum speed/height	-
Load	Up to 6,614lb (3,000kg)
Hermes 8P8 1990	
Status	Project
Span	-
Length	41ft 8in (12.7m); 61ft 0in (18.6m) with HRM
Gross Wing area	-
Gross weight	48,502lb (22,000kg)
Engine	None
Maximum speed/height	56mi (90km) transfer orbit
Load	Up to 6,614lb (3,000kg)

RIGHT Patrick Baudry was the first spationaut to be involved in the design of Hermes. Aerospatiale promoted him as the 'manned spaceflight advisor' in this picture.
Aerospatiale

ABOVE Dassault studied in depth the spacesuits that could be worn by spationauts during EVA around Hermes and MTFF. Here a CATIA computer simulation of space-suited astronaut.
© Dassault Aviation

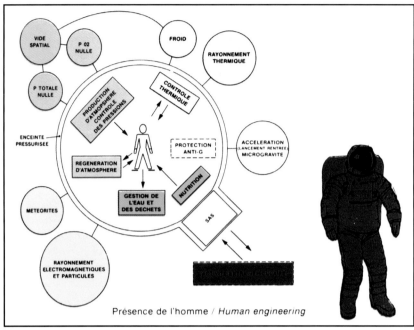

ABOVE In this drawing, Dassault indicated all the features needed in a spacesuit (a pressure suit = enceinte pressurisée) in blue = supplying an atmosphere; in white = thermal protection of the astronaut; in green = feeding the astronaut and dealing with waste; in yellow = protecting the astronaut against radiation and micro-meteorites. In red: the result, a spacesuit. © Dassault Aviation

The project schedule was now to have the first flightworthy Hermes in 1995 followed by the second machine the next year. Subsonic free-flight tests (Approach and Landing Tests: ALT) would occur in 1996. The first space flight (unmanned) would be launched in 1998 to be followed by the first manned mission in 1999. It was expected that qualification and experimental flights would follow, probably reduced to one per annum instead of the original two per year planned. All in all, the operational life of Hermes was not expected to begin until…2015. The two Hermes spaceplanes had 34 operational missions scheduled for them.

In October, Wolfgang Wild the Director-General of the Deutsche Agentur für Raumfahrtangelegenheiten (DARA: the German Agency for Space Affairs—created in 1989 to take over the space activities previously managed by DFVLR) announced that he did not believe it possible to afford the European Long Term Programme and asked for Columbus and Hermes to be delayed, while cutting their budgets by 15–20%. In consequence of this declaration (which was confirmed by a letter to ESA asking the same) ESA agreed to push the decision about going to Phase 2 until February of the next year.

In November 1990 EuroHermespace was founded as an industrial consortium of Aerospatiale, Dassault Aviation, Deutsche Aerospace (Daimler Benz AG had absorbed MBB in 1989 and would absorb Dornier and MTU by 1992) and Alenia. EuroHermespace took over the role of prime contractor from Aerospatiale with Philippe Couillard as its CEO. This French-law company was set up at the request of ESA and CNES which wanted to concentrate the industrial responsibilities and Europeanise the prime contractor of Hermes.

This was not well received by industrialists in other countries, who felt 'excluded.'

Of the creation of EuroHermespace, Philippe Couillard recounted in *Rêve d'Hermes*,

'The Hermes programme should be carried out like other space projects, therefore using proved methods. EuroHermespace should first have the qualities expected from any programme leader: ability to synthesise, ability to decide, recognised authority among industrial partners. I therefore advised use of the Société Anonyme constitution, rather than the Groupement d'Intérêt Economique for two main reasons:

i Delegation of responsibility to the CEO from the shareholders of a Société Anonyme is stronger and larger than to the gérant (managing director) of a GIE.

ii Besides, the personnel are recruited by the Société Anonyme under a single statute while in a GIE they remain employed by their original company.

Hermes SRS	
Status	Proposal to NASA
Span	-
Length	41ft 8in (12.7m); 61ft 0in (18.6m) with HRM
Gross Wing area	-
Gross weight	40,786lb (18,500kg)
Engine	None
Maximum speed/height	-
Load	6 × astronauts (+ 2 crew)

Dassault Maia	
Status	Proposal
Span	11ft 0in (3.36m)
Length	17ft 6in (5.34m)
Gross Wing area	-
Gross weight	3,748lb (1,700kg)
Engine	-
Maximum speed/height	-
Load	None

'Obviously, duplication of jobs disappears in such an organisation. The hierarchy was constituted by a delicate proportioning between the shareholding companies and each posting corresponded to an actual function.

'As DASA and Alenia were not familiar with the very French Société Anonyme with Président Directeur Général, the form selected was the Société à Directoire et Conseil de Surveillance (with Board of Directors and Supervisory Board). This has the advantage of not concentrating power on a single nationality. If the General Manager is French, then the Director of the Supervisory Board can be of any other nationality.'

He also revealed a little anecdote. 'I can't tell you why we called this company EuroHermespace, which is a remarkably long and difficult name to pronounce.

Initially we thought of "Euro-Hermes" but Hermes the World-renowned luxury goods manufacturer wanted to keep this name for its own use, as it already had branches like Hermes-Italy and Hermes-Japan. So, we went for Hermespace, like we already had Arianespace. There was no longer any ambiguity with consumer goods. But our German partners wanted to see a reference to Europe, which was essential to them. Those two constraints led us to this fourteen-letter word.'

In parallel, the two French partners, Aerospatiale and Dassault formed Hermespace France while in the Assemblée Nationale, French Parliament's lower house, asked, through senator Paul Torre, 'Is a shuttle still needed?'[92]

On 7/8 February 1991, a special ESA Council was held at Santa Margherita in Italy and a new schedule was validated:

- MTFF launch was delayed from 1998 to 2001.
- The European Manned Space Infrastructure (that is a French Space Station) was postponed (and would eventually be cancelled).
- Hermes maiden flight was pushed from 1998 to 2000. The first manned flight occurring in 2001.

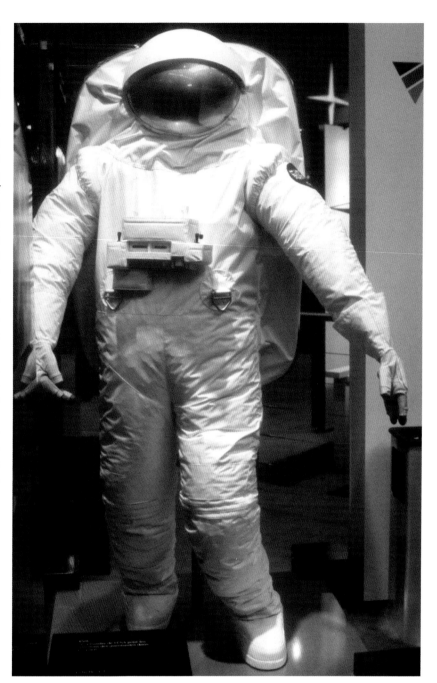

ABOVE Photographed at the 1991 Paris Air Show, a full-size model of Dassault's spacesuit. *JC Carbonel*

- Ariane 5 would be developed and would be able to lift Hermes with its full payload due to the slip in the Hermes schedule.

A major reduction in European ambition was to build only one Hermes spaceplane instead of two, as had been planned so far. '*We must not go too far in the cost-killing if we don't want to miss the programme's main objective, which is to guarantee future European autonomy in regard manned spaceflights*,'[93] wrote Philippe Couillard, who added, '*This was a headlong rush; in just a few months the development [of Hermes] was stretched from eight to twelve years. Of course, the short-term yearly budgets were reduced, and the cost spike was delayed and on the whole, annual spending was limited, at constant economic conditions, to the level agreed at The Hague. That was the goal and that was obtained.*

ABOVE CNES and ESA had prepared for ground support equipment for Hermes. Here the Crew Training Complex. *ESA*

'But stretching a programme has a cost. There are fixed costs and pushing a programme from eight to twelve years increases the global cost. The global cost increase from The Hague was nearly 40% which became an unfavourable argument against the programme as a whole.'

In March, ESA reviewed the reply to its request for proposals of the previous year.

In June, at the Paris Air Show, a full-scale mockup of Hermes with HRM attached was exhibited. To many visitors it must have been a strong signal that Hermes would soon be a reality. Actually, it was a swan song and the beginning of the end.

Maia Enters…and Leaves

Among the first significant project evolutions was a proposal by Dassault to build and fly a sub-scale demonstrator called Maia. (This was not an acronym but the name of Hermes's mother in Greek mythology.) The company saw this project as very important because computer power at the time did not allow a complete simulation of Hermes's re-entry and the Europeans had no first-hand experience in the aero-thermo-dynamics of it.

Bruno Revellin-Falcoz justified Maia in *Air & Cosmos*. 'Considering the importance of aerodynamics and aerothermal technology, of the good knowledge of the airflow along the spaceplane and of the flight qualities required, it appeared to us during the studies that it could be important to fly a sub-scale demonstrator before flying a crewed spaceplane.'[94]

Michel Rigault confirms, 'Dassault and MBB proposed Maia [to acquire hands-on experience on re-entry questions]. MBB was involved at first because it wanted to have a demonstrator for the HORUS second stage of Sänger. After Maia was rejected by the commission under Jean-Jacques Dordain, future ESA general manager from 2003 to 2015, it was also recommended that the first flight of Hermes be unmanned. In 2015, the IXV performed a flight very similar to what we had in mind for Maia. Its shape was derived from a Dassault proposal for a simplified shuttle craft (project Launac).'[95]

The idea was to launch Maia in an Ariane 4 (in the same flight planned for the SPOT-3 observation satellite… which dictated the mass and volume available for Maia) and study its re-entry. Maia would be a small, 1.5-tonne spacecraft specifically designed to test re-entry with an underside built of titanium and covered by silicate shingles (SiO_2 or SiC/Al_2O_3) designed to support temperatures up to 1,400°C. Re-entry was automatically programmed and final recovery would be by parachute.

Studies of Maia were prepared by Dassault, in association with the German company MBB in 1986 and resulted in two vehicles: Maia-A (to validate the design tools and the overall configuration) and Maia-B (to validate the precise shape and its thermal protection).

The initial idea to piggy-back on the SPOT launch proved full of difficulties so Maia was finally rescheduled to benefit from its own launch (a launch shared with SPOT would have complicated the recovery, which was originally expected in the Atlantic Ocean, near Kourou, but the SPOT orbit was not very practical for that, requiring a cross-range of 1,802 miles [2,900km]).

A non-recoverable Maia was considered but this would have not allowed an after-flight examination of the shingles, which was an important intention of the Maia concept. This non-recoverable Maia was quickly passed over in favour of a larger ⅓ scale Maia (the original Maia was 23% of Hermes size to fit inside the Ariane fairing: *'What we can put under the fairing and bring back to Earth,'* said Bruno Revellin-Falcoz). In the meantime, Hermes had been shortened (5L3 shape) with a fuselage reduced by 6ft 0 in (1.82m) in length. This study was completed in November 1986 and presented to the CNES Maia evaluation board which asked that all previous experience available be taken into account (US Shuttle and French ballistic missiles) and that a maximum of tests (wind tunnel, heat stress) be carried on the ground. Alain de Leffe confirmed that a lot of wind tunnel testing was carried out, not only for Maia but for the whole Hermes programme: *'We worked with a lot of different wind tunnels including those managed by the military Direction Générale à l'Armement. We sometimes ran into difficulties with DGA which found we were occupying their wind tunnels too often. So, I told them 'you don't have a monopoly on wind.'*[96]

While work continued, CNES began questioning the justification of Maia. It was smaller than Hermes, so its aero-thermo-dynamics could not be completely extrapolated; its all-titanium undersurface did not represent the way Hermes was to be built; many components of Maia would need to be procured outside Europe (and ESA worked on very stringent 'industrial return on investment' for each country funding it: every item bought outside the European partners compromised the equilibrium between them); and, probably most importantly, there was fear of the consequences of a failure of a Maia flight—could it delay or even stop the whole Hermes programme?

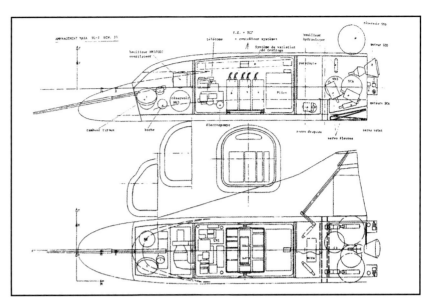

ABOVE Interior details of the Maia demonstrator proposed by Dassault.
© Dassault Aviation

In May 1987 CNES selected what was probably perceived as a safe and economical way: cancel Maia. Interestingly, ESA objected. It naturally required an emphasis on ground tests and on prior experience (pointing out that the European experience was very limited in comparison with what the US had obtained with X-15, ASSET and PRIME programmes[97], and the Russians had flown their BOR demonstrators before either had started on its own shuttle programme). But it also pointed out that while safety concerns required that the first flight of Hermes would be unmanned and completely automated this flight was not an alternative to Maia: Maia could guide the development of Hermes. If the first flight of Hermes encountered problems, it would be too late to correct them.

A Véhicule d'Essais Hypersonique (Hypersonic Test Vehicle) was proposed, but CNES killed the idea. CNES was still concerned that a demonstrator could 'dilute' the programme and its funding, and refused to present the VEH proposal to the 1987 ministerial level Council of ESA. This position generated mixed opinions among the programme engineers: many feared that the failure of a demonstrator could kill the whole Hermes programme, being proof that the political will to support Hermes was very thin indeed. The termination of Maia was indeed an early sign of mismanagement of the project which went against Dassault's policy of 'small steps' which it had demonstrated successfully in fighter design. In some ways Hermes became a technological and political gamble: let's design an all-new machine, with no previous hands-on experience, and hope everything goes right from the first flight.

Philippe Couillard was very critical of Maia which he called an 'internal attack' against Hermes, considering it presented a multiple risk on the project:

- To see ESA only deciding (and voting on) the side-project which was Maia.
- To have Maia eat a consequent part of the budget allocated to Hermes.

But this may just have been in defence of Hermes, which had become his own project by that time.

The Final Hermes

To define and test Hermes, it was determined that eight models would be needed including the two flightworthy space-rated vehicles:

i   Maquette d'Aménagement 1 (Layout Mockup). This full scale, wooden model would be used to check the integration of all systems within the spaceplane frame, the ergonomics of the crew compartment and simulate operation of the vehicle with astronauts. Assembled in Aerospatiale Building B03 at Toulouse, this was actually being built when the programme was terminated.

ii Maquette d'Aménagement 2 was a second mockup, built in metal designed to define assembly procedures of the various sections of Hermes. This was to be built in the Batiment d'Integration Hermes en Europe (BIHE = Hermes Integration Building in Europe).

iii Banc d'Integration Systèmes (System Integration Bench): linked to Aerospatiale Data Centre, this model would represent the electronics aboard Hermes and would serve to validate hardware and software integration including fly-by-wire controls. This was to be housed in the BIHE.

iv Maquette d'Identification (MI: Identification Model). This mockup represented all the materials, electrical and data wiring, etc, and would be used to test the electromagnetic environment of Hermes. It was to be built in the BIHE.

v   Maquette Structurelle et Thermique (Thermal and Structural Mockup): Full-scale representation of Hermes used for thermal, electromagnetic tests. This model was also intended to be used to define the attachment of Hermes to its Airbus transport aircraft (Hermes Carrier Aircraft: HCA).

vi Cellule d'Essais Statiques (CES: Static Test Airframe): to be installed at Aerospatiale's Toulouse Blagnac factory.

vii Avion 1 (Airframe 1): The first space-rated airframe. It would be used for Approach and Landing Tests (ALT) and also for HCA experimental flights. Subsonic test flights launched from the Airbus HCA were planned for 1996.

viii Avion 2 (Airframe 2): Second space-rated airframe would be used to test the resistance of the machine to various climatic conditions and electromagnetic disturbances. Those tests were obviously required because the climatic conditions at Kourou are different and somewhat harsher to any machine than those encountered in Europe.

The Cockpit:

At some point it had been suggested that the windscreen of Hermes be eliminated for heat stress reasons. A test was carried out using a Dassault Mirage III fitted with blanked-out windscreen. The pilots disliked it and the windscreen was reinstated. It was later found out that the thermal stress would be less than originally estimated. De Leffe explained to the author, 'The windscreen was not a hot area. Re-entry was at a 40° pitch backward so the windscreen was not in the lee of the wind. Besides, today there are types of glass which are very resistant to heat.'[98]

Ejection seats were provided for the three crew members. It was not possible to buy Russian seats directly but acquiring the technology to licence-build them appeared possible. So an Association des Vols Habités (Society for Manned Flights) was set up by Dassault, Energia, Molnia and Zvezda with French *Spationaut* Jean-Loup Chrétien as president. Henri Lacaze remembers 'We *worked a lot on the question of the cockpit. We were required to call in ergonomists but we could have done the cockpit as we wished—on landing, Hermes was flown like any other aircraft.*'[99]

Michel Rigault also remembers the input of the pilots. '*During the call for proposals period (before the 31 March 1985), Dassault had, as advisor, Spationaut Jean-Loup Chrétien, who was known by the company from his work as State test pilot during the Mirage 2000 programme. Later we were also advised by Jean Coureau (Dassault company test pilot who had flown the Mirage 2000 for its maiden flight on 10 March 1978) on the internal size of the cockpit. A test was undertaken aboard a Dassault Falcon 2000 business jet with Jean Coureau fitted with the "suit of light," the high-altitude pressure suit of the Armée de l'Air which had been developed*

BELOW In this rendering of the heatwaves engulfing the Hermes shape, the hottest sections are in blue-violet. © Dassault Aviation

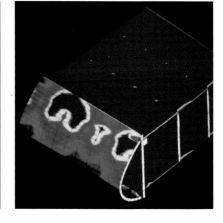

BELOW A similar Dassault depiction of a section of the leading edge of Hermes's wings. Again, blue-violet represents the hottest areas. © Dassault Aviation

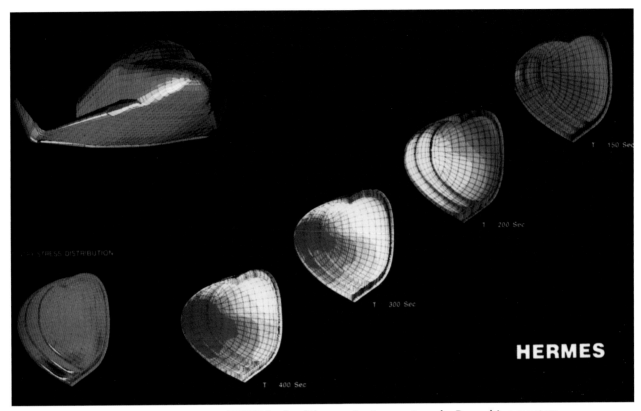

ABOVE Study of the re-entry temperatures by Dassault's computers.
© Dassault Aviation

for the rocket-equipped Mirage III interceptors (which could reach 78,740 to 82,020ft (24,000 to 25 000m). The size of the Falcon cockpit had been appreciated sufficiently to give the spationauts access to the seats in case of pressure loss. This has determined the size of the fuselage of Hermes. Spationauts were obviously involved later, in the definition of the man-machine interfaces, like on all aircraft.'[100]

Alain de Leffe also had his anecdote about pilot's reaction to Hermes. 'The hypersonic flight sequence was quite easy; the major difficulty lies in the thermal protection. It's around Mach 2 that keeping control was difficult. We could not use the Variable Stability Mirage; the flight envelope of the Mirage was just not wide enough to simulate the flight of Hermes. The fall rate was just too high so we did it all in simulator. I remember astronaut Michel Tognoni getting out of the simulator and complaining to me, 'Alain, an iron would fly better than that!'[101]

RIGHT ONERA worked with Dassault on thermal tests. Here is a picture of one of the models used for this purpose.
© Dassault Aviation

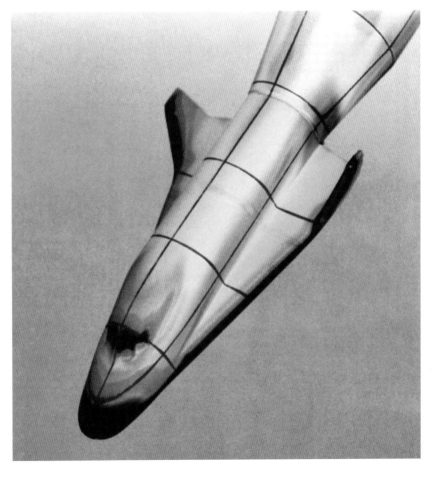

ABOVE Some emphasis was given at the time (1987 for this picture) to the fact that many of the studies were carried out using computer simulations. Note the computer set up for this ONERA technician. © Dassault Aviation

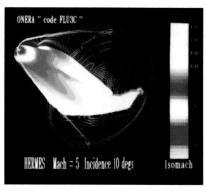

ABOVE And here is the result: an impression of the spaceplane flying at Mach 5 with a pitch of 10°.
© Dassault Aviation

Thermal Protection:[102]

Hermes thermal protection was *'of a more advanced technology than the Shuttle 356lb/sq ft (15kg/sq m) versus 949 lb/sq ft (40kg/sq m)'* using three different materials:

i Shingles of ceramic-ceramic material supplied by SEP for the hot structure (nose, leading edge, winglets).

ii Other shingles for 'mild' areas (front and undersurfaces).

iii Panels of 'Protecalor' isolation material on the cold surfaces.

Carbon-carbon materials were also considered for specific locations such as the nose 'bullet'.

Avionics and Computers:

Most of the flight phases of Hermes would be automated. During the launch and up to orbit, the crew intervention would be very limited—its role just to monitor the screens and press the red button to eject if something went amiss. Re-entry would be equally automated with the pilots intervening only in case of problem. Only when in orbit, would the astronauts gain some autonomy from the computers.

On the ground, two mission management computers would provide guidance, while on board four guidance, navigation and control computers would help fly the spacecraft. Alain de Leffe remembered, *'The opposition between the launcher people and the aviation people manifested itself in a comical way when they were designing the avionics. Avionics for a rocket launcher is quite simple, avionics for a spaceplane is much, and much more complex…more like the avionics of a Rafale fighter plane.'*[103]

BELOW A model of Hermes flying piggy-back on an Airbus A320 for transport to Kourou. This photograph was taken during the 1985 Paris Air Show (the Aerospatiale styling of Hermes is obvious). Aerospatiale participated in Airbus, in Arianespace and was keen to display the synergies between its different products. *JC Carbonel*

ABOVE This illustration comes from a 1986 brochure but must have been prepared earlier, because Hermes is still drawn with a single central fin. *Aerospatiale*

Transport to Kourou:

Various modes were considered to transport Hermes to Kourou: aircraft (as seen with the US Shuttle Orbiter and the Russian Buran) or by ship (as was done for Ariane). Various aircraft were considered for the Hermes Carrier Aircraft (HCA): Airbus A300, Airbus A310, Airbus A340 (considered safer because it has four-engines). Hermes could be either attached directly to the HCA or carried inside a large fairing.

The HCA was also intended to be used for subsonic Approach and Landing Tests which were to be carried using the first airframe. Dassault test pilots were planned to carry these tests.

In the end, the programme was terminated with no decision regarding the HCA. Hermes would be carried by ship. This was not considered a problem, because of the small number of annual flights planned.

Typical Mission:

It was expected that a mission campaign would take 40 days. The spaceplane would be carried to Kourou by aircraft while the preparation of the flight would begin on Day –37. The crew would arrive on Day –12 and the payload would be integrated on Day –9. Hermes would be installed atop Ariane 5 on Day –3 (on launch pad ELA-3). Almeria Airport, in southern Spain, was the 'nominal' landing site for manned flights.

BELOW This is how Aerospatiale intended to fit Hermes atop an Airbus A320. *Aerospatiale*

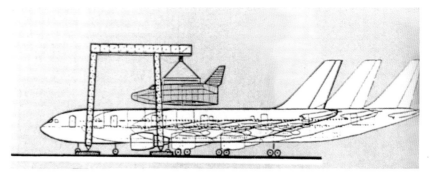

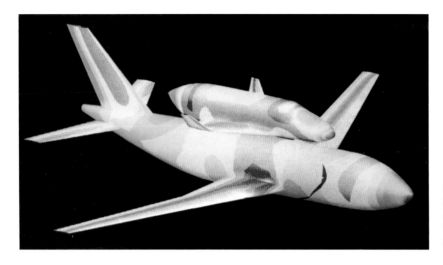

LEFT Dassault also studied the transport of Hermes in a piggy-back configuration. In this 1989 illustration, the rear section of Hermes is faired by a cone to reduce drag. © Dassault Aviation

[73] Quoted in *Air & Cosmos,* 26 October 1985m.
[74] Technically, 'shingles' are bolted to the airframe structure while 'tiles' are just attached to it. Hermes constantly referred to 'shingles' but 'tiles' were also considered.
[75] Mail exchanges with the author, October 2020.
[76] Before the reunification, Bonn was the location of West German government offices.
[77] *Rêve d'Hermes*.
[78] Phone call with the author, November 2020..
[79] Quoted in Assemblée Nationale 1ere Séance, 4th June 1986.
[80] Quoted in Assemblée Nationale 1ere Séance, 25th June 1986.
[81] Both schedules: *Air & Cosmos,* 5 July 1986.
[82] Phone call to the author, November 2020.
[83] Mail exchanges with the author, October 2020.
[84] *Rêve d'Hermes,* 1993.
[85] Henri Lacaze, interview with the author, 8 February 2019.
[86] Ibid.
[87] *Rêve d'Hermes,* 1993.
[88] *Rêve d'Hermès,* 1993.
[89] *Rêve d'Hermes,* 1993.
[90] Phone call to the author, November 2020.
[91] Phone interview with the author, 25 November 2020.
[92] French Sénat, session of 4 December 1990.
[93] *Les Echos*, 18 June 1991.
[94] *Air & Cosmos,* 26 October 1985.
[95] Mail exchange with the author, October 2020.
[96] Phone interview with the author, 25 November 2020.
[97] ASSET and PRIME were programmes designed to test re-entry shapes and materials. ASSET under the auspices of the USAF encompassed four flights between 1963 and 1965. It was intended to examine re-entry heating and structural resistance of a spacecraft originally tied in to the X-20 Dyna-Soar programme and was continued after the cancellation of the X-20. PRIME (Precision Recovery Including Manoeuvring Entry) was designed to test the re-entry shape of the X-24 lifting body. Three flights were carried out in 1966. Both ASSET and PRIME were part of the Spacecraft Technology and Advanced Re-entry Tests (START) programme. In Europe similar programmes were undertaken: France with VERAS (see Chapter 5) and Germany with Bumerang (see Chapter 2). Neither was flown into space.
[98] Phone interview with the author, 25 November 2020.
[99] Interview by the author, 8 February 2019.
[100] Mail exchange with the author, October 2020.
[101] Phone interview with the author, 25 November 2020.
[102] *Air & Cosmos,* 30 November 1985.
[103] Phone interview with the author, 25 November 2020.

Chapter Nine
Competing with Hermes

ABOVE To promote its design, MBB built a large model said to be 10 metres (30ft) long. *MBB*

While France and ESA were busy developing Hermes, British and German companies felt they could contribute their own national designs. Although discussion and financing was labelled 'European' under the heading of ESA, everyone was fiercely national and wanted to promote their 'own' project. In the end, one should keep in mind that Hermes fell; in part because it was perceived as 'too French' for some people—with the consequence that even today, European astronauts depend on other countries to get into space. Disunity creates weakness.

British HOTOL

In **1982** British Aerospace began studying an alternative launcher, with the idea of drastically reducing the cost of each launch. '*Convinced that total reliance on expendable launchers can only be seriously detrimental to Europe's interest, British Aerospace has undertaken far-reaching studies to identify and design the optimum commercial launch vehicle,*' proclaimed the pamphlet distributed at the 1987 Paris Air Show. The shuttle was entering service, so BAe decided to set its sights on cutting the costs of the shuttle five-fold. Interestingly, Rolls-Royce was at that time working on a new type of engine which could just be the solution to BAe's conceptual space plane.

In **1984** the project 'HOTOL' for HOrizontal Take-Off and Landing, was revealed. BAe had seemingly toyed with the idea of creating an SSTO (single-stage-to-orbit), '*A vertical take-off version [of a reusable launcher] will always suffer from the unfortunate fact that the Earth's gravitational field is about 10% too high to make a single-stage-to-orbit rocket practical, even with the highest performance propellants of the foreseeable future.*'[104] This was an entirely automated vehicle optimised for satellite launching: BAe was a major satellite designer and assembler so it was only logical that it offered a launcher. BAe was never keen to have people aboard HOTOL and Robert Parkinson, manager of future launch systems and chief promoter of HOTOL, was quick to point out that manning HOTOL would add $1.5 billion to the development cost—contrary to the objective of reducing costs by all means.

But before launching satellites at a much reduced cost, a huge investment, estimated at £4 billion, was required to build just four machines. Each launch was estimated to cost $10 million and each machine was intended to fly between twice a month and once a week and be able to fly 120 times

ABOVE HOTOL at take-off. The take-off carriage was a characteristic of HOTOL. This illustration comes from a 1985 BAe brochure. *BAE Systems*

BAe HOTOL 1984	
Status	Project
Span	91ft 10in 28.00m)
Length	203ft 5in (62.00m)
Gross Wing area	–
Gross weight	551,150lb (250,000kg)
Engine	4 × RB545 Swallow 165,230lb st (735kN)
Maximum speed/height	–
Load	15,432 to 17,637lb (7,000 to 8.000kg) in 186mi (300km) orbit

before retiring. (In 1986, BAe claimed that the minimal turn-around time was 'two days'). Thirty-five different configurations were studied before the project was sufficiently stabilised to be shown outside the company.

With the project revealed, it appeared in direct concurrence with the French Hermes programme, but this could have played in its favour as a few European countries were not happy to see France, already perceived as the space leader of Europe with the Ariane launcher, heavily promote its next space project, Hermes. But then it has been claimed that German government willingness to fund MBB's Sänger hypersonic project also stemmed from the appearance of HOTOL in the perceived competition for the next big European space programme.

In January **1985** ESA elected to initiate development of Ariane 5, while just taking a moderate interest in the BAe proposal, merely asking to be kept informed of the project's progress.

In any event, Rolls-Royce, probably anticipating significant development costs, was not willing to commit itself alone to the development of the engine named RB545 Swallow and asked for involvement of the Royal Aircraft Establishment. The USA was, indeed interested and in November 1985, technical advisors to Margaret Thatcher and Ronald Reagan met to discuss the project. Politically, the British Government was reticent toward any American involvement in the project, fearing that the British invention, which could give Britain a leading role in Europe, could become diluted among various American space projects. Finally, the British Government agreed to finance a preliminary study for two years, with the intention of arriving at a future ESA decision meeting with a working proof-of-concept.

At that time, BAe and Rolls-Royce expected that it would take twelve years to bring the project to fruition. They made a curious move, trying to convince France to abandon Ariane 5 and Hermes in favour of HOTOL. Peter Conchie, British Aerospace Director of Business Development declared during the Paris Air Show, '*I think we have to get the French to embrace this programme for it to go. [They] spend more than any other European nation in this area and always have.*'[105] Amusingly, one year later, in April 1986, after admitting that if HOTOL was 'Europeanised' France would get 'the lion's share' of work, he would say

BELOW Three launch trajectories for HOTOL are shown: unassisted take-off in which the carriage did not include booster; normal take-off in which the carriage has integrated boosters; and rejected take-off in which the procedure has to be aborted during the initial run. *Author's collection*

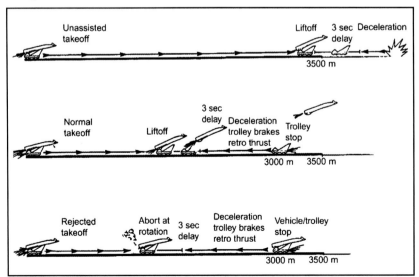

exactly the opposite, using the same words. '*Traditionally France spends most money within Europe on launch vehicles. It is not easy to say how they can embrace HOTOL.*'[106] He even tried to frighten Europe into American dependence if it did not make the right choice. '*We don't have any solution to the problem of how we get the HOTOL programme endorsed by ESA. It is only a matter of time before the US catches up – all we can do is give Europe a little grace.*'[107] Interesting talk when, one year earlier in March 1985, the press had announced that Rolls-Royce had concluded a technological exchange agreement with the American company Rocketdyne.

Promotion of HOTOL to ESA was certainly handicapped by the fact that its core asset, the Rolls-Royce RB545 Swallow needed '*further definition*'[108] in the words of Rolls-Royce's technical director Gordon Lewis. And, perversely, what was defined was classified. What was explained was that RB545, although a rocket engine normally running on LH_2/LO_2 propellants, could use gaseous oxygen while in the Earth's atmosphere. According to Gordon Lewis this could be achieved through '*the judicious exploitation of turbomachinery*'. The trick, he said, was in the design.

According to Henri Lacaze, '*No one really believed in this project. It should have been driven by a miraculous engine, which probably was not so much. In the end it never was in competition with Hermes.*'[109]

In his souvenir *Rêve d'Hermes*, Philippe Couillard was even more caustic, saying, '*On the technical side, the British were not very talkative, notably regarding the air-breathing engine and the intakes which were the most critical aspects of this type of vehicle. Actually, this project's goal was to show there are alternatives to Hermes which, in their view, had two major defects: being French-inspired and not being advanced enough, technologically speaking. If, for the engineers, it was obvious HOTOL was, on the contrary, too advanced and could not be developed in parallel with Ariane and Hermes, for the political delegations of ESA member-states, it*

ABOVE **Model of HOTOL exhibited on BAe's stand at the 1985 Paris Air Show**. *JC Carbonel*

BELOW **Four-view drawing of HOTOL as originally proposed.** *NASA*

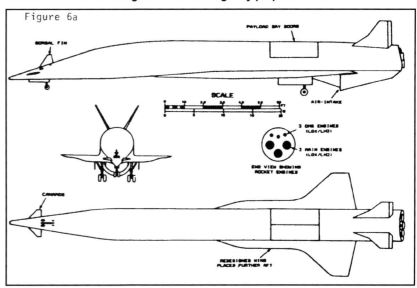

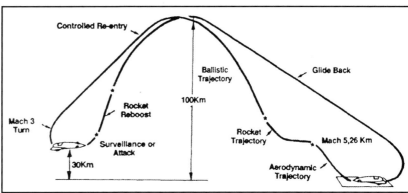

ABOVE **HOTOL was considered for both military and civilian missions. For the latter, it would have been in direct competition with Ariane/Hermes and Sänger II.** *NASA*

generated doubts as to the way to go in the matter of manned spaceflight.'

In 1985, the British National Space Centre (BNSC) published its own estimates of the cost, £5 billion, and the delay, twenty years, needed to complete the project. Two major development uncertainties emerged: first, up to that time, only the ascent phase had been studied, so there was great concern regarding the re-entry. Neither aerodynamic braking nor thermal stress appeared particularly forthcoming, due to the small wings of HOTOL. Second, the very heavy engine, at the extreme rear of the aircraft. generated a serious problem of weight distribution—again during the critical re-entry phase when the fuel tanks would be exhausted, and therefore the weight of the aircraft minimal but concentrated around the engine.

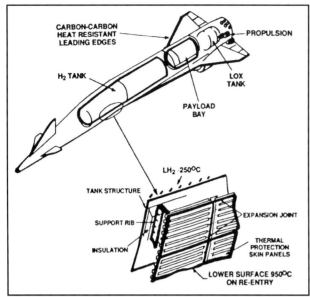

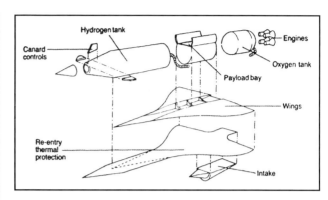

ABOVE Structural details of HOTOL. Most of the fuselage was taken up by fuel tanks. NASA

LEFT Drawing illustrating the interior arrangement of HOTOL. In the foreground, detail of the thermal protection skin panels. NASA

ABOVE HOTOL accelerating through the atmosphere. In this 1989 brochure, some design evolutions are evident—notably removal of the rear fins. *BAE Systems*

ABOVE Another illustration of HOTOL taking of from a conventional runway. Note the absence of the rear fins. *Author's collection*

Geoffrey Pattie, British Minister of Industry presented a more detailed file to ESA in **1986**. The idea was to propose the project for Europeanisation at the October ESA meeting. The development schedule would have been:

- Early studies (Phase A) from January 1988 to December 1989.
- Second study phase (Phase B1) from January 1990 to December 1991.
- First flight in 1998-1999.
- Operational flights from 2000-2001.[110]

He said, *'These studies will build on work already funded by the two firms [British Aerospace and Rolls-Royce] and in particular establish if there are any insuperable difficulties with the concept and to provide credible design, performance and cost data on which to explore the technology with our partners in the European Space Agency.'*[111]

To support this plan, the project was expected to be declassified during the year. Two years went by then, in February **1988**, the British Government indicated that Rolls-Royce expected to have finalised its 'propulsion concept' inside one year and that BAe gave themselves three years of industry-funded research to fully design the spaceplane and present it for 'Europeanisation.' By that time HOTOL's design had been revised with an increase of wing sweep to 54° while the wing itself had been pushed back to the extreme rear of the aircraft. The fuselage was stretched to achieve the best compromise between aerodynamic drag and structural weight. A droop of 5° was given to the nose section. The foreplanes were made jettisonable. In October, the UK Government withdrew its support (and financing) for the project which appeared burdened by unsolvable problems.

In **1989**, with Ariane 5 having been selected officially by ESA, the market for HOTOL was evaporating and Rolls-Royce elected to get out of the project, estimating it would never recoup the development costs of the Swallow engine…and the aerodynamic re-entry questions remained unsolved. BAe continued alone promoting a modified design: jettisonable, canard foreplanes were intended to improve transonic and low supersonic flight characteristics.

Wings acquired a straight leading edge and were enlarged to increase stability in hypersonic and re-entry flight.

Additionally, the wing was moved forward to reduce the length of the front fuselage. Wind tunnel experiments revealed a heat of up to 1,750°K (1,475°C) at the nose and 1,300°K (1,026°C) on the undersurface. Finding partners was difficult considering that, contrary to earlier promise by the British Government, the engine details were still classified. All was known was that the engine remained fed by the outside atmosphere up to Mach 5 and 19 miles (30km) altitude, then switched to a LOX/LH$_2$ rocket engine. In **1992**, redundancies were announced at British Aerospace and among them, the whole HOTOL team.

HOTOL was a 249ft (76m) long (interestingly, dimensions were given in Metric units in all the documents examined by the author), 196.8-ton

ABOVE On this wind tunnel model one can see that the intake for the engine has been changed to a wedge type. *Author's collection*

(200-tonne) fully automated aircraft which took off on a ground trolley after a 7,545 ft (2,300 m) run at a speed of 290 knots (335mph [540km/h]). The trolley was equipped with supplementary propulsion for ground manoeuvre; laser steering (not explained); and a parachute braking system. The idea behind this trolley was to be able to lighten the undercarriage, which, not being used at take-off, could be designed for landing only —when the HOTOL would be much lighter, its fuel exhausted (fuel represented about 50% of the take-off weight of HOTOL) and its load delivered to orbit.

BAe HOTOL 1989	
Status	Project
Span	65ft 7in (20.00m)
Length	249ft 4in (76,00m)
Gross Wing area	-
Gross weight	440,925lb (200,000kg)*
Engine	4 × RB545 Swallow 165,230lb st (735kN) LO$_2$/LH$_2$
Maximum speed/height	-
Load	15,432 to 17,637lb (7,000 to 8.000kg) in 186mi (300km) orbit

RIGHT Different evolutions of HOTOL presented side by side. *Author's collection*

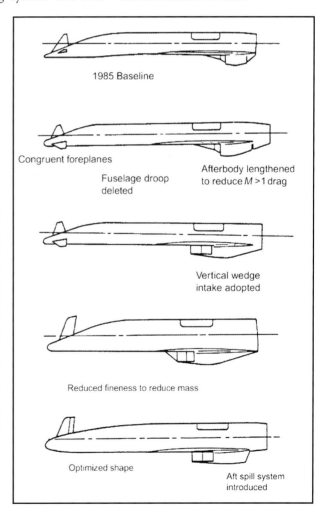

ABOVE HOTOL in orbit deploying its payload. This art depicts a rather late evolution of the design with a single nose fin and a wedge intake. *Author's collection*

* 606,271lb (275 000kg) quoted in *Air & Cosmos*, July/August 1989.

ABOVE Artist's impression of HOTOL being launched from an Antonov An-225 Mriya. *BAE Systems via Ron Miller*

Consideration was also given to take-off noise. While HOTOL would probably be less noisy than the shuttle, a relatively secluded area would be required for launching. Robert Parkinson claimed that a launch site could be found using '*a good atlas and some thought*,' but obviously not near an existing rocket launch site, so away from Kourou. The discussion also demonstrated that taking off from '*airport-style runaway*' certainly did not imply take-off from any existing airport, in stark contrast to many other Aerospace Transporters. Vertical acceleration at that time was 1.15G and the climb attitude was 24°. Mach 1 was reached after 2 minutes of flight and Mach 5 after 9 minutes. At that point, at an altitude of about 16 miles (26km) '*various problems preclude air-breathing and dictate a ballistic trajectory on main engine.*'

Orbital velocity was achieved at 56 miles (90km), after which the engine was cut and HOTOL continued on a ballistic trajectory until it reached its operating altitude of about 186 miles (300km). An Orbital Manoeuvring System would then be used to achieve a precise orbit.

Once in orbit, the cargo bay doors would be opened and the payload extracted. With the load in its own orbit, HOTOL would be de-orbited using OMS to bring it back to about 43 miles (70km). Re-entry would be at an incidence of about 80° to bleed-off speed. Hypersonic glide would begin only at 16 miles (25 km) altitude. Re-entry temperatures were deemed to be lower than the American shuttle '*because of [HOTOL's] large wing and low mass.*'

In consequence, high-temperature alloys would be used for the undersurfaces and ordinary titanium/ René 41 nickel sandwich for the uppersurfaces. Carbon-silicon carbide material would be used in the areas of high temperature, up to 1,750°K. Skin panels designed to withstand 700°K would be one foot by three feet (30cm by 91cm) and would not require periodic replacement (as do the Shuttle tiles). Titanium aluminides were also considered for the skin panels, both materials having limitations in term of resistance to heat stress.

Dispensing with tiles was seen as easing between-flight maintenance and therefore favouring high turn-around rates. '*The high hypersonic lift-to-drag ratio of HOTOL during re-entry (more than twice that of the Shuttle) gives it a high cross-range capability sufficient, say, for a landing in Europe from an Equatorial orbit. Indeed the Kourou site need be used only as a re-fuelling stop on the way from Europe to orbit.*' The final approach was also deemed '*gentler*' than the Shuttle's with an angle of 16° and a touchdown speed of 170 knots (196 mph [315 km/h]). A roll of 5,905ft (1,800m) was expected at landing. Total duration of a standard mission was given as 50 hours. Basically, HOTOL was a fully-automated spacecraft, with only one machine out of the six (previously four) intended to be built being man-rated.

The idea of using HOTOL as passenger transport was also briefly hinted at, with dreams of a '*one hour passenger flight from Europe to Australia.*' In this guise, HOTOL was advertised as the 'transatmospheric Skyliner for the 21st Century.'[112] It was announced that the passenger transport version would begin flying in 2010.

Wings acquired a straight leading edge and were enlarged to increase stability in hypersonic and re-entry flight.

Additionally, the wing was moved forward to reduce the length of the front fuselage. Wind tunnel experiments revealed a heat of up to 1,750°K (1,475°C) at the nose and 1,300°K (1,026°C) on the undersurface. Finding partners was difficult considering that, contrary to earlier promise by the British Government, the engine details were still classified. All was known was that the engine remained fed by the outside atmosphere up to Mach 5 and 19 miles (30km) altitude, then switched to a LOX/LH$_2$ rocket engine. In **1992**, redundancies were announced at British Aerospace and among them, the whole HOTOL team.

HOTOL was a 249ft (76m) long (interestingly, dimensions were given in Metric units in all the documents examined by the author), 196.8-ton (200-tonne) fully automated aircraft which took off on a ground trolley after a 7,545 ft (2,300 m) run at a speed of 290 knots (335mph [540km/h]). The trolley was equipped with supplementary propulsion for ground manoeuvre; laser steering (not explained); and a parachute braking system. The idea behind this trolley was to be able to lighten the undercarriage, which, not being used at take-off, could be designed for landing only —when the HOTOL would be much lighter, its fuel exhausted (fuel represented about 50% of the take-off weight of HOTOL) and its load delivered to orbit.

ABOVE On this wind tunnel model one can see that the intake for the engine has been changed to a wedge type. *Author's collection*

BAe HOTOL 1989	
Status	Project
Span	65ft 7in (20.00m)
Length	249ft 4in (76,00m)
Gross Wing area	-
Gross weight	440,925lb (200,000kg)*
Engine	4 × RB545 Swallow 165,230lb st (735kN) LO$_2$/LH$_2$
Maximum speed/height	-
Load	15,432 to 17,637lb (7,000 to 8.000kg) in 186mi (300km) orbit

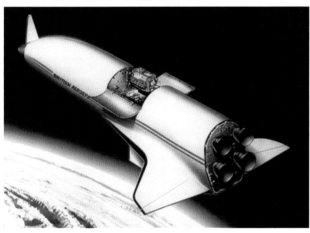

ABOVE HOTOL in orbit deploying its payload. This art depicts a rather late evolution of the design with a single nose fin and a wedge intake. *Author's collection*

* 606,271lb (275 000kg) quoted in *Air & Cosmos*, July/August 1989.

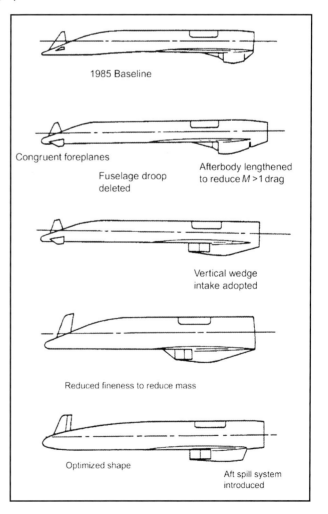

RIGHT Different evolutions of HOTOL presented side by side. *Author's collection*

1985 Baseline

Congruent foreplanes
Fuselage droop deleted
Afterbody lengthened to reduce *M* >1 drag

Vertical wedge intake adopted

Reduced fineness to reduce mass

Optimized shape
Aft spill system introduced

ABOVE Artist's impression of HOTOL being launched from an Antonov An-225 Mriya. *BAE Systems via Ron Miller*

Consideration was also given to take-off noise. While HOTOL would probably be less noisy than the shuttle, a relatively secluded area would be required for launching. Robert Parkinson claimed that a launch site could be found using '*a good atlas and some thought,*' but obviously not near an existing rocket launch site, so away from Kourou. The discussion also demonstrated that taking off from '*airport-style runaway*' certainly did not imply take-off from any existing airport, in stark contrast to many other Aerospace Transporters. Vertical acceleration at that time was 1.15G and the climb attitude was 24°. Mach 1 was reached after 2 minutes of flight and Mach 5 after 9 minutes. At that point, at an altitude of about 16 miles (26km) '*various problems preclude air-breathing and dictate a ballistic trajectory on main engine.*'

Orbital velocity was achieved at 56 miles (90km), after which the engine was cut and HOTOL continued on a ballistic trajectory until it reached its operating altitude of about 186 miles (300km). An Orbital Manoeuvring System would then be used to achieve a precise orbit.

Once in orbit, the cargo bay doors would be opened and the payload extracted. With the load in its own orbit, HOTOL would be de-orbited using OMS to bring it back to about 43 miles (70km). Re-entry would be at an incidence of about 80° to bleed-off speed. Hypersonic glide would begin only at 16 miles (25 km) altitude. Re-entry temperatures were deemed to be lower than the American shuttle '*because of [HOTOL's] large wing and low mass.*'

In consequence, high-temperature alloys would be used for the undersurfaces and ordinary titanium/ René 41 nickel sandwich for the uppersurfaces. Carbon-silicon carbide material would be used in the areas of high temperature, up to 1,750°K. Skin panels designed to withstand 700°K would be one foot by three feet (30cm by 91cm) and would not require periodic replacement (as do the Shuttle tiles). Titanium aluminides were also considered for the skin panels, both materials having limitations in term of resistance to heat stress.

Dispensing with tiles was seen as easing between-flight maintenance and therefore favouring high turn-around rates. '*The high hypersonic lift-to-drag ratio of HOTOL during re-entry (more than twice that of the Shuttle) gives it a high cross-range capability sufficient, say, for a landing in Europe from an Equatorial orbit. Indeed the Kourou site need be used only as a re-fuelling stop on the way from Europe to orbit.*' The final approach was also deemed '*gentler*' than the Shuttle's with an angle of 16° and a touchdown speed of 170 knots (196 mph [315 km/h]). A roll of 5,905ft (1,800m) was expected at landing. Total duration of a standard mission was given as 50 hours. Basically, HOTOL was a fully-automated spacecraft, with only one machine out of the six (previously four) intended to be built being man-rated.

The idea of using HOTOL as passenger transport was also briefly hinted at, with dreams of a '*one hour passenger flight from Europe to Australia.*' In this guise, HOTOL was advertised as the '*transatmospheric Skyliner for the 21st Century.*'[112] It was announced that the passenger transport version would begin flying in 2010.

BAe HOTOL II	
Status	Project
Span	70ft 10in (21.60m)
Length	119ft 7in (36.45m)
Gross Wing area	-
Gross weight	552,150lb (250,000kg)
Engine	4 × RD-0120 rocket 1,762,500lb st (7,840kN) LO_2/LH_2
Maximum speed/height	-
Load	15,432 to 17,637lb (7,000 to 8.000kg) in 186mi (300km) orbit

British HOTOL 2

BAe tried to salvage what it could by completely redesigning the project, now with standard rocket engines. In 1989, M Q Hassan had suggested using an Antonov An-225 to air-launch HOTOL, effectively replacing the still-classified, and of unknown quality, HOTOL air-breathing engine. The new, compact design would then be carried skywards atop an An-225 Mriya (the aircraft used to transport the Soviet Buran orbiter). This new 'Interim HOTOL' (I-HOTOL or HOTOL 2) was proposed in 1991, without success, although the Soviet Ministry of Aviation agreed to collaborate in the project. A tail mockup was built to test the effects of rocket exhaust from RD-0120 engines for Interim HOTOL. Various wind tunnel experiments were carried out.

David Ashford's designs

David Ashford developed Aerospace Transporter projects while working at Hawker Siddeley Aviation (see Chapter 2) and claimed that they 'could and should have been built then.' He was the author of the 1965 study, *Boost-Glide Vehicles for Long Range Transport* in which he assessed the feasibility of 'everyday,' commercially-competitive travel between the Earth and space stations. He concluded that aerospace transporter-type vehicles were possible and would be *'reasonably economical, given cheap hydrogen fuel and a market large enough to support about 50 aircraft, each making five antipodal flights per day.'*[113] In 1981, he became an exponent of 'space tourism' which he thought could be achieved by reducing the cost of space access by a magnitude of 1,000, *'using 1960's technology.'*

Much less publicised than HOTOL, a 'Space Transporter'-type project was put forward in 1984 by Mr Ashford, then working for Bristol Aerospace (Dynamics) Ltd. The proposed vehicle was called Spacecab II and was a less ambitious variant of an earlier design named Spacebus—a 'spaceliner' intended to bring 50 passengers to an orbiting

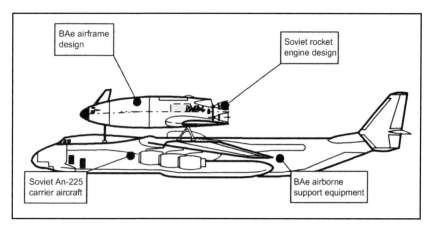

ABOVE Sketch showing the Interim HOTOL piggy-back on the Antonov An-225. *BAE Systems*

BELOW Launch trajectory of Interim HOTOL. Note that in this version, the spaceplane launched from the Antonov is not yet the 'Interim HOTOL'. *BAE Systems*

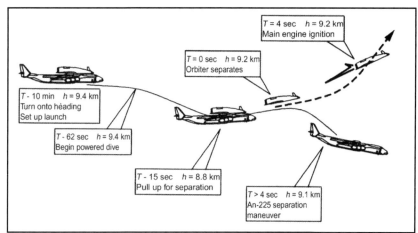

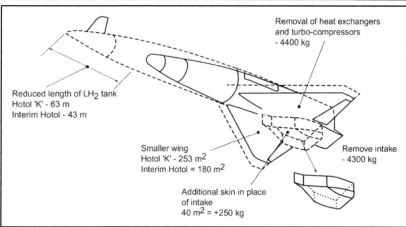

ABOVE Drawing explaining how HOTOL became 'Interim HOTOL'. *BAE Systems*

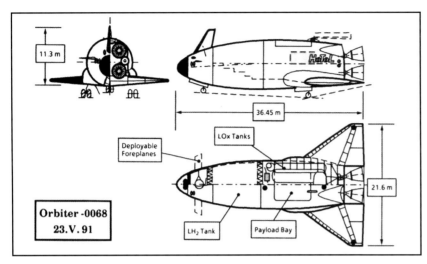

Spacecab 1st stage	
Status	Project
Span	92ft 10in (28.30m)
Length	212ft 6in (64.77m)
Gross Wing area	218.5 tons (222 tonnes)
Gross weight	–
Engine	4 × RR Olympus turbojets
	2 × Viking IV rockets
Maximum speed/height	Mach 4
Load	40.4 tons (41 tonnes)
Spacecab 2nd stage	
Status	Project
Span	53ft 6in (16.30m)
Length	55ft 0in (16.76m)
Gross Wing area	–
Gross weight	40.4 tons (41 tonnes)
Engine	6 × HM7 rockets
Maximum speed/height	–
Load	2,205lb (1,000kg)
	(including two astronauts)

ABOVE **Three-view drawing of the revised Interim HOTOL.** *BAE Systems*

BELOW **David Ashford's Spacecab Spacecab proposal from 1985. The first stage was much inspired by Concorde, but the second stage, the real spaceplane, was both all-new and not as well detailed as the first stage.** *NASA*

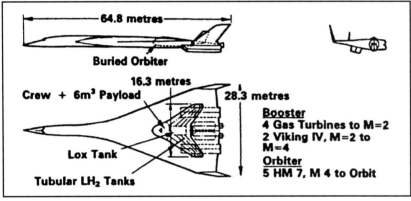

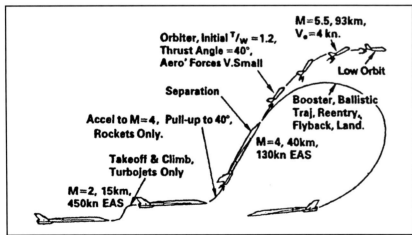

ABOVE **Intended flight path for launch of Spacecab II.** *NASA*

'Space Hotel.' The development of Spacebus involved two preliminary steps: a rocket-powered research aircraft; then a 'Spacecab,' or partially-recoverable, small shuttle. By 1984 the design had evolved and Spacecab II had become a fully-recoverable vehicle for six passengers and two pilots. The vehicle was described as '*not unlike the 1960's Eurospace Aerospace Transporter in appearance*.'[114] The first booster stage used four Olympus turbojets (like those of Concorde) for take-off, acceleration up to Mach 2, return to base and landing. Two complete Ariane second-stage engines were buried in the rear section of the fuselage to accelerate the complete vehicle up to Mach 4, the separation speed. The second orbiter stage was buried in the centre-rear fuselage. It was propelled by no fewer than six Ariane third-stage engines. Two astronauts and 212cu ft (6.0cu metres) of equipment (or six passengers) could be put into orbit.

Designer David Ashford planned various uses for his Spacecab, ranging from tourism to military operations. But in his mind, the main employment would be the re-supply of the US Space Station and the European module *Columbus*, thus entering into direct competition, in terms of use, with Hermes. Basic to Mr Ashford's spacecraft philosophy was the belief that costs could be greatly reduced by designing a fully-reusable vehicle. Economy would come not from the saved expense of having to build from scratch for each flight, but from the confidence and security which the continuous reuse of the same vehicle could bring. '*We suggest that this search for safety using expendable launchers has been the driving factor behind the high development cost of piloted spacecraft and explains why it has cost so much to qualify a piece of equipment for use on a piloted spacecraft than on an aeroplane.*'[115]

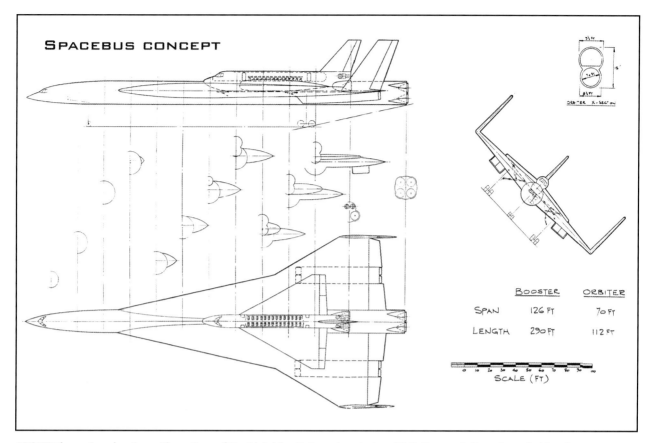

ABOVE Three-view drawing with sections of David Ashford's Spacebus design. While Spacecab II was intended for six passengers, Spacebus was for 50. bristolspaceplanes.com

Besides, the following features were intended to reduce drastically the development and operating costs:

- Re-purposing basically Concorde and Ariane technology, the vehicle could use existing equipment and could be developed as an aircraft, not as a spacecraft.
- Complete reusability.
- Burying the orbiter allows it to be optimised for re-entry (contrary to Hermes, which had severe constraints imposed by its launch atop Ariane).

The development approach based on aircraft logic instead of spacecraft logic was intended to bring down costs to the point where Spacecab could be developed by Britain alone, without need for Europeanisation.

In 1991, Mr Ashford set up Bristol Spaceplanes Ltd to develop and promote his designs, which have included different spaceplanes. The company remains active today, but its current schemes go well beyond the scope of this book. In fact, one year earlier Mr Ashford had co-written with Mr Patrick Collins a book titled *Your Spaceflight Manual*, in which he detailed the then state of his projects, focusing on Spacebus and Spacecab (actually the fully recoverable Spacecab II). He also described at length how a space hotel could be built and what life aboard would be like. However, he gave credit to the US Company ETCO for this design.

The book is intriguing, as it is obviously intended to promote Mr Ashford's projects to the general public. Its introduction ends with an adress to the reader, 'Since democratic governments and business managers respond to public opinion, you can help to eliminate the chicken-and-egg situation outlined above [people are interested in space tourism, but nobody wants to fund it because of the high costs] by writing to the airlines, travel companies and politicians to ask what they are doing about space tourism. If you want a space holiday, start asking (and saving) for it.'

While in this book he referenced the Dassault Aerospace Transporter (Transporteur Aérospatial, see Chapter 4) as representative of state-of-the-art 1960 technology (going so far as to illustrate it in Air France livery) on which his research was based, none of the then-current state-supported projects found grace in his eyes. It is interesting to describe what the failings of those designs were, in his view, as, in parallel, he also advertised the advantages of his designs:

- Hermes: *It perpetuates the cost and safety problems of vehicles with expendable stages, so it is unsuitable for space tourism. It is similar to the Boeing X-20 Dyna Soar cancelled in 1963. Thus Hermes is 20 years late [...]*
- Sänger: *Sänger is not unlike Spacebus, but it lacks the design features necessary for early development. For example, the booster has jet engines*

167

only, which will have to be extremely advanced to achieve the required separation speed of Mach 7. Spacebus uses off-the-shelf rocket engines for the difficult, high-speed part of the boost phase. The Sänger project lacks the ski-jump separation of the Spacebus, so the orbiter has to fly out of the atmosphere using athmospheric lift, exposed to high airloads and to severe aerodynamic heating. The Spacebus orbiter avoids exposure to both on the way up. Noneless, of all the current official projects, Sänger is the most suitable for tourism.

■ HOTOL: *British Aerospace proposal for an unpiloted, single-stage-to-orbit satellite launcher. As it is designed for cargo-carrying, it is aimed at a different market from Spacecab and Spacebus, whose main purpose is transporting people.*

In 1993 Bristol Spaceplanes Ltd received a study contract from ESA regarding the feasability of Spacecab. It concluded that the vehicle could be developped using available technology.

Spacebus 1st stage	
Status	Project
Span	126ft 0in (38.40m)
Length	290ft 0in (88.39m)
Gross Wing area	–
Gross weight	393.6 tons (400 tonnes)
Engine	4 × 'new design' turbo-ramjets
	4 × HM60 or 4 × J2S rockets
Maximum speed/height	Mach 6
Load	108.3 tons (110 tonnes)
Spacebus 2nd stage	
Status	Project
Span	70ft 0in (21.34m)
Length	112ft 0in (34.13m)
Gross Wing area	–
Gross weight	108.3 tons (110 tonnes)
Engine	2 × unspecified LOX/LH$_2$ rockets
Maximum speed/height	17,500mph (28,160km/h)
Load	11,905lb (5,400kg)

German Sänger II

From April **1973**, Germany continued its space effort with the ART-Programme (*Allgemeine Rückkehr Technologie*: Advanced Re-entry Technology) using DM15 million and 130,000 man-hours in research on aerothermodynamics and material technologies. This programme which absorbed the Bumerang research (see Chapter 2) was intended to develop a hypersonic aircraft during the latter part of the 'seventies, with a first flight planned around 1980-82 and a full scale re-entry test in 1983.

Germany's Sänger II proposal was originated by Messerchmitt-Bölkow-Blohm in **1984,** continuing the Junkers Raum Transporter project of the 'sixties, with financing from the *Bundes Ministerium für Förschung und Technologie* (BMFT: German Federal Ministry for Research and Technology). Actually, in its status report brochure MBB stated that it was a *'German tradition and history'* going back to 1943. Initially working with its own funds, MBB designed a two-stage, fully-recoverable vehicle giving access to orbit to ten astronauts or a heavier load.

Sänger was actually the first stage: a 'Jumbo-Jet'-sized hypersonic which could be associated with either of two different second stages: HORUS (Hypersonic, Orbital, Research and Utilisation System manned spacecraft,

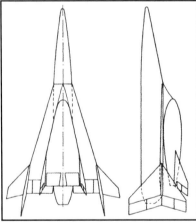

RIGHT MBB design for a two-stage-to-orbit spaceplane which fits 'somewhere' in the Sänger genealogy. *NASA*

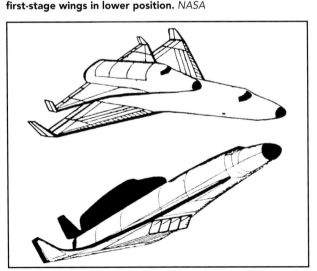

BELOW Two drawings of Sänger II, early design, with the first-stage wings in lower position. *NASA*

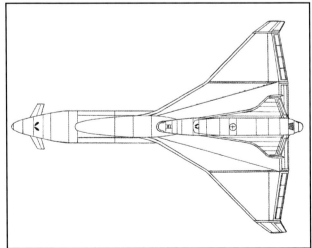

BELOW Plan view of an early Sänger II design, this time with the first-stage wings in upper position. *Via Ron Miller*

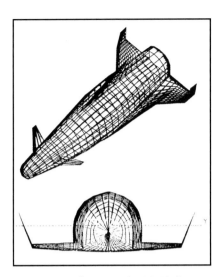

ABOVE Line drawing of HORUS the spaceplane intended as the second stage of Sänger II. *NASA*

ABOVE The same design depicted taking off from a runway on the cover of the magazin *Fusion*. *Author's collection*

similar in concept to Hermes); or CARGUS (CARGo Upper Stage, an unmanned freighter module with a 22,046lb (10,000kg)-to-orbit capacity, non-recoverable). Consideration was also given to using Sänger solo, as a long-range, hypersonic airliner for 230 business-class passengers. The cruising speed was given Mach 4.4 with a range of 6,835 miles (11,000km)—'for example, Frankfurt to Los Angeles in less than 3 hours,' claimed MBB. This option was never really detailed and appeared to be more of a promotional gimmick than a real project (as had been the case with the Nord Mistral, which had also been promoted as a SST in its time).

Sänger's initial design was indeed very aircraft-looking with a pronounced cylindrical fuselage having an apparently conventional windscreen, mated to a high delta wing with downturned tips. On its back was carried a kind of mini-shuttlecraft. The machine was propelled by six TRA-400 turbojets using liquid hydrogen as fuel (illustrations showed four; but some documents indicated five). This design, however, was soon superseded by a more ambitious creation in which wing, fuselage and engines were completely integrated in the mothership (first stage).

The second stage (HORUS/CARGUS) used a single ATC-500 H_2/O_2 rocket engine. This engine was presented as being *'much more thrust-efficient than Ariane V's HM-60 engine.'* The use of these fuels (7.9 tons [8 tonnes] of H_2 and 29.5 tons [30 tonnes] of O_2) meant that nearly half the fuselage was devoted to fuel tanks.

MBB Sänger II (1984-86)	
Status	Project
Span	-
Length	-
Gross Wing area	-
Gross weight	Under 881,850lb (400,000kg)
Engine	6 × TRA-400 LH₂ turbo-ramjets 89,921lb st (400kN) each
Maximum speed/height	-
Load	-
MBB HORUS (1984-86)*	
Status	Project
Span	51ft 2in (15.60m)
Length	97ft 5in (29.70m)
Gross Wing area	-
Gross weight	50,927lb (23,100kg) (net)
Engine	1 × H1 pressure rocket
Maximum speed/height	-
Load	2 × crew + 10 × astronauts in orbit at 248mi (400km) and 4,409 to 8,818lb (2,000 to 4,000kg) load

* Sänger report, September 1986.

FRENCH SECRET PROJECTS: FRENCH AND EUROPEAN SPACEPLANE DESIGNS 1964-1994

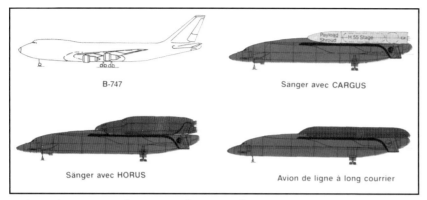

ABOVE Alternative configurations of Sänger II from a French brochure. The 'avion de ligne long courrier' refers to a suborbital (very) long range airliner. *Author's collection*

BELOW Sectioned drawing of the HORUS orbiter. Soute = Cargo bay. Note that half of the available space inside the HORUS shape is taken by a liquid hydrogen tank. *Author's collection*

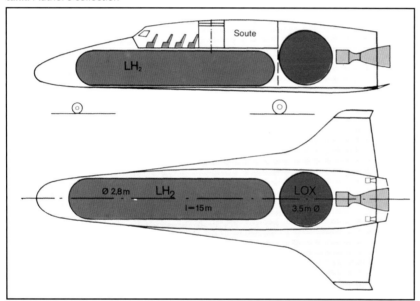

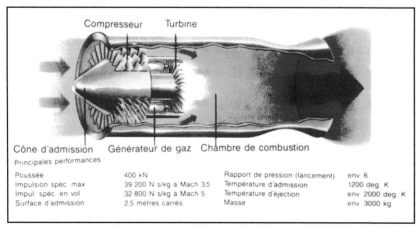

ABOVE The TRA 400 was the engine intended for Sänger II. In this drawing it looks more turbo than ramjet. Legend: Compresseur = Compressor; Turbine = Turbine; Cône d'admission = Entry cone; Générateur de gaz = Gas generator; Chambre de combustion = Combustion chamber. *Author's collection*

A typical mission would have seen the complete vehicle taking off from Germany (MBB, in its pamphlet distributed at the 1989 Paris Air Show, insisted that among the advantages of Sänger was the 'direct access to all orbits from Europe') like a conventional aircraft, and reaching the Equator while accelerating to Mach 7. At an altitude of 18.6miles (30km) above the Equator, HORUS would separate and begin its climb to orbit, reaching an acceleration of 26ft/s (8m/s) at 50miles (80km) at which altitude the rocket engine would cut, the spaceplane then coasting on its trajectory, eventually reaching apogee at 250 miles (400km).

From 1984 to 1986 about 20,000 man/hours were invested in the early design of Sänger II. The project was presented in 1986 to the ESA Council. Shrewdly it was labelled as the successor, rather than the competitor, to Hermes. MBB hoped to have Sänger II operational between 2004 and 2008 if it received the go-ahead (and the corresponding funding) in 1987. Clearly such proximity of timescale (1999-2000 for Hermes, 2004-2008 for Sänger) meant that Sänger would have been a direct competitor to Hermes, at least in budgetary allocations. Besides, it was announced that to place a ton in orbit would be *'less than 20%'*[116] of what it would cost with Ariane 5 + Hermes.

MBB CARGUS (1984-86)	
Status	Project
Span	-
Length	107ft 7in (32.80m)
Gross Wing area	-
Gross weight	25,353lb (11,500kg) (net)
Engine	-
Maximum speed/height	-
Load	13,228 to 26,455lb (6,000 to 12,000kg) in orbit at 248mi (400km)

MBB Sänger II 1st stage (1986)	
Status	Project
Span	136ft 2in (41.50m)
Length	277ft 3in (84.50m)
Gross Wing area	-
Gross weight	328,475lb (149,000kg) (net)
Engine	6 × TRA-400 LH$_2$ turbojets 89,921lb st (400kN) each
Maximum speed/height	-
Load	211,645lb (96,000kg)

MBB HORUS (1986)

Status	Project
Span	51ft 2in (15.60m)
Length	97ft 5in (29.70m)
Gross Wing area	–
Gross weight	50,927lb (23,100kg) (net)
Engine	1 × ATC-50 rocket LH_2/LOX 119,820lb st (533kN)
Maximum speed/height	–
Load	10 × astronauts in orbit at 248mi (400km); 7,275lb (3,300kg)

MBB CARGUS (1986)

Status	Project
Span	–
Length	107ft 7in (32.80m)
Gross Wing area	–
Gross weight	25,353lb (11,500kg) (net)
Engine	1 × ATC-50 rocket LH_2/LOX 119,820lb st (533kN)
Maximum speed/height	–
Load	11,023 to 30,865lb (5,000 to 14,000kg) in orbit at 248mi (400km)

Sänger was also selected by the German Federal Ministry of Research and Technology (BMFT) as reference concept for a National Hypersonic Technology Program.

According to Michel Rigault, '*For the Germans, Ariane 5, a CNES project (whose pre-eminence was difficult to accept), would have to be replaced at short term by Sänger. And the fact that most of the innovative technologies aboard Hermes were contracted to French companies was difficult for them to accept.*'[117] Philippe Couillard recognised this concurrence, saying, '*In the spirit of its designers, this project is not in competition with Hermes because it fits in a long term, post-Hermes, perspective. It would use a large part of the technologies developed for Hermes and would add to them advancement in air-breathing propulsion. Its main goal is to show to European partners that after the French-initiated Hermes, Germany will be ready to offer ESA new challenges. Yet, one must recognise that on a political level, this discourse was summed up in a competition between Hermes and Sänger. On a financial level, financing Sänger preliminary work was really in competition with Hermes in the German space budget.*'[118]

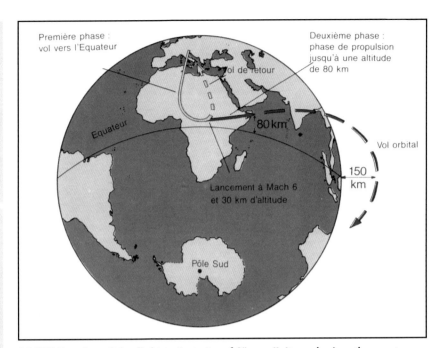

ABOVE Drawing of the flight trajectories of Sänger II. It emphasises the great advantage of a winged first stage: launch into an equatorial orbit from a base in Europe. Legend: 1ere phase vol vers l'équateur = 1st phase: flying toward the equator; Lancement à Mach 6 et 30km d'altitude = Separation at Mach 6 and 18 miles height; Deuxième phase: phase de propulsion jusqu'à une altitude de 80 km = Second phase: propulsion sequence up to 49 miles; Vol orbital 150 km = Orbital flight at 93 miles. *Author's collection*

MBB Hypersonic Passenger Plane (1986)

Status	Project
Span	136ft 2in (41.50m)
Length	277ft 3in (84.50m)
Gross Wing area	266ft 9in (81.30m)
Gross weight	–
Engine	–
Maximum speed/height	Mach 4.4
Load	230 pax

By **1988** the project had evolved with a completely new shape with soft, curved contours for both the mothership and the HORUS (and in a lesser approach to CARGUS). The crew had no direct vision ports: the view outside was relayed via video camera. Performance was slightly revised, with a more limited payload for HORUS and a slightly increased payload for CARGUS. The question of heat was detailed: the mothership was deemed to be buildable using titanium alloys without new technologies involving active cooling, if the separation speed was kept under Mach 6.8 (most of the flight time would be a cruise at Mach 4.4. with just a short burst of acceleration prior to separation).

Titanium alloys were assessed as being '*of major interest for this kind of application due to their favourable high-temperature strength-to-density characteristics*,'[119] with the ability to sustain temperatures up to 800°C. However, '*a substantial increase of the stage separation speed above Mach 6.8 (for example, with scramjet propulsion) would undoubtedly result in significantly higher technology requirements of the structural design […] compared with single-stage-to-orbit concepts (NASP,*

MBB Sänger II 1st stage (1988)*

Status	Project
Span	136ft 2in(41.50m)
Length	–
Gross Wing area	749,575lb (340,000kg) (1st + 2nd stage) 571,000lb (259,000kg)
Gross weight	–
Engine	5 × turbo-ramjets 67,441lb st (300kN) (LH_2 fuel) each
Maximum speed/height	Mach 6.8
Load	–

* 1988 data: *Propulsion Systems Optimisation For Two-Stage Winged Launcher Systems 1988.*

ABOVE **The main engines of Sänger were to be turbo-ramjets so MBB did some bench testing of ramjets, as seen here.** *MBB*

HOTOL) *the materials and structure technology demands for Sänger are eminently reduced*.'[120] To sustain the flight profile thus determined, engine designers expected to use turbo-ramjets, operating as turbojets at low supersonic speeds and as ramjets for hypersonic speed up to Mach 6.8. Flaps were to be used to direct airflow to the appropriate engine type. Engine designers were also acutely conscious of the heat question, especially around the nozzle of the first-stage turbo-ramjet. They intended actively to cool these parts by circulating the liquid hydrogen fuel in them. A sub-scale ramjet demonstrator was planned to be built in 1988.

At the **1989** Paris Air Show, Dietrich Koelle, MBB Space Transport Director announced that the Sänger II could become operational around 2005-2006, but was countered by H Hertrich, spokesperson of the German Ministry of Research and Technology who said it was unlikely that a commercial hypersonic transport system would enter service in the foreseeable future. Indeed, Dietrich Koelle admitted that hypersonic transport would not be economically feasible unless the price of the hydrogen fuel ($3.3/kg in the USA in 1989) could be halved.

The key problem to be solved was the engine: two-thirds of the budget allocated to Sänger was for engine design.

The matter of temperature was not the same for HORUS although its temperature requirements were said to be lower than the American Shuttle and Hermes. A technology called 'metallic multiwall' was developed for it but the materials required – like cobalt/nickel alloys – demanded the development of new tools and methods. Regarding the propulsion system of HORUS, it was changed from two 157,375lb st (700kN) rocket engines[121] to a single 269,775lb st (1,200kN) unit with two 9,000lb st (40kN) orbital manoeuvring engines.

That year Germany instituted a Hypersonic Technology Programme of which Sänger was the flagship. Phase 1 was the conceptual phase and was funded until end of 1992. A year later, the overall design had slightly evolved with HORUS being more integrated into the overall shape of the first stage, now labelled EHTV: European Hypersonic Transport Vehicle. A detailed separation sequence had been calculated: '*After the pull-up manoeuvre when the first stage has reached an altitude of some 37 km at a trajectory angle of 8 deg (v = Mach 6.6),*

MBB HORUS (1988)	
Status	Project
Span	-
Length	-
Gross Wing area	-
Gross weight	193,345lb (87,700kg)
Engine	1 × rocket (LOX/LH$_2$) 269,775lb st (1,200kN)
Maximum speed/height	-
Load	2 × crew + 4 × astronauts in orbit at 249mi (400km) 7,275 lb (3,300kg)

MBB CARGUS (1988)	
Status	Project
Span	-
Length	-
Gross Wing area	-
Gross weight	25,353lb (11,500kg) (net)
Engine	1 × rocket (LH$_2$/LOX) 44,961lb st (200kN)
Maximum speed/height	-
Load	11,023 to 33,069lb (5,000 to 15,000kg) in orbit at 882mi (400km)

MBB Sänger II EHTV (1989)*	
Status	Project
Span	135ft 10in (41.40m)
Length	277ft 3in (84.50m)
Gross Wing area	-
Gross weight	749,575lb (340,000kg) (1st + 2nd stage) 537,925lb (244,000kg)
Engine	6 × turbo-ramjets 41,589lb st (185kN) (LH2 fuel) each
Maximum speed/height	Mach 6.8
Load	-

* 1989 data: Sänger progress report 1989.

the second stage is lifted mechanically to an 8 deg angle of attack while the first stage lift is reduced to zero. At the same time, the orbit control engines of the upper stage (2 × 8,992 lbf [40kN]) are ignited in order to balance the increased drag of the upper stage […] The angle of attack of HORUS [increases] from 8 to 12 deg and the main engine thrust build up to the full 269,770 lbf (1,200 kN) level […] The time from ignition to full thrust is about 4 sec, so that the distance between the stages is large enough to avoid interference with the engine exhaust. HORUS is then accelerated into a 186-mile (300-km) phasing orbit with subsequent adaptation to the space station orbit at 288 miles (463km) altitude.'[122] HORUS was reworked and reduced in size while CARGUS was also reworked around Ariane 5 core stage.

Titanium alloy with thermal production was definitively selected for the fuselage. Temperatures were expected to be lower than 600°C for the undersurface and 400°C for the uppersurface. Work continued regarding the wing construction. All this was supported by various wind tunnel tests.

Even French representatives were worried by this project and queried the reaction of their government to the announcement of Sänger II. The French Ministry of Research and Technology replied, 'The decision to pursue this project could happen in 1992, after the study period. This project could only be carried out in co-operation. France is also interested in those technologies.'[123] Indirectly one could deduce that France was willing to participate in Sänger II… but would it have got the funds?

ABOVE Seen from above, Sänger shows clearly that it is composed of two vehicles: a winged first stage launcher and the second, a spaceplane named HORUS. *MBB*

BELOW The two main variants of Sänger II differed by reason of their second stages: a recoverable, winged spaceplane to carry astronauts named HORUS; and a non-recoverable, rocket-powered booster to put satellites into orbit named CARGUS. *MBB*

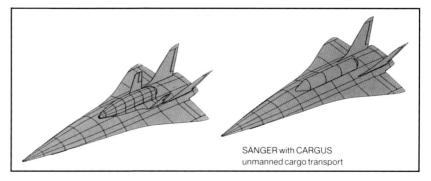

MBB HORUS 3C (1989)	
Status	Project
Span	51ft 6in (15.70m)
Length	92ft 10in (28.30m)
Gross Wing area	2,207sq ft (205.0sq m)
Gross weight	211,645lb (96,000kg)
Engine	1 × rocket ATC-1200 (LOX/LH$_2$) 269,775lb st (1,200kN)
Maximum speed/height	–
Load	2 × crew + 4 × astronauts in orbit at 882mi (400km) 7,275lb (3,300kg)

MBB CARGUS (1989)	
Status	Project
Span	–
Length	–
Gross Wing area	–
Gross weight	–
Engine	1 × rocket HM-60 Vulcain (LH$_2$/LOX) 240,545lb st (1,070kN)
Maximum speed/height	–
Load	11,023 to 33,069lb (5,000 to 15,000kg) in orbit at 882mi (400km)

RIGHT Sänger II was exhibited as a small model at the 1987 Paris Air Show. *JC Carbonel*

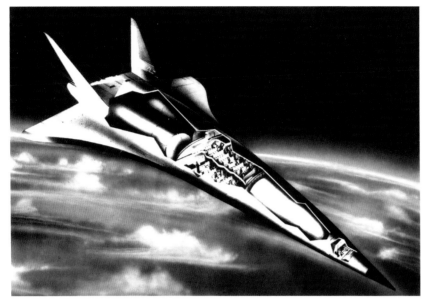

ABOVE Illustration of the supersonic transport variant of Sänger II.
Author's collection

RIGHT Rare! Hermes and Sänger II 'flying' side by side on a MBB leaflet distributed at the 1989 Paris Air Show. MBB

BELOW For Sänger II, MBB compared its size to the Boeing 747 'Jumbo Jet'.
Author's collection

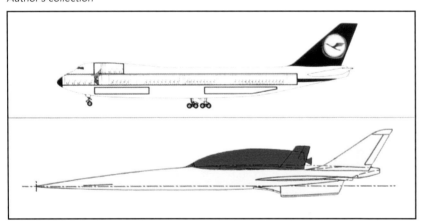

MBB HYTEX

Status	Demonstrator project
Span	About 30ft 6in (9.30m)
Length	About 75ft 6in (23.00m)
Gross Wing area	–
Gross weight	–
Engine	Turbo-ramjet × 1?
Maximum speed/height	Mach 5.5 at 98,435 ft (30,000m)
Load	–

MBB HYTEX R-A3

Status	Demonstrator project
Span	12ft 9in (3.89m)
Length	50ft 7in (15.42m)
Gross Wing area	–
Gross weight	–
Engine	1 × Turbo-ramjet
Maximum speed/height	Mach 5
Load	Various test equipment

1990 saw the abandonment of CARGUS, it being replaced by an unmanned variant of HORUS named HORUS-C, the manned HORUS being re-labelled HORUS-M. HORUS-M was fitted with a docking adapter for the then-future International Space Station.

That same year the German Ministry of Research awarded a contract to MBB to design a conceptual hypersonic, sub-scale proof-of-concept named Hypersonic Experimental Aircraft Technology Demonstrator (HYTEX) which was conceived in **1991**. First flight of the 30ft 6in (9.30m)-span aircraft was planned for 1998. The aircraft was designed for a maximum speed of Mach 5.5 and would have tested the turbo-ramjet engine intended for Sänger. It was also intended to use the same type of material (titanium alloys with insulation covering) as on Sänger. The manned aircraft design was still too costly and the HYTEX evolved into an unmanned HYTEX R-A3. In 1994, co-operation with Sweden, Norway, France and Russia was evaluated. Consideration was then given to adapting the Russian Raduga D2 missile with a turbo-ramjet. One example of Raduga was acquired in **1995** from Russia but modification to the missile, and by consequence test-flight, did not occur.

A 33ft (10m)-long display model of Sänger II was built in 1991 to be exhibited at air shows.

Between 1990 and **1992**, experiments were carried out regarding ramjet (notably the 'sharktooth' flameholder designed in co-operation with the Swedish company Volvo, intake and nozzle designs) and thermal protection systems. Various alternatives were considered, manned and unmanned, for the technology demonstrator.

By that time Sänger was the 'technical reference concept' of the German Hypersonics Technology Programme but it was noted that *'the developed technologies are applicable to a wide range of potential future space transport systems.'*[124] The whole concept was then evaluated by TAB (*Büro für Technikfolgenabschätzung am Bundestag*: Office for Technological Evaluation of the National Parliament of the Federal Republic of Germany) in April 1991 (*'Low cost to Orbit'*[125]) and then again in October 1992. TAB proposed three options:

i Option I was a moratorium on development of hypersonic technology. This *'would effectively interrupt the technological work. On the other hand resources would have been spared if later, a policy against a "progressive" use of space was adopted.'*

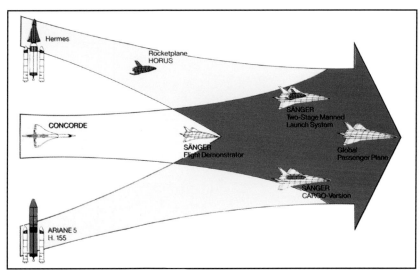

ABOVE MBB wanted to show Sänger II in a broader dimension by tying it to other programmes: Concorde, Ariane 5, Hermes. *Author's collection*

BELOW Details of the first stage structure. *MBB*

ABOVE Bench test of the hypersonic intake for the Sänger engines. *MBB*

BELOW Test sample of thermal protection material. *MBB*

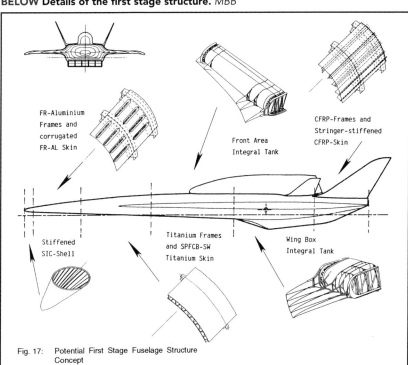

BELOW Test sample of the leading edge of Sänger II wing. MBB produced a few hardware samples to test ramjets, scramjets, and thermal protection material. *MBB*

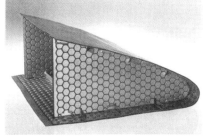

BELOW Wind tunnel model of Sänger II with various alternative components. This picture suggests that nose canards (on the left) were contemplated. *MBB*

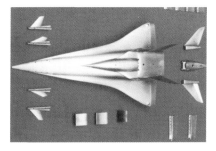

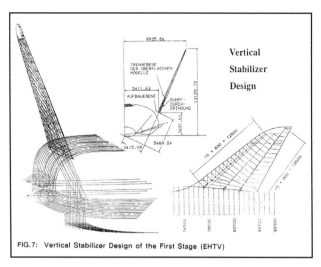

ABOVE **Design of the vertical stabilizers.** *MBB*

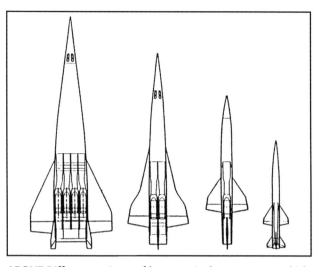

ABOVE **Different variants of hypersonic demonstrators which were intended to prove the feasibility of Sänger II.** *MBB*

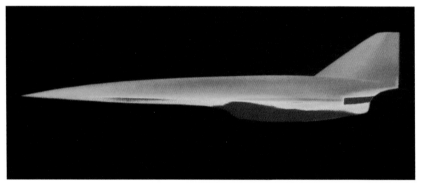

ABOVE **This picture made the cover of the Hytex brochure— but was not captioned. It does look like a temperature gradient illustration but strangely the 'hot' parts are only shown in the engine area and just behind it, while the nose section remains apparently 'cool'.** *MBB*

No decision was taken; the project slowly atrophied until 1995 and was not continued afterwards. What had been achieved was transferred to the ESA programme FESTIP (Chapter 12). Henri Lacaze summed up the Sänger II proposal, '*The people at the German Research Ministry had not much faith in the project and were not willing to invest in a programme which could end in a failure. HERMES was more realistic so it was not difficult to take them aboard the HERMES project. However, they were very keen to do things with the USA, hence they favoured the Columbus project. Besides, they felt, probably rightly, that it was more interesting to be part of the International Space Station rather than build smaller, limited-life, stations as was planned in the projects associated with HERMES.*'[127]

ii Option II: '*The second option would be to continue until 1995 the hypersonic technology programme based on the technical reference concept Sänger, following the present previsions of BFMT, notwithstanding modifications and complements.*'

iii Option III: '*The third option follows the detractors of the hypersonic transport programme which consider premature any decision on this project […] Besides, studies show that it is not yet possible to compare the different concepts.*'[126]

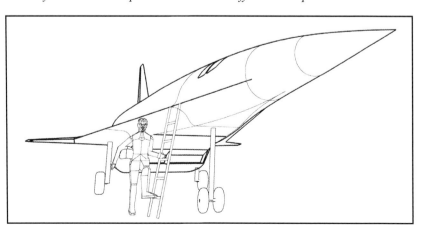

LEFT **Line drawing from the same brochure. For once, a human figure gives an impression of size to a project.** *MBB*

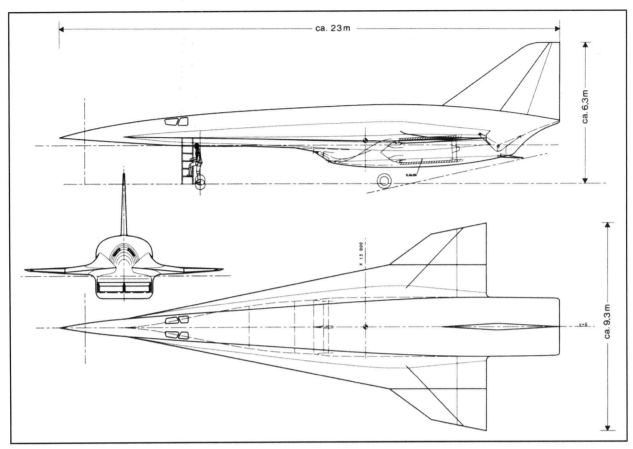

ABOVE Three-view drawing of the HYTEX demonstrator. The aircraft would have been of similar size to a General Dynamics F-111 'Aardvark'. *MBB*

ESA: PLATO PLATform Orbiter concept

This project, designed jointly by two engineers, one British, Stephen Ransom and one German, Rainer Hoffmann – and which should not be confused with the current PLATO Planetary Transits and Oscillations of stars – was a late-'eighties design for an unmanned, winged re-entry vehicle designed to be launched by Ariane 4 and retrieved through a conventional landing at a site in Europe. It was intended to carry science experiments into orbit and test material, structures, navigation and control technologies for spacecraft.

RIGHT Representation of the grid system around the HYTEX as used for the computer simulations. *MBB*

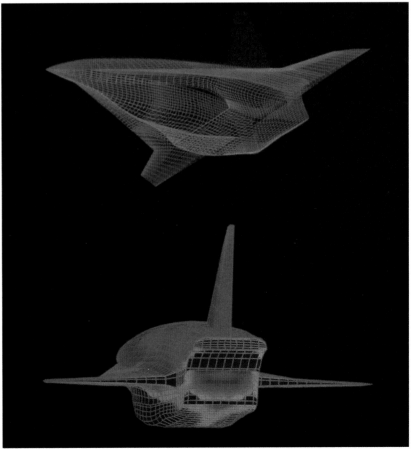

PLATO	
Status	Project
Span	21ft 0in (6.40m)
Length	39ft 4in (12.00m) (Rect)
	46ft 7in (14.20m) (Delta)
Gross Wing area	729sq ft (67.7sq m)
Gross weight	11,023lb (5,000kg)
Engine	–
Maximum speed/height	Mach 20
Load	1,764 to 2,205lb
	(800 to 1,000kg)

Simplicity in design and minimisation of costs were key factors in the development of PLATO.

The impetus for PLATO came from the accident to space shuttle *Challenger* in 1986. At that time, ESA had a programme called EURECA (European Retrievable Carrier) which was entirely dependent on the shuttle for its operation. EURECA remained a single vehicle launched on 31 July 1992 by Space Shuttle *Atlantis* (mission STS-46) and was retrieved one year later on 1 July 1993 aboard Space Shuttle *Endeavour* (mission STS-57), but in the late 'eighties there was a real risk that the whole program could be stranded if no shuttle was available. So, PLATO was designed to carry the EURECA experiments and bring them back to Earth. Actually, PLATO went further than that in proposing to use the EURECA systems (power, attitude and orbit controls, communications and data handling systems) to pilot PLATO's return flight.

The EURECA unit was scaled to one-half so as to fit into the winged airframe of PLATO, while deleting the equipment required to dock with a shuttle. During the launch phase, the vehicle was mounted axially on the Vehicle Equipment Bay adapter of Ariane 4. Two different configurations were studied: a rectangular shape and a delta shape. The designers did not choose between those and presented them side by side in their report. A quite unusual feature was the installation of the undercarriage in the upturned wing tips so the spaceplane needed to rotate 180° for landing, actually landing on its back. The designers had some concern that wheels, and especially tyres able to survive the rigours of space flight, would not be ready in time, so they elected to use sprung skids integrated in the tips of the fins.

When the launch configuration was established, the maximum weight for PLATO was estimated at 11,023lb (5,000 kg). For the return trip, the cross-range capability was computed at 1,243 miles (2,000 km) which was deemed sufficient for a recovery of PLATO from a 25.5° orbit (intended for EURECA project) at landing sites located in southern Europe. Heating was taken into account in studying the re-entry but the solution to be implemented (tiles?) was not explicitly given in the PLATO description.

The final section of the flight, through Earth's atmosphere was studied using data from NASA and DFVLR for super and hypersonic flight sequence, while subsonic flight was actually tested, using a $1/20$ model in a wind tunnel at Aachen.

---

[104] *Spaceflight,* Vol 26 No 11, November 1984.

[105] Quoted in *Hermes Europe's Dream of Independent Manned Spaceflight,* by Luc van den Abeelen, (Springer 2017).

[106] Quoted in *Hotol Versus Hermes, Spaceflight,* Vol 28 No 4, April 1986.

[107] Ibid 106.

[108] *Flight International,* March 1986.

[109] Interview by the author, 8 February 2019.

[110] *Air & Cosmos,* 5 July 1986.

[111] *Spaceflight,* March 1986.

[112] All the citations are from the advertising pamphlet distributed at the Paris Air Show.

[113] https://www.cambridge.org.

[114] *Spaceflight,* Vol 27, January 1985.

[115] *Your Spaceflight Manual.* David Ashford and Patrick Collins, Headline book publishing, 1990.

[116] Sanger II status report, September 1986.

[117] Mail exchange with the author, October 2020.

[118] *Rêve d'Hermes,* 1993.

[119] *Structural Requirements and Basic Design Concept For a Two-Stage Winged Launcher System* (Sänger), 1988.

[120] Ibid.

[121] This reference to 'two' rocket engines appears in the *Propulsion Systems Optimization for a Two-Stage Winged Launcher System* document of 1988, but not in earlier documents in the possession of the author.

[122] Sänger progress report 1989.

[123] Assemblée Nationale *Questions to the Government,* 9th Legislature 1989.

[124] Sänger progress report, 1992.

[125] *Billinger in den Orbit?* TAG Bericht AB-001.

[126] *Technikfolgen-Abschätzung zum Raumtransportsystem "Sänger,"* TAG Bericht AB014, Germany, October 1992.

[127] Interview by the author, 8 February 2019.

Chapter Ten
Co-operation Around the World

ABOVE A rare model of the Hyperplane. The blue/white scheme, large underside intake and cockpit faired into the fuselage make it look like the 'SX-1 Swordfish', mechanical hero of Edgar P Jacobs's graphic novel *Secret of the Swordfish*, published in 1950.
Via H Lacaze

While Europe was engaged into the design and construction of Hermes, many agencies knocked at the doors of the main contractor Aerospatiale, eager to share in its experience with spaceplanes, and benefit from European knowledge on the subject.

India: Hyperplane

In 1989, Air Commodore (Ret) Raghavan Gopalaswami Managing Director of Bharat Dynamics Ltd (BDL), a company belonging to the Ministry of Defence of India, telegraphed Henri Lacaze at the Direction des Etudes Avancées (Advanced Studies Directorate) of Aerospatiale following an earlier meeting (which could have taken place in April). BDL had, by that time, a long history of co-operation with Aerospatiale. When incorporated in 1970, it had acquired a licence to produce the SS.11 anti-tank missile (called locally the 1st generation Anti-Tank Guided Missile [ATGM]). Then it went on to licence-produce the 2nd generation ATGMs, one of them being the French MILAN (produced under the auspices of Euromissile, a consortium created by Aerospatiale and MBB for the production of MILAN and Roland missiles) and the second, the Russian Konkurs. Both weapons were upgraded during their life, maintaining the contact with their original designers.

Raghavan Gopalaswami's telegram referred to 'Hyperplane', a hypersonic, air-breathing spaceplane whose design had been submitted to Aerospatiale for advice and about which Henri Lacaze had '*expressed doubts regarding the engine design*' in the earlier meeting.

This telegram predated by only a few days an official letter announcing the visit to France, in the week beginning 12 June, of Dr V S Arunachalam, scientific adviser to the Defence Minister of the Republic of India and secretary of the Department of Defence Research & Development of the Government of India and Dr A P J Abdul Kalam, Director of the Defence Research & Development Laboratory (DRDL) of the Ministry of Defence of the Republic of India. The two Indians expected to meet the President of Aerospatiale, Henri Martre, and sign with him a Memorandum of Understanding (MoU) regarding the collaboration between BDL and Aerospatiale on the concept validation of Hyperplane. Aerospatiale did not want to get too involved in this project and agreed only to collaborate on conceptual propulsion principles, even if further MoUs could deal with '*the ultimate goal of the joint collaboration effort between Aerospatiale and DRDL [which] shall be the joint development of a future single-stage-to-orbit launch vehicle and the potential joint commercial utilisation of this launch vehicle.*'[128]

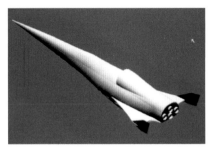

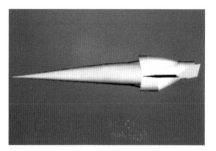

ABOVE Three computer portrayals of the Indian SSTO Hyperplane. The Indian company Bharat Dynamics Ltd approached Aerospatiale during 1989 in the hope of getting help with developing the propulsion system of Hyperplane.
Bharat Dynamics Ltd via H Lacaze

Aerospatiale's Division des Systèmes Stratégiques et Spatiaux was actually only interested in the propulsion studies which could be included in the scope of other conceptual studies like the STS-2000 (a future SST project). DRDL would have liked to have the MoU signed before December 1989 because of the Indian internal political agenda but this was not to be. In February 1991 Aerospatiale informed the Indian Embassy in Paris that such a MoU would require political approval at Government level (Ministry of Defence, Ministry of Industry) – which might be complex and of long duration – and proposed to limit co-operation to *'punctual exchange of information'* without a formal frame like an MoU.

Raghavan Gopalaswami revived the question in May 1991, but Aerospatiale maintained that formalising the co-operation was not advisable for the time being, pointing out that it was itself engaged in other co-operative works on the same subject with Dassault, SEP, SNECMA and ONERA.

Hyperplane description:

Hyperplane was to take-off from a conventional runway and reach orbit at a fraction (between a tenth and a thirtieth) of the cost of rocket launch and without the need for *'massive, vulnerable launch pads and ground support'.*

Highly advanced technologies were to be used in its design:

- *Revolutionary scramjet and turbo ram-rocket propulsion.*
- *Advanced structural materials: carbon-carbon, titanium and engine materials.*
- *Artificial Intelligence, expert systems, robotics for flight management.*
- *Supercomputers and advanced CAD/CAM for design and development.*[129]

Hyperplane missions were triple:

i Satellite launcher.

ii Mass missions in space involving hundreds of thousands of tonnes per year applied to solar power, space manufacturing, sub-orbital transport, mining, colonisation and nuclear waste disposal into the sun.

iii Transport over transcontinental distances.

The brochure recognised that, '*The crux of the matter would be the transatmospheric engine [...] A separate and major programme needs to be launched in this area, with international collaboration to cut costs and time.'*

After the failure to formalise collaboration with Aerospatiale, BDL continued on its own. Furthermore, in 1998 Air Cdre Gopalaswami had revealed the development of a further spaceplane named 'Avatar'. The project was commissioned by the Indian Defence Research and Development Organisation in 2001.

Hyperplane	
Status	Project
Span	51ft 6in (15.70m)
Length	190ft 3in (58.00m)
Gross Wing area	1,889sq ft (175.5sq m)
Gross weight	–
Engine	Air-breathing
Maximum speed/height	–
Load	–

BELOW From this drawing, it would seem that the key to the propulsion system of Hyperplane was to extract oxygen from surrounding air during atmospheric flight.
Bharat Dynamics Ltd via H Lacaze

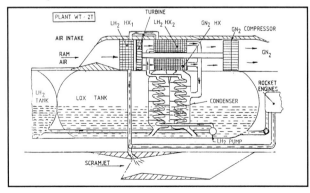

BELOW Three-view drawing and perspective grid shape of Hyperplane. *Bharat Dynamics Ltd via H Lacaze*

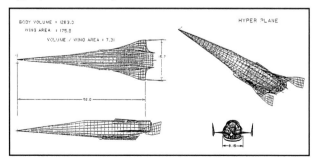

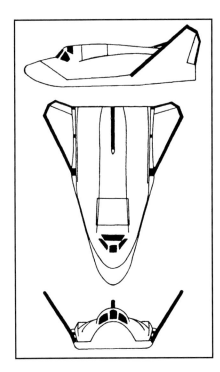

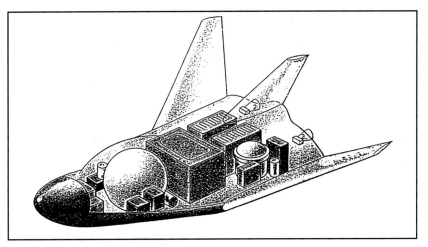

ABOVE Cutaway of a BOR-4 vehicle which was presented to Aerospatiale engineers by Russian counterparts during their 1991–92 meetings. *Aerospatiale via H Lacaze*

LEFT NASA drawing of the BOR-4 *Bespilotnyi Orbital'nyi Raketoplan 4* (Unpiloted Orbital Rocketplane 4). The BOR vehicles did fly re-entry trajectories and were the basis of the Soviet/Russian experience in spaceplanes. *NASA*

Russia: Molniya Mini-Shuttle

It seems that in 1992, there were talks with Russia about co-operation on spaceplanes. France and Russia had always collaborated on space projects: General Jean-Loup Chretien was the first Frenchman in space, flying aboard Soyuz T-6 in 1982. In 1992, he was training for pilot qualification on Buran, the Soviet Shuttle. (The Soviet Union had dissolved on 26 December 1991 into its component – now autonomous – republics).

So it was no surprise that EuroHermespace and Dr Gleb E Lozino-Lozinsky, General Director of NPO Molniya, began sharing information on building a spaceplane together. In 1990 NPO Molniya had announced that it was studying a mini-shuttle also named *Molniya* which would be launched from the back of an Antonov An-225 Mriya. It had claimed that '*with a technological effort on materials, equipment and engines, it was possible to develop a fully-reusable space system with a shuttle launched from the six-engine An-225.*'[130] NPO Molniya put forward the reusability, the agility (independency from fixed launch sites) of its concept. The *Molniya* spacecraft was designed to put 6.9 to 7.9 tons (7 to 8 tonnes) in orbit, claiming that it was sufficient to cover 90 to 95% of the current needs in term of orbited mass.

In 1991 NPO Molniya contacted CNES, to pass it a small brochure describing the proposed mini-shuttle, or as CNES put it 'Soviet new generation orbital plane.' This design, referred to by the umbrella-name 'MAKS', was communicated 'exclusively' to CNES which indicated at the beginning of its report that NPO Molniya had not yet revealed publicly this project in the public domain. The missions imagined for this system – in which the mini-shuttle still took off from an An-225 Mriya, but now carrying with it a large fuel tank very much like the US Shuttle – were the same as those initially detailed for Hermes:

BELOW Drawings submitted by NPO Molniya to Aerospatiale. They are reconstructions of HERMES shape by Molniya. Caption in Cyrillic text reads *orbitalnii samolet Hermes* (orbital aeroplane Hermes). *Aerospatiale via H Lacaze*

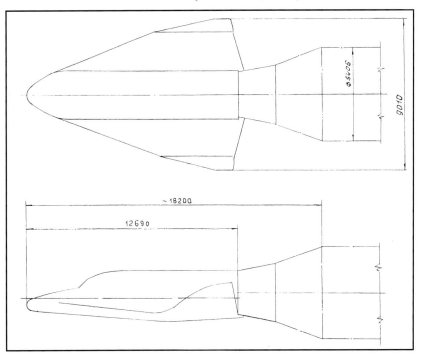

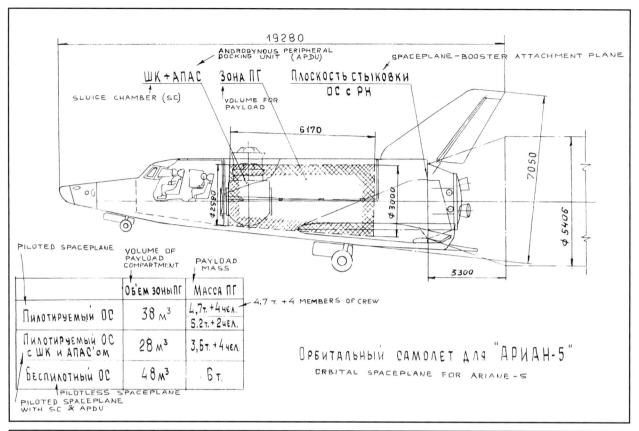

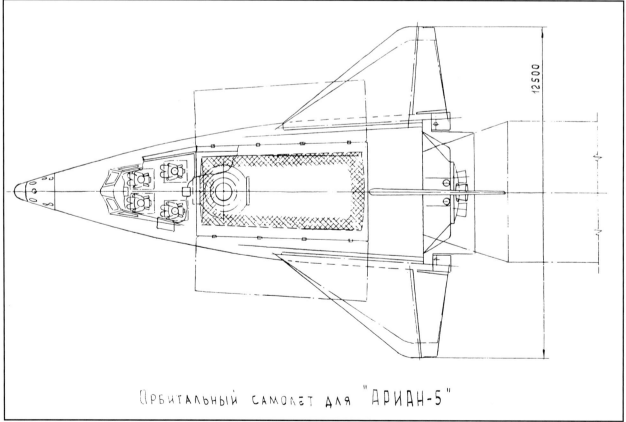

ABOVE NPO Molniya offered Aerospatiale the MAKS spaceplane for launch atop Ariane. Here is a sectioned view of the MAKS adapted for this type of launch. Caption in Cyrillic text reads *orbitalnii samolet alya Arian-5* (orbital aeroplane for Ariane-5). *Aerospatiale via H Lacaze*

- Servicing manned space stations.
- Servicing unmanned space factories.
- Using the vehicle as a space laboratory or for limited production of special materials and biological and pharmaceutical products.
- Astronomical or ecological research.
- Deployment and retrieval of satellites in low orbit.

The aircraft had a launch mass of 53,259lb (24,158kg) with a payload of 13,228lb (6,000kg) in a 1,377cu ft (39cu m) bay. An unusual feature was positioning of the pressurised crew cabin at the rear together with a deployable airlock and docking ring.

Another aspect was variable geometry, with the dihedral of the wings which could be adapted to the mission sequence (high dihedral at take-off to minimise wind effect; low dihedral during subsonic approach to improve the flying characteristics). CNES was obviously impressed by this design, remarking, '*The variable-geometry wing for a vehicle re-entering Earth atmosphere in a glide is a clever choice, off the beaten track. It allows better optimisation of the aerodynamic features of the vehicle and flight qualities in a wide range of Mach numbers and flight altitudes*.'

CNES also noted that high wing loading did not appear to trouble the Russian designers who based their experience on both wind tunnel measurement and experiments with the BOR-4 experimental re-entry vehicle. It wondered, '*Could it be that in designing ASH [Avion Spatial Hermes] we overestimated the influence of wing loading on the intensity of re-entry thermal flow and landing speed?*' CNES also noted the shape of the nose, asking, '*would it be necessary to re-think [our study of the nose]?*'

Finally, the position of the crew cabin was appreciated, pointing out that the pressurised volume was reduced. If it became necessary to pressurise the cargo bay, then its access would be facilitated. Forward vision was achieved through TV cameras, which was deemed acceptable. Crew escape solution was the installation of ejector seats. CNES also noted with

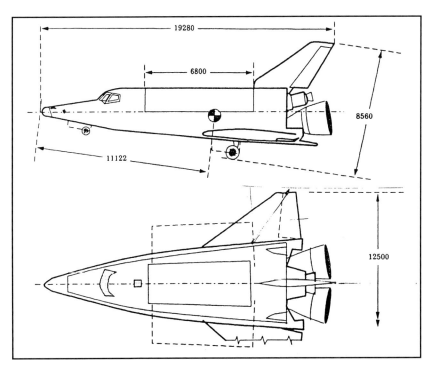

ABOVE Drawings of MAKS orbiter. This is the version intended to be launched from an Antonov An-225 Mriya. *Aerospatiale via H Lacaze*

interest the philosophy of the role of the crew, which was, '*Piloting until rolling on the landing strip and stopping is entirely automated. The crew would be high-level technicians in charge of the mission and adapting the vehicle to the necessities of the time rather than joystick virtuosos who must heroically save the vehicle and its passengers from a menacing peril*.'

In 1992, the proposal made to EuroHermespace was to adapt *Molniya* to be launched atop Ariane 5. In this configuration *Molniya* had its payload reduced to 10,362lb (4,700kg) to and from orbit. The payload brought back from orbit was the same as the payload put into orbit, a feature very distinctive from Hermes which was quite limited in the mass it could bring back to Earth. The unpressurised cargo bay was 20ft 3in (6.17m) but part of this length was taken by an airlock with an 'androgynous' docking unit, suggesting it could dock with both the US and Soviet Space Stations. (The androgynous part refers to the Apollo-Soyuz mission in which neither country wanted to have a 'female' docking ring…obliging both spacecraft to have an extra tunnel able to link with both Apollo and Soyuz. Later docking devices were designed to be androgynous, so they could all dock together).

A weight table for the proposed vehicle indicated a gross mass (launch mass) of 50,706lb (23,000kg), well within the ability of Ariane 5 with a crew of four and the payload already mentioned. However, wing surface was not indicated by the Russians and may have been a source of problems, as it had been a severe limitation to Hermes, notably regarding the payload brought back to Earth. An interesting point revealed by this weight table was the weight of each spacesuit: 35lb (16kg).

Although co-operation with Russia was often mentioned at political level, nothing came out of the talks with Molniya.

Molniya	
Status	Project
Span	41ft 0in (12.50m)
Length	56ft 8in (17.28m)
Gross Wing area	807sq ft (75.00sq m) (launch) 936sq ft (87.00sq m) (landing)
Gross weight	53,259lb (24,158kg)
Engine	Two small rockets
Maximum speed/height	-
Load	13,228lb (6,000kg) + 4 × crewmembers

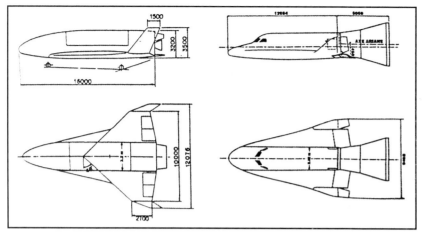

ABOVE When in dialogue with ESA, the Japanese Space Agency was keen to compare its HOPE Spaceplane to HERMES—as in this 1991 document. *ESA via H Lacaze*

BELOW Sectioned drawing of HOPE which shows the similarities to HERMES. A major difference appears to be the lack of an airlock/docking ring—a question which long vexed the design of HERMES. *JASDA/JAXA*

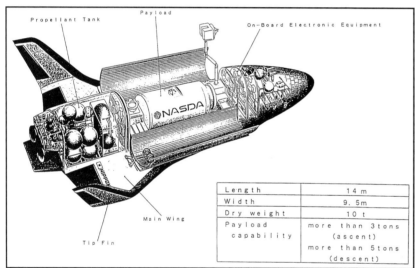

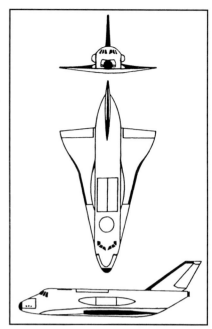

ABOVE This large variant of HOPE appearing in a NASA report showed that in 1987, HOPE was a quite big vehicle, somewhat larger than HERMES. *NASA*

Japan: HOPE/HOPE-X

From 1985, Japan conducted a programme to design a recoverable spaceplane named HOPE (H-2 Orbiting PlanE). The name suggests that in the same way Hermes was an offshoot of the Ariane launcher, HOPE was an offshoot of the H-II launcher. It was intended to stay a few days in low Earth orbit, re-entering the atmosphere to land horizontally on conventional runways. The main work was to design thermal protection. Carbon/polyimide (C/Pi) and carbon-carbon materials were finally selected. After a half-decade of research, NASDA wanted to go into flying experiments and approached ESA to share experience…and budgets.

In June 1991, a joint meeting between ESA and NASDA was held in Tokyo. HOPE and Hermes were compared and NASDA described its own programme which encompassed the OREX re-entry vehicle, intended to test specifically the thermal protection systems; and the ALEX flight demonstrator.

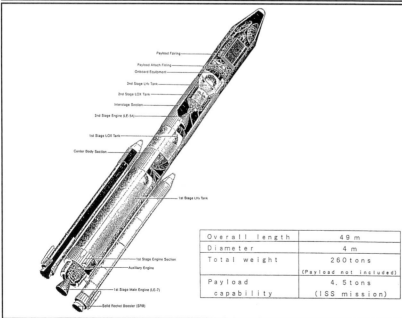

LEFT The H-II rocket also had many similarities to Ariane. *JASDA/JAXA*

CHAPTER TEN   CO-OPERATION AROUND THE WORLD

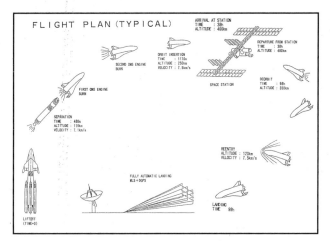

ABOVE Trajectories proposed for HOPE. *JASDA/JAXA*

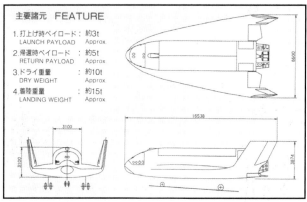

ABOVE Three-view drawing of Kawasaki's HOPE proposal. *Kawasaki*

BELOW Artist's impression of the spaceplane HOPE in its element. This is a Kawasaki proposal. *Kawasaki*

In February 1994, with Hermes now cancelled, ESA wanted even more to co-operate with Japan, feeling that Japan and Europe shared a common situation *vis à vis* the two original space powers, USA and Russia. Henri Lacaze had his views on the Japan-Europe collaboration. '*At diplomatic level, there was action, but nothing came out of it at the technical level. But we were very impressed by the Japanese step-by-step method.*'[131] OREX was flown on the first H-II mission in early 1994 but fell into the sea and could not be recovered. The later HYFLEX flew a successful re-entry in 1996 but was again lost at sea.

Like Hermes, HOPE suffered a contraction of budget, leading to a reduced model called HOPE-X (for HOPE demonstrator). Although some common work was done with CNES during the early 21st Century, HOPE was cancelled in 2003 while NASDA was reorganised into JAXA.

RIGHT Low-speed tests of HOPE. The test vehicle was air-dropped from a helicopter. *Kawasaki*

185

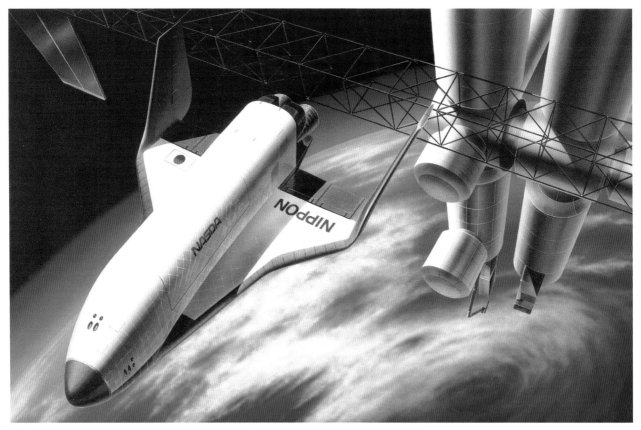

ABOVE AND LEFT These other artist's impressions depict Mitsubishi's vision of the HOPE project.
Mitsubishi Heavy Industries

HOPE	
Status	Project
Span	28ft 3in (8.60m)
Length	54ft 3in (16.53 m)
Gross Wing area	-
Gross weight	33,069lb (15,000kg) (at landing)
Engine	None
Maximum speed/height	-
Load	6,614lb (3,000kg) (to orbit) 11,02lb (5,000kg) (from orbit)

BELOW Three-view drawing of Mitsubishi's HOPE. *Mitsubishi Heavy Industries*

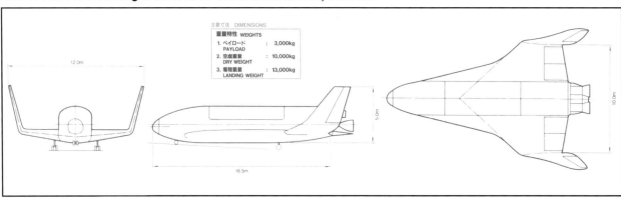

RIGHT HOPE on the H-II rocket. *JASDA/JAXA*

BELOW While the Europeans had elected not to conduct re-entry testing, JASDA wanted to do some testing with a capsule-type vehicle called OREX. *ESA*

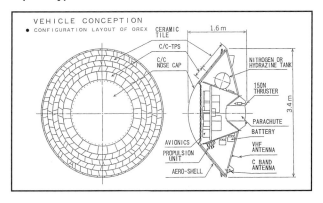

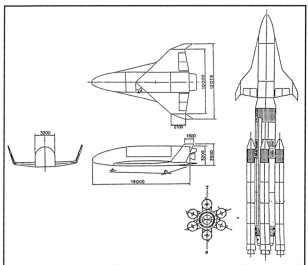

RIGHT OREX was intended to test the thermal protection for HOPE. It was flown successfully in 1994 but was lost during recovery at sea. *ESA*

BELOW Another research vehicle planned at that time was HYFLEX. *JASDA/JAXA*

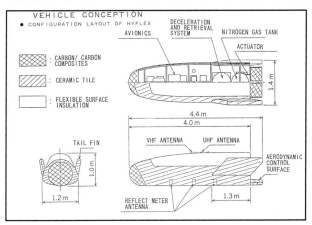

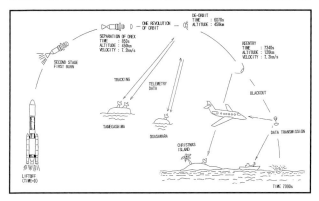

RIGHT HYFLEX was flown in 1996; it executed a successful re-entry, but was also lost in the sea during recovery. *JASDA/JAXA*

BELOW ALEX was a third experimental vehicle intended to test the low speed capability of HOPE. *ESA*

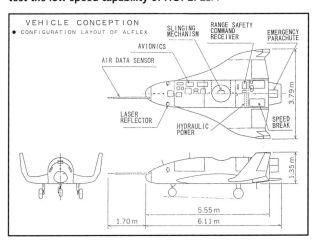

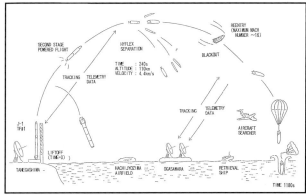

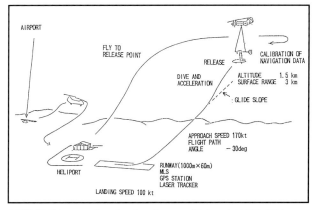

ABOVE This was the plan for the test flight of ALEX. *ESA*

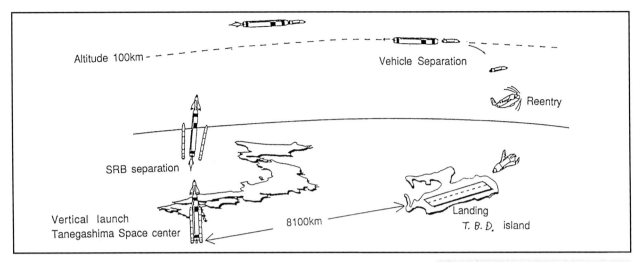

ABOVE This would have been the intended flight plan of HOPE-X. *JASDA/JAXA*
BELOW After HERMES was cancelled, JASDA was left alone and opted for a reduced demonstrator called HOPE-X—which was not flown either. *JASDA/JAXA*

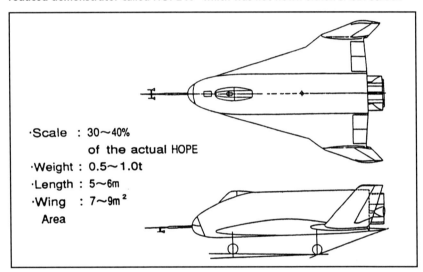

ALEX	
Status	Demonstrator project
Span	-
Length	16 to 20ft (5 to 6m)
Gross Wing area	75 to 86sq ft (7 to 9sq m)
Gross weight	1,100 to 2,200lb (500 to 1,000kg)
Engine	None
Maximum speed/height	Dropped from about 4,920ft (1,500m)
Load	None

Christofferus	
Status	Study
Span	32ft 6in (9.90m)
Length	32ft 6in (9,90m)
Gross Wing area	-
Gross weight	13,228lb (6,000kg)
Engine	None?
Maximum speed/height	N/A
Load	5 × astronauts

Switzerland: Christofferus

The Christofferus project was presented in 1992 during a workshop about FESTIP (Future European Space Transport Investigations Programme, actually opened in 1996) by Professor Hans-Rheinhard Meyer-Piening as a continuation of Hermes development engineering. It was, in fact, a science project carried out by Prof Meyer-Piening's students. Christofferus was a small spaceplane designed to be launched either atop Ariane 4 or in the cargo bay of the American Space Shuttle. Its uses were as follows:

■ Fast astronaut rescue (for example if astronauts became ill or injured during a space mission, Christofferus could be launched to retrieve and bring them quickly to Earth while the mission continued).

■ EVA shelter.

■ Shelter for space station crew.

■ Space tug to geosynchronous orbit.

■ Communications relay.

■ Space taxi (for example to help space assembly tasks when constructing space stations or space factories).

■ Space garbage monitoring & control.

■ Event survey platform.

Christofferus was diminutive, less than 33ft (10 metres) long, with the ability to carry up to five astronauts, but also to return to Earth automatically. It was fitted with folding wings so that it could fit into the Ariane 4 shroud. Landing would be at about 150 knots (173mph [278km/h]), on three retractable skids. Being just a university project, it was not financed.

[128] Draft MoU-1, 30 June 1989.

[129] Hyperplane brochure.

[130] Reported in *Air & Cosmos*, 11 October 1990.

[131] Interview by the author, 8 February 2019.

Chapter Eleven
End of Hermes

ABOVE Photograph of the full-scale mock-up exhibited at the 1991 Paris Air Show. *UTA via Xavier Truchet*

Following the ESA Council meeting in Santa Margherita (Chapter 8), Hermes entered a period of intense budgetary discussion. The German DARA had proposed to delay the Hermes-*Columbus*-Ariane 5 schedule, which ESA accepted. In March 1991, ESA received the responses of the industry in reply to the request for proposals issued the previous year… and the cost of building Hermes escalated by 30%, half of the increase being caused by technical problems and the other half by the recurring delays in the project.

Time was not in favour of space projects in 1991. That year the ISS aka 'Space Station *Freedom*' was going through troubled times and was nearly cancelled outright by the US Congress. This was reflected in the press. In the Paris Air Show Special Issue of *Aviation Magazine*, Philippe Couillard tried to defend the role of man in space, declaring, '*Because we build ever more complex mechanisms [we need men]. With automatics, there is always the risk of a minor glitch which will stop the whole system. And the more complex the stations and systems we build, the more we will need men [in space]*.' But the title of the whole magazine article was *Hermes, pourquoi faire?* (Hermes, what for?)

However, there were also rumours of good news. Just before that year's Air Show, there was a hint that UK might, in the end, join Hermes. Philippe Couillard was hopeful in this respect, observing, '*So far, I am attentive. The flight of a British astronaut aboard a Soyuz was effected last month; maybe it will induce a few modifications in Britain's view of manned spaceflight*.'[132] Helen Sharman was the first British astronaut. She flew aboard Soyuz TM-12/11 and boarded the *Mir* Space Station, having taken off from Baikonur Cosmodrome on 18 May 1991. But the UK did not put a pound into Hermes.

By November this rumour had evaporated and the Delegate to Space at the French Ministry of Transport and Space could only take note that, '*In the foreseeable future, one cannot realistically hope to have the UK attain a level of participation in ESA more coherent with its part in the European GNP*.'[133] In the same interview he did not give much chance to Hermes in the next round of budgetary decisions. Asked about the space priorities of the French Government he replied, '*To begin with, we must complete Ariane 5, that's our first priority. Then we must acquire the whole orbital infrastructure; that means access for Europe to the technology of manned spaceflight*.' Neither Hermes by name, nor even a spaceplane was mentioned.

Hermes's configuration was by now identified as **8R1** but did not differ greatly from previous designs. A few detail aspects were renunciations—it was now admitted that some elements would have to be procured outside Europe: GPS receivers, fuel cells and water separators, tanks, Freon pumps, intra-vehicular activity (IVA) suits associated with ejection seats, biomedical units and others. Some problems remained unsolved: The use of Ariane 5 'Mk1' for the first launches meant that constraints on mass still applied: 49,383lb (22,400kg) at launch and 33,069lb (15,000kg) at landing were the limits.

The second *Revue de Definition Préliminaire* noted various points which needed further study:

■ The weight margin had dropped to 4% which was deemed insufficient.

ABOVE Close-up of the nose landing gear of Hermes on another stand in 1991. *JC Carbonel*

BELOW As could be seen, it differed somewhat from this 1989 study
© *Dassault Aviation*

ABOVE From a 1993 GIFAS brochure, this wind tunnel model appears to be quite small. For high speed testing, small models could be used. *GIFAS*

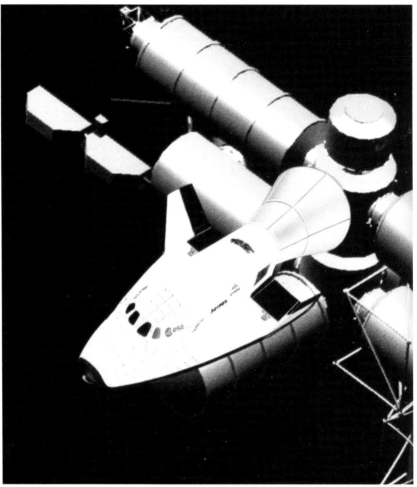

ABOVE Never to come to pass: an 8R1 Hermes docked to a large space station. *Author's collection*

- Spacionic architecture needed improved failure tolerance.
- Reliability of unmanned flights should be increased.
- Protection against space debris should be examined.
- Spaceplane HRM separation before re-entry showed unacceptable risk.
- Reliability, availability, maintainability, safety (RAMS) analyses showed numerous issues.

Another 'special' council, held at Darmstadt on 30 August and 1 September, saw the schedule again redefined with the first manned flight delayed one more year, to take place in 2002. Germany and Italy supported the scenario. France was in total opposition to this plan and threatened to terminate finance of *Columbus*. This would have probably resulted in killing both Hermes and *Columbus*, heralding the end of all European man-in-space ambitions.

At the end of the year, on 18-20 November in Munich, a ministerial-level Council was held and which boldly decided—not to decide, and to push the final decision back until the next ministerial-level Council to be held the following year. That was, in itself, a major decision, because the programme would have to be approved each year by the governments, which many in the industry considered to be incompatible with such a long term venture. Many were also surprised by the approval by France of the new delay. An aspect which emerged, possibly as a consequence of the Munich Council, was a schism inside the French Government itself, in that Science and Telecommunication Minister Paul Quilès continuously

Hermes 8R1 (1991)

Status	Project
Span	30ft 10in (9.40m)
Length	41ft 8in (12.70m); 61ft 7in (18.76m) with HSP
Gross Wing area	900sq ft (83.6sq m)
Gross weight	52,644lb (23,879kg)
Engine	None
Maximum speed/height	-
Load	-

supported the spaceplane, while his colleague Defence Minister Pierre Joxe began questioning the use of Hermes. At the same time, the industrial partners wondered about the creation of EuroHermespace. Was it still worth it?

Finally, on 8 January 1992, they decided at a shareholders' meeting to go with it and the company was registered in Toulouse on 10 January and began its activities on 1 February. To integrate the smaller contributors, three seats were offered to them on the Surveillance Board: ETCA (Belgium), CASA (Spain) and a third seat that would have been allocated to Contraves (Switzerland), Saab-Ericsson Space (Sweden), Fokker (Netherlands), CRI (Denmark) and ORS (Austria) on a two-year rotation.

Meanwhile, the engineers kept working. On the reference design Hermes, a 'suspended cabin' was introduced as a reply to criticism relating to the resistance of Hermes to space debris. It was found that this solution facilitated the manufacturing and integration of the cabin in regard to the structure of Hermes. Another advance in design concerned the ergonomics of the cabin: a full-size mockup was tested aboard the 'Zero-G' Caravelle during parabolic flights. This validated the basic organisation of the cabin.

Dassault, with SEP and MAN, produced a sample of 'hot structure'— that is, a panel representing the future leading edges, winglets and undersurfaces. This sample was trialled by *Ottobrunn Industrieanlagen-Betriebsgesellschaft mbH* (IABG: Industry Testing Organisation). A 60% model of the nose was supplied by Aerospatiale to Dassault which sent it to IABG. After Germany, it went to Spain to be tested in Almeria's solar

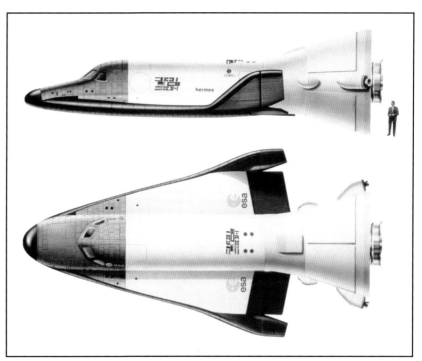

ABOVE This two-view drawing illustrates a configuration very close to the final Hermes version. *ESA*

BELOW David Ducros was CNES resident artist at that time and he painted quite a few illustrations of the final HERMES. *ESA*

furnace. The leading edge section was later sent to NASA Ames Research Center in the USA.

ESA began discussions with NASDA (National Space Development Agency of Japan[134]). Simultaneously, ESA had also begun talks with Russia (at that time part of the Commonwealth of Independent States) on a future joint space station which could be served by Hermes. This led to investigations of the feasibility of bringing Russian industries into Hermes. Among the acts of co-operation with Russian industry would be the use of a BOR-4 vehicle to represent the nose of Hermes. This section would also receive elements of Hermes's thermal protection. However, a budget was never made available for this experiment.

ABOVE After the Paris Air Show, the full-size mockup went to ESTEC to study how to handle a spaceplane…on Earth. *ESA*

In 1992, Hermes was still being refined with:

- Efforts to reduce the heat stress on the windscreen by reclining the windows or repositioning them.
- Efforts to reduce the acoustic stress (at take-off), turning the wingtips inward.
- Reducing the size of the heat-exchangers to save weight.

In March 1992, in reaction to the stretching of the programme decided at the Munich conference, the industrial partners proposed to concentrate on proving their know-how in a relatively short term.

According to Philippe Couillard in *Rêve d'Hermes*, 'To reply to the will of Europeans to master the manned spaceflight technologies while replying to the budgetary constraints, it was necessary to split our objectives. The first one was to develop a vehicle able to be launched with Ariane 5 and to perform atmospheric re-entry. The second one was to complete the programme by adding the orbital capacity, manoeuvring in orbit and space rendez-vous. The first objective was in relation to the basic abilities of any spaceplane: launch and re-entry. The second objective deals with the fitting of the spaceplane for its mission. Being closely linked to the station to be serviced, the second objective could be delayed because the European space station was itself delayed and the agreements to visit the US and Russian space station were not yet signed.* This line of thought resulted in 'Hermes X-2000' being proposed. It would be a smaller, unmanned vehicle foregoing any attempt at an operational use. It was to be a demonstrator, to prove the capacity of European industry to have a spacecraft into space and then to recover it by gliding through the atmosphere.

Hermes X-2000's mission was composed of four phases:

- Atmospheric launch phase.
- Exo-atmospheric launch phase which ended just short of an insertion into orbit
- Pseudo-ballistic phase (there was no orbital flight planned).
- Re-entry phase and landing.

The 'pseudo-ballistic' phase was perceived as the most delicate as it was a precondition of a good re-entry.

Outwardly, Hermes X-2000 would have had the appearance of the manned Hermes but much simplified by deletion of life support equipment and all the other accessories required by a human crew. The HRM would also be dispensed with, being just an adaptor to fit the spaceplane onto the Ariane launcher. This induced a serious reduction in weight to 18 tonnes, perhaps just 16 tonnes, which made it fully compatible with Ariane 5 Mk1 capacity of 22.5 tonnes.

To design the X-2000, Dassault offered the 'project Launac' which Michel Rigault described to the author. *'The idea emerged after a workshop held in the Château de Launac, near Toulouse. I was not there but Pierre Perrier, the designer of the 1964 TAS, was there and was probably at the origin of the Launac design. This was among the proposals to reduce the cost of the Hermes shuttle late in the programme. The overall shape was similar to what flew in 2015 as the IXV. The major difference was that the curves of the leading edges were relatively smaller, first because it was much bigger—about the size of Hermes fuselage and therefore had to bear lower thermal flows. Secondly Dassault, to keep costs down, offered to do the leading edges in ablative materials, which could be replaced easily and cheaply. They would allow smaller curves and therefore better aerodynamic properties. This allowed a better cross-range ability. This had been the solution retained for the TAS orbiter in 1964.*

'When CNES initiated the Pré-X study [in 2004], I proposed in the name of Dassault a shape based on the Launac which, after a comparison with the Aerospatiale proposal, was retained by CNES. The follow-up ESA programme IXV used the same shape as Pré-X. One could claim that IXV was the Maia of Launac. A difference between Launac and IXV was in the final step of the recovery: Launac would have opened its parachute in subsonic flight while IXV opened its one in low supersonic flight. Flight simulation at Dassault had shown that rear rudders allowed the vehicle to be controlled in subsonic flight. On IXV those rudders were ablative skids outside the vehicle to reduce costs, while those of Launac were directly extrapolated from Hermes technology. Internally at Dassault, we had designed variants of Launac able to land on a runway through a simple extendable wing.'[135]

After the successful flight of the unmanned machine, in a second step, the manned Hermes would take over. In some ways this was a return to the aborted Dassault Maia.

To split the programme into as many segments as could be palatable to the budgetary reviewers, Step 2 would only cover the validation flights of the manned Hermes. Its evolution into an operational vehicle would be Step 3. This programme was submitted to ESA but, being unsolicited, was not well received. Then in July 1992, there was even study of a reduced-scale X-2000, but this generated no interest.

With the programme so obviously in decline, another problem appeared: human resources—engineers were leaving the team, asking to be reassigned to other projects. Noticing this trend, any effort to reduce the scope of X-2000 was stopped as it was unrealistic to hope to relaunch Step 2 after Step1. The important thing was to carry on with what had been budgeted and hold-off dissolution of the teams.

On 3 July German Minister for Research and Technology Heinz Riesenhuber met his French counterpart, Hubert Curien. The outcome of this meeting was the unofficial renunciation by France of the X-2000. Yet on 19 August, at a reception in honour of the Franco-Russian Antares mission, Curien continued officially to support Hermes.

In September a further ESA meeting defined a new programme which only endeavoured to *'provide Europe in the medium term with a manned transport system for low earth orbit, either in co-operation with other spacefaring nations or as an autonomous undertaking.'*[136] That same month German Minister Heinz Riesenhuber announced that Germany was getting out of Hermes and of the MTFF. He actually proposed pushing the Europeanisation of Sänger II (see Chapter 9) by including it under the Future European Space Transport Investigations Programme (FESTIP) which was about to be launched (see Chapter 12).

BELOW Drawing of the Hermes X-2000. In blue, the equipment required for an orbital flight. Legend: Piles Lithium = Lithium batteries; Equipement des mesures technologiques = Technological measurement recorders: PCDU Pile Lithium = Power Conditioning Distribution Unit (for) Lithium batteries. *Author's archives*

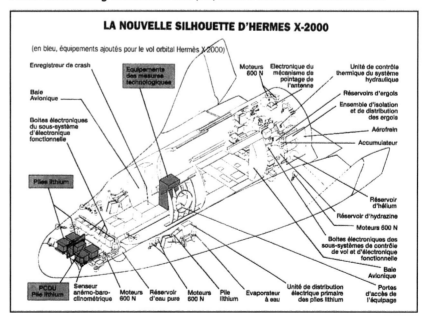

ABOVE AND OPPOSITE With an illustration lacking for the Launac project, here is the nearest approach in shape: the IXV. It flew successfully in February 2015 and performed a controlled re-entry before landing in the Pacific Ocean. It is shown at the Paris Air Show, in June of the same year. The chequered vehicle is a model used for study of the return flight's splashdown. *JC Carbonel*

Hermes X-2000 (1991-92)	
Status	Project
Span	30ft 10in (9.40m)
Length	41ft 8in (12.70m); 61ft 7in (18.76m) with HSP
Gross Wing area	900sq ft (83.6sq m)
Gross weight	–
Engine	None
Maximum speed/height	–
Load	None

At the same moment, Hubert Curien talked about reorienting Hermes toward the Russians. He proposed to have Hermes replace the old Soyuz spacecraft by having it link the future *Mir-2* space station with Earth. Industry clamoured for the immediate construction of Hermes 2000.

In October the French *Comité National d'Evaluation de la Recherche* (National Committee for the Evaluation of [scientific] Research) issued a report stating that Hermes had no chance of being financed, and claiming that it was outdated, coming after the American and Russian shuttles. If one followed this report the future was seen to be more in hypersonic research like the German Sänger II or the French *Programmes d'Études sur la Propulsion Hypersonique Avancée* (PREPHA: Programmes for the Study of Advanced Hypersonic Propulsion) (see Chapter 12).

French députés were worried about the outcome of the approaching ESA Meeting. *Député* Robert Galley of the Gaullist party RPR said, '*We have received a document from EuroHermespace, dated 26 October, which reduces the programme to a multitude of limited operations, without interest for France and without coherence […] The name Hermes has disappeared […] Mr Prime Minister, why should France's position fall in line with Germany, which favours co-operation of its industry with the USA?*'[137]

On 9/10 November 1992 Granada (Spain) hosted the now annual ministerial-level ESA Council Meeting, that year presided over by France's Hubert Curien. Although a possible co-operation with Russia on a Hermes crew transport vehicle was mentioned in the conclusion, Hermes and Hermes 2000 were killed, in effect, at this meeting.

EuroHermespace was closed down and put into a dormant state on 30 June 1993.

Naturally, all the companies involved had to reorient their activities and for Dassault it was especially difficult as the end of Hermes pronounced the end of its effort to break into space. Even Kourou had to suffer from the abandonment of Hermes as the regional development programme PHEDRE (Partenariat Hermes de Développement Régional) fell over. It had been planned to improve the regional infrastructure (including hotel, roads, port and airport) but this programme was greatly reduced.

French députés balked at the abandonment of Hermes and accused the government of giving in to American and German pressures. Communist Senatrix Danielle Bidard-Reydet said, '*We can't help but see behind the stopping of Hermes the result of American political pressure and a new proof of the submissiveness of the [French] Government to German injunctions.*'[138] At the Assemblée Nationale, Député Jean de Gaulle of the right-wing party UMP was no less critical, pointing out the '*technological incoherence*' as the '*complementarity between Ariane 5, Columbus and Hermes appeared obvious*' and ended, '*Mr Prime Minister […] why did you renounce so suddenly and so surreptitiously the great challenge of space conquest for France and Europe?*'

Philippe Couillard was very disappointed, ending with, '*France gave up at Munich and renounced Hermes X-2000 which was presented in June 1992 even before Granada. It is true it did not lead to a manned spaceflight, yet it would have been a magnificent demonstration of know-how and a significant step forward. But to obtain this result, the French Government and CNES should have shown their motivation more than they did. They should have fought to pursue their European ambitions. They did not do it. But was there any will to do it?*'

To other questions about the future of Hermes by Député Alain Cousin (also from RPR), Jean-Pierre Brard (Communist Party) the Government replied in January 1993, '*Studies will be funded regarding two complementary concepts: a spaceplane used as crew transport to the Space Station or as rescue vehicle; and an automated vehicle to carry freight to the* Freedom *Space Station.*'[139] However, only the second part was to be developed, in the form of Automated Transfer Vehicle (ATV) which first flew on 9 March 2008.

In 1993 it was announced that the Russians would join the ISS with their own modules. In the end, the first Russian module was attached to the ISS in 1998 while the European *Columbus* module only joined the ISS in 2008. Even in 1993, French députés kept asking the Government to re-launch Hermes, notably stressing the abandonment of sovereignty which the killing of the spaceplane represented—but to no avail. After such a decision it would have been impossible to go back. The European spaceplane was definitively dead.

In 2001 the French Senate published a report on national space policy which interviewed many French authorities on space subjects. On Hermes, Claude Allègre, then Minister of Education Research and Technology was unambiguous, declaring, 'Hermes was cancelled. And I would say that was fortunate. You know my position about manned flights. Manned flight was not something unthinkable in the beginning, but today, with the progress made in automation technology, I no longer see, personally, what its use could be, either technologically or scientifically, when compared to the considerable cost. And I say it clearly: one of the first wishes of the French Government is to continue its ambitious space policy with an increased European independency'.

Not all participants in this report agreed. Pierre Ducout, President of Parliamentary Group on Space thought differently, saying 'There was in the Hermes project, which Patrick Baudry supported, a rendez-vous other than just mastering a new technology [it was] the opportunity to get into the small club of true space powers, those that brought humanity from dreams into reality [...] after the cancellation of Hermes, France remained, so to speak, on the doorstep of the space kingdom [...] Behind the manned flights, there is, in my opinion another danger. It is not just investments, it is also a question of the pertinence of science. Research may appear to some like a luxury rather than a fundamental question pertaining to our common progress. Today we note that the question of man in space is debated and to some

people, it may appear like an ideal symbol against which to fight. That's why in defending manned space flight, it is also defending science, research—even if science is not entirely based on the presence of man in space'.

Christian Poncelet, President of the French Senate also suggested that the cancellation of Hermes was a mistake. 'The question of the manned spaceflight – to which Europe partially renounced when abandoning the Hermes shuttle project – is essential to popular support for space. In this domain Europe must take care to maintain its rank and not leave a monopoly to the US power'. Astrophysicist Jacques Blamont pointed out that Hermes cost 1 billion francs 'just for show' which would have been better employed in the French military budget.

Philippe Couillard himself was interviewed for this report and put much faith in the needs of the ISS for support vehicles 'We are currently developing an automated supply vehicle, to be launched by Ariane 5 to service the Space Station: it is the ATV (Automatic Transfer Vehicle) which offers some interesting technical challenges [...] But the ISS is also the participation of Europe in other vehicles, the rescue vehicle, which the lifeboat of the station and to which we could participate and even the crew transport vehicle'.

M Salomon, honorary professor at the Conservatoire National des Arts et Métiers was, perhaps, the person who came closest to identifying the real trouble with Hermes when he declared, 'While talking about Hermes, I had students from CNES who worked on the project and, coming from CNES, they were convinced that Hermes could be done. Then after one and half years, progressively, they discovered that the decision-making process, for reasons that had nothing to do with science or technology, was such that the affair would get cancelled'.

---

[132] *Air & Cosmos*, 1 June 1991.

[133] *Aviation Magazine*, 1 November 1991.

[134] NASDA was merged into the Japan Aerospace Exploration Agency (JAXA) in 2003.

[135] Mail exchange with the author, October 2020.

[136] Quoted in *Hermes Europe's Dream of Independent Manned Spaceflight*, by Luc van den Abeleen, Springer 2017.

[137] Assemblée Nationale, 2nd session, 4 November 1992.

[138] Sénat session of 25 November 1992.

[139] Assemblée Nationale, Questions au Gouvernement 9th Legislature.

Chapter Twelve
Around and Beyond Hermes

ABOVE Artist's impression of STAR-H in the high atmosphere. The powerplant section is kept undefined as the type and number of engines had not been decided when this was painted.
© Dassault Aviation

Hermes was a catalyst for many projects. First was the question of the European Space Stations. There was obviously the module *Columbus*, which was to be attached to the US Space Station but there was also the Man-Tended Free-Flyer, an autonomous space station which could dock with either the US Space Station or with Hermes for re-supplying.

Then designs for an advanced air-breathing launcher were put forward: for Dassault Star-H and early Dornier EARL proposals the orbiter stage was just Hermes or a very close cousin of it. From the late 'eighties Aerospatiale ran many studies to assess the interest and feasibility of SSTO and TSTO launchers using air-breathing propulsion or rocket engines. Those studies produced four different designs in the years 1992–1993:

- Taranis [Celtic god]: rocket-propelled TSTO 1990
- STS-2000: air-breathing TSTO & SSTO 1990
- Oriflamme [battle flag]: air-breathing SSTO 1991
- Radiance: air-breathing TSTO 1993

After Hermes, ESA launched the programme FESTIP (Future European Space Transport Investigation Programme) to study future, recoverable launchers; while in France, PREPHA attempted to design an advanced hypersonic ramjet for an SSTO vehicle. This programme – which echoed ramjet research of the 'sixties (see Chapter 3) – continued well beyond the scope of this story. In the enthusiasm of the times, Aérospatiale's Henri Lacaze, former Hermes project manager, was moved to write about what a Moon base could be.

A – Companion to Hermes: the MTFF

The MTFF

MTFF (Man-Tended Free-Flyer) was part of the *Columbus* project:

- A Man-Tended Free-Flyer (MTFF), as a space station element.
- An Attached Pressurised Module (APM), as a manned space station component.
- An unmanned Polar Platform (PPF) for remote sensing and data return.

An Attached Pressurised Module was finally built and flown in 2008, aboard Space Shuttle *Atlantis*. It therefore falls outside the scope of this story. The Polar Platform was unmanned and for this reason, will not be detailed either.

The MTFF was developed by MBB-ERNO. This was an autonomous space station which was intended to remain in orbit for thirty years, being serviced occasionally by astronauts, either by docking with the US Space Station or by having Hermes docking into it (MTFF docking system was also compatible with the Rockwell Shuttle orbiter). While the *Columbus* module permanently attached to the US Space Station was more-or-less independent, the MTFF was heavily dependent on Hermes…and reciprocally, servicing the MTFF was a major function of Hermes.

As a space laboratory, MTFF was intended for micro-gravity material science, fluid physics and life science projects. Structurally it was built around two different modules:

i A pressurised module (PM-2). This provided a shirt-sleeve environment in which astronauts could operate. 24ft 7in (7.50 m) long, it comprised two elements, each 13ft 1in (4.00m) in diameter. Its primary structure was based on the earlier Spacelab project. (Spacelab was a manned pressure module installed inside the US orbiter cargo bay. Built by the German company VFW-Fokker and the Italian company Aeritalia, it allowed astronauts to perform scientific experiments in a shirt-sleeve environment. Its total length was 22ft 10in (6.96m) and diameter 13ft 6in (4.12m). It was flown 25 times). The interior of PM-2 was configured for '1G operations'—that is with a dedicated floor and ceiling with numerous racks on the sides. Outside PM-2 were mounted radiators to evacuate excess heat of both PM-2 and RM. They were also expected to provide PM-2 with protection against micro-meteorites and space debris.

ii An unpressurised resource module (RM) onto which were attached two solar panels providing an output of 20kW. The main thrusters used for

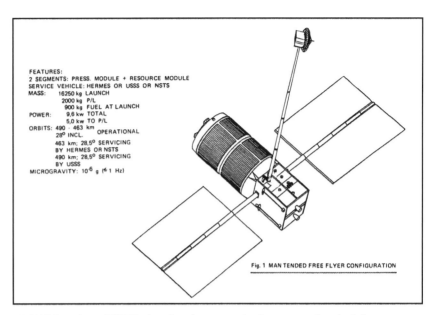

ABOVE Drawing of MTFF, showing the pressurised segments (on the left, meant to be cylindrical), the Resource Module on the right, the antenna and the solar array. *MBB-ERNO*

BELOW MRFF was part of the *Columbus* project, here depicted as a model. *ESA*

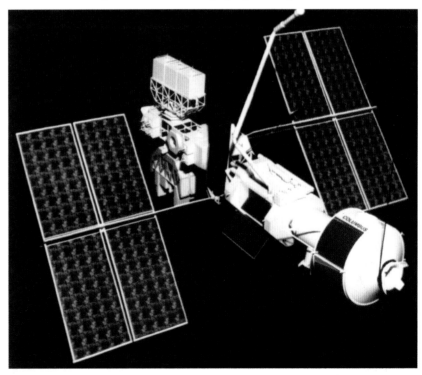

197

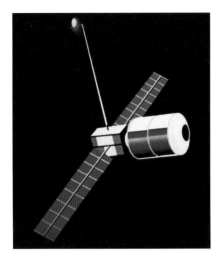

ABOVE A slightly stylised artist's impression of the MTFF from a 1987 ESA brochure. *ESA*

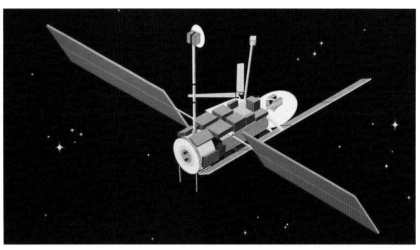

ABOVE From the same ESA brochure, a view of the Polar Platform. *ESA*

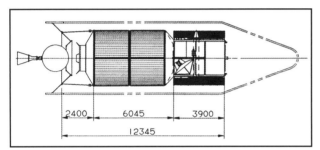

ABOVE The MTFF could fit in the fairing of Ariane 5 and therefore could be orbited with a single launch. *MBB-ERNO*

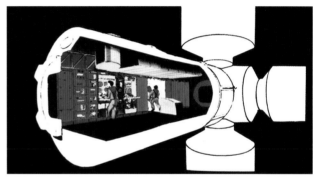

ABOVE The Attached Pressurized Module was the only *Columbus* design actually built and flown—ten years later after this conceptual drawing. *ESA*

LEFT An early Hermes docked to a somewhat fanciful MTFF using its upper docking ring is shown in this artistic rendering. *Aerospatiale*

BELOW In 1987 it was thought that the MTFF could be serviced by Hermes using its upper docking ring. *MBB-ERNO*

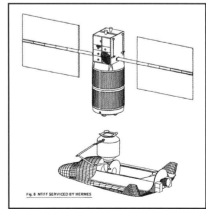

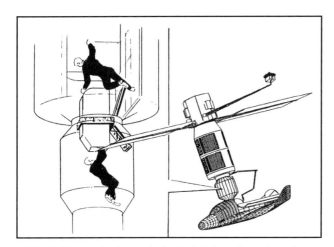

ABOVE This required a Logistic Module interface since the Hermes docking ring was smaller than the MTFF docking ring. *MBB-ERNO*

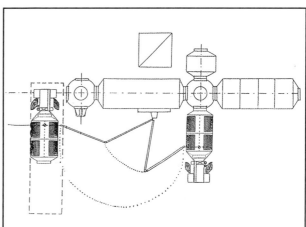

ABOVE That was because the MTFF had been configured to dock primarily with the US Space Station which used larger docking rings. *MBB-ERNO*

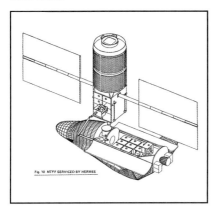

ABOVE For the bi-yearly maintenance and servicing, the MTFF could berth with Hermes using a telescopic mast. This way there was sufficient clearance above the Hermes airlock to allow astronauts easy egress during EVAs. *MBB-ERNO*

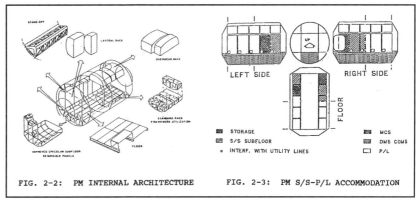

ABOVE A fairly basic drawing of the MTFF's interior. *MBB-ERNO*

BELOW By the end of 1987, Hermes had evolved, with a larger docking ring at the rear, so the Logistic Module could be dispensed with. *MBB-ERNO*

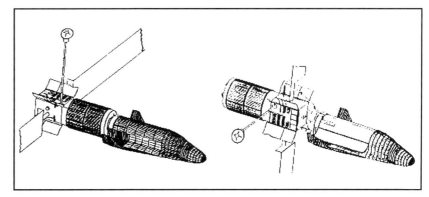

orbit transfer manoeuvres were also integrated into RM (MTFF was to follow an orbit more or less similar to the ISS but obviously, during a thirty-year lifespan, orbit decay had to be accounted for and manoeuvres would be necessary during docking). Docking systems were provided at both ends of the space station, but only that on PM-2 had an airlock for astronauts. The system at the extremity of RM was described a 'berthing interface,' into which Hermes could dock but could not transfer material inside the module. Various grapples were provided to allow astronauts to easily perform extra-vehicular activities (EVA) around it.

MTFF was to be orbited with just one Ariane 5 launch, with a base payload of 4,409lb (2,000 kg) and a fuel load of 1,984lb (900kg). Obviously, both could be increased (maximum payload was calculated to be 11,023lb [5,000kg]) with each servicing operation. Intervals between each servicing operation were intended to be between 30 and 180 days. This was for mission requirements; pure servicing, for example, maintenance or refuelling, was only needed every two years.

The 30-year life requirement induced design constraints such as easy accessibility for replacement of key elements like batteries and computers. Experiments were also to be designed as self-contained modules to be attached to racks. The ease with which astronaut

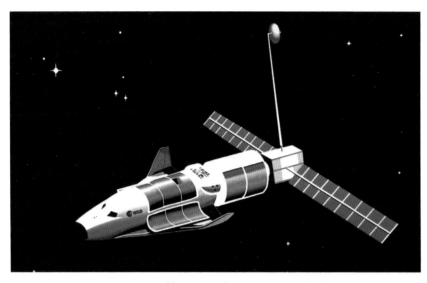

ABOVE An artist's impression of late 1987 showing Hermes docked with MTFF using the rear airlock. *ESA*

BELOW In this late 1987 document, models were used to illustrate the launch of MTFF followed by the launch of Hermes and the docking of the two. *ESA*

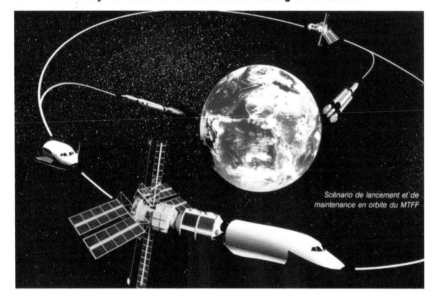

crewmembers could transfer those modules from and to the payload bay was a key element in the design of Hermes's bay, airlocks and docking systems. Equipment on RM could not be accessed from inside so they were designed to fit standardised orbital replacement units (ORU) which could be manoeuvred by Hermes's robotic arm. This was intended to reduce EVAs as much as possible. Solar arrays and radiators could not be made to fit a standard ORU because of their shape and would require EVA during servicing. During rendezvous with ISS, it was intended to retract solar arrays and antennas before docking.

As we have seen (Chapter 8), there was much debate about the size of Hermes's airlock. So for MTFF, it was decided to have the PM-2 airlock adapted to ISS size. For docking with Hermes, a Logistic Module (LM), which had two different port sizes, was to be carried inside Hermes and thus would serve as adapter between its hatch and the MTFF hatch. The reason for this ungainly arrangement was not explained. Dedicated Hermes maintenance flights would have been required every two years. For this operation, the MTFF would be attached to Hermes by a telescopic mast allowing sufficient clearance for astronauts to get out of the Hermes airlock for EVAs. The telescopic mast could be rotated, giving Hermes's robotic arm access to all sides of the MTFF.

BELOW Two pictures (one a cutaway) depicting a latter-day HRM Hermes docked to the MTFF. MTFF lasted until 1991. *ESA both pictures*

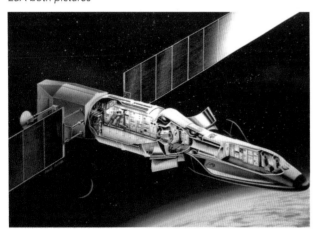
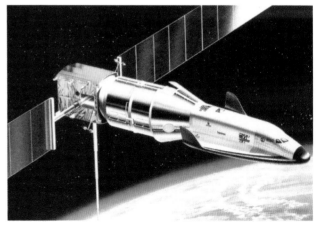

Servicing and resupply operations evolved with Hermes: in 1988 docking was at the rear and in line with Hermes, while a year earlier it had been above Hermes (in T configuration) because Hermes dock was on top of the spaceplane.

B – Beyond Hermes: the Fully-Recoverable Launcher

B1 Dassault

Dassault Star-H

In 1986 Dassault was contracted by CNES to assess the trends of future space transport systems. Two other studies were also called-for: a similar one for Aerospatiale (which led to the STS-2000 proposal; see below) and an engine-oriented study for SNECMA, SEP and ONERA. These studies eventually contributed to establishing the basis of a national programme of hypersonic air-breathing propulsion, especially scramjets: PREPHA (see below). In this context, Michel Rigault proposed a two-stage-to-orbit, air-breathing launcher named STAR-H (*Système de Transport Aérobie Récupérable à décollage et atterissage Horizontaux*. Recoverable Air-breathing Transport System with Horizontal take-off and landing).

The selected STAR-H configuration was based on the following elements:[140]

- Low-wing first stage.
- Second stage in the upper position (readers may remember that during the Aerospace Transporter studies, a few proposals had been rejected because they did not allow for dropping separation of the second stage).
- A single propulsive pod under the fuselage.

The vehicle was composed of three elements:

i A large delta air-breathing first stage; fully recoverable
ii A second-stage booster; not recoverable
iii Hermes orbiter

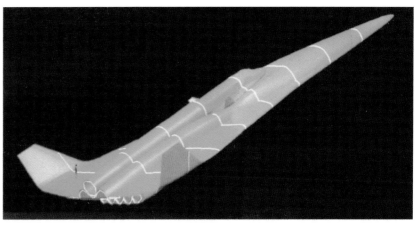

ABOVE Computer-generated view of STAR-H, possibly the document from which the artwork was created. © *Dassault Aviation*

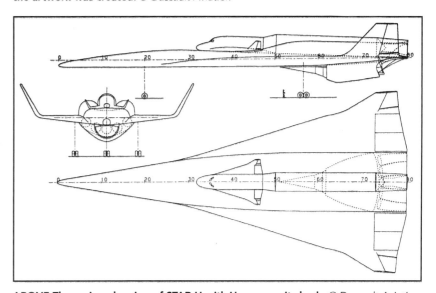

ABOVE Three-view drawing of STAR-H with Hermes on its back. © *Dassault Aviation*

BELOW Two other views of STAR-H, this time showing the underside of the vehicle and engine pod with five nozzles and the side view without the second stage and Hermes. The numbers are the temperatures in Celsius during the hottest part of the flight. © *Dassault Aviation*

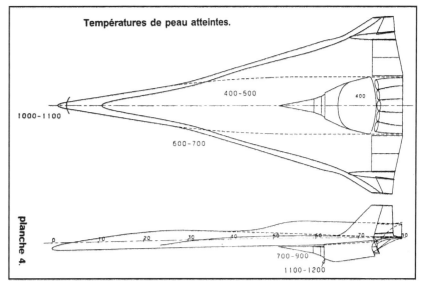

In his study Michel Rigault acknowledged the origins of the vehicle in the Transporteur Aérospatial (TAS) study carried out during the 'sixties (see Chapter 3)[141] in part due to the lack of actual flying references or contemporary studies. The study intended to be the most realistic possible, using technologies due to mature during the development of Ariane 5 and Hermes. Advantages of the aircraft-type first stage were recognised: lower weight as the oxidiser fuel is replaced by the oxygen in the air. The use of wings leads to both a wide cross-range ability and an easy recovery. Finally, being aircraft-like, the vehicle should have operating costs much lower than a space vehicle's. But the drawbacks were also recognised: notably the added weight of the wings; the heavier air-breathing engines; and the trajectory which led to more severe conditions in terms of temperature, dynamic pressure and total pressure.

Some variants were considered:

- An assisted take-off device (trolley). This was not considered useful for two-stage variants but could be interesting for a SSTO where take-off mass would be very different to landing mass. Michel Rigault indicated that this was done in reply to a CNES query initiated by the presentation of HOTOL at an ESA meeting.

- Use of methane instead of liquid hydrogen. Liquid hydrogen requires large, bulky and drag-inducing tanks which could be much reduced by the use of methane. However, methane is heavier than hydrogen and has a lower specific impulse resulting in a heavier vehicle. In addition, the use of methane limited the maximum velocity reached by the air-breathing stage, which would require a larger second stage. A last advantage of methane was its easier handling during refuelling on the ground…but naturally this advantage was lost if liquid hydrogen was still used for the second stage. Therefore, the study concluded that methane was not attractive for a TSTO.

- Recoverable second stage. A fully-recoverable second stage, integrating the entire cryogenic booster, would retain the overall shape of Hermes but with the size of the Rockwell Space Shuttle Orbiter. Thus, a second, non-recoverable stage was selected, echoing the solution retained in 1964 for the TAS. The engine for this stage would be extrapolated from the HM-60 Vulcain developed for Ariane 5.

The main mission of STAR-H would be resupply and servicing of a space station (so identical to Hermes's mission of the times). STAR-H would take-off from a conventional airfield (although Kourou was retained for the study, as it allowed a supersonic acceleration over the sea) after a ground run of 6,562ft (2,000m) and reaching a take-off speed of 240 knots (276mph [444km/h]). This performance allowed the use of standard 9,843ft (3,000m) runways and was consistent with tyre limitations. Flight duration for the first stage would be one hour.

The following points describe the vehicle in detail:

- Aerodynamics. The question of aerodynamics was claimed to be similar to supersonic aircraft and

BELOW Comparison of the various configurations investigated by Dassault: low-wing, shoulder-wing, engines in wing pods, a single propulsion pod below the fuselage. © Dassault Aviation

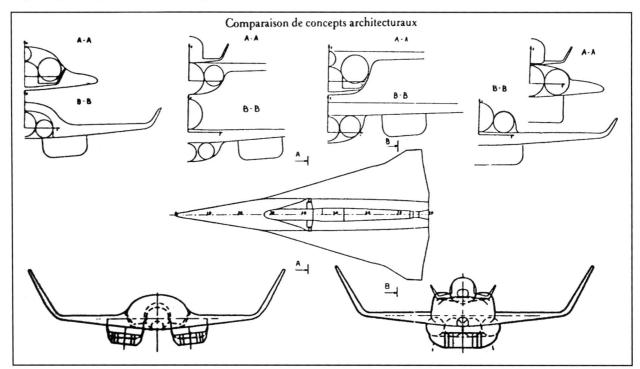

evaluated using CFD codes currently used for combat aircraft and Hermes. Said Michel Rigault, '*The integration of propulsion and mainly the air inlet was analysed in more depth due to the high impact on overall propulsion performance and the difficulty of coping with both the transonic and high Mach number constraints.*'[142]

■ The separation phase has been simulated with preliminary design tools, accurate enough due to the relatively low dynamic pressure during this phase. The upper position of the second stage is unavoidable, taking into account the priority given to the propulsion system on the lower surface. In the past this separation configuration had raised a few eyebrows but for Michel Rigault, '*While unusual, this position had already been used on previous experimental flights (Leduc ramjet test aircraft on Languedoc carrier; and Space Shuttle Orbiter above B747 for the ALT missions). The lower position proposed for the TAS was consistent with the in-flight experience of Dassault on supersonic separation of bombs under the Mirage IV and the smaller size of the propulsion system due to the lower Mach number.*'[143]

■ Propulsion. Various types of engines were analysed among the ones studied by SNECMA, SEP and ONERA: turbo-rocket, twin-duct turbojet-ramjet, turbo-rocket expander, rocket-ramjet. The turbo-rocket expander was a new type of engine which functioned as follows: '*The engine operates in turbo mode from take-off to about Mach 3, then in ramjet mode to reach a maximum Mach number of 6 at the end of the air-breathing acceleration phase.*'[144] Selection of the engine type had a great influence on such parameters as propellant mass, dry (structural) mass, wing area and number of required engines. Two types the turbo-rocket-expander and the rocket-ramjet were found to present the best opportunities, with the rocket-ramjet being too heavy.

■ The turbo-ramjet was not studied in detail, considering that its parameters would be similar to those of the turbo-rocker-expander. In the end, the requirement was estimated as '*four or six*' turbo-rocket-ramjets of which the ramjets light-up at Mach 3.5 to 4.0, fuelled by liquid hydrogen (+ liquid oxygen). (The number of engines was arrived at by parametric calculus which gave a result of 6.3, requiring a scaling effect on the engine). Overall available thrust was found to be critical to the design, due to its limited capacity for acceleration in the transonic region. Another crucial aspect was the capture area of the air intakes, as it determined the thrust at the higher Mach speed, and therefore, the maximum speed (which, in itself, determined the speed and altitude of separation, which influenced the volume of fuel needed to raise the upper stage into orbit). The synthesis of the study presented at the 1991 IAF conference showed that the turbo-rocket-expander had a slight advantage in weight. The higher weight of the turbo-rocket-ramjet powered vehicle generated some concern regarding the ability of the tyres to withstand the take-off stress.

■ Heat protection. The first stage was basically comparable to a large aircraft (the size was similar to a Boeing 747) but its flight envelope was not. Heat-resistant materials were to be used for the skin of the vehicle. For the structure, titanium and composites were retained. In chosen areas, high-temperature

STAR-H (1990)	
Status	Project
Span	131ft (40m)
Length	262ft (80m)
Gross Wing area	13,100sq ft (1,217sq m)
Gross weight	881,850lb (400,000kg)
Engine	Undecided. 1990 studies concluded in favour of turbo-rocket-ramjets but later 1991 studies favoured turbo-rocket-expanders
Maximum speed/height	Mach 6
Load	264,550lb (120,000kg) [222,675lb (101,000kg) for 2nd stage booster and 41,888lb (19,000kg) for orbiter]

BELOW Computer simulation of the separation of the winged first stage and the rocket second stage with Hermes. © *Dassault Aviation*

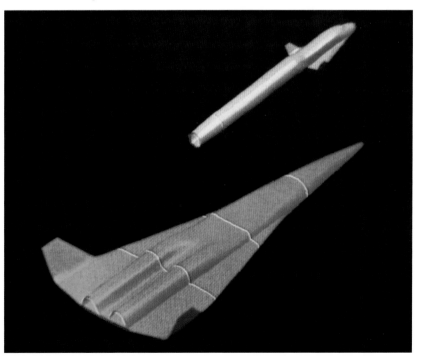

materials, like those developed for Hermes, were selected: carbon-carbon or ceramic-matrix composites. In addition, liquid hydrogen fuel was circulated in some areas as an active coolant. It was stressed that this solution had been used on other Mach 3 aircraft, including North American B-70 and Lockheed SR-71.

- Propellant storage. This was found to require more severe specifications than for either aircraft (because of the type of fuel used) or rockets (horizontal position of the tanks, variation of acceleration orientation, centre-of-gravity location requirements, external temperature).
- Power supply. About 1,000 kW was found to be required; a cryogenic APU was selected.
- Weight. With a dry weight of 147.6 tons (150 tonnes) the first stage received 127.9 tons (130 tonnes) of fuel. The second stage booster would have a dry weight of 10.8 tons (11 tonnes) and contain 88.6 tons (90 tonnes). As pointed out by Michel Rigault, *'To launch an orbiter of 18.7 tons (19 tonnes), the take-off weight is 393.7 tons (400 tonnes), compared with the 738 tons (750 tonnes) of Ariane 5.'*
- Safety of operation. It was recognised that because the vehicle was re-usable, its loss would have much more impact than in the case of a single-shot launcher. This would elevate the safety requirement to the level of manned flights. Catastrophic failure probability needs to be reduced through system redundancies. An escape system for the crew needed to be designed. However, one thing had to be taken into consideration early on: the failure of a single engine should never lead to the destruction of the vehicle. For this reason, back-up landing sites were defined. The study took the approach that the normal landing configuration was the first stage carrying on its back the second stage and the orbiter. The difference with take-off was that the fuel for the second stage had to be either jettisoned in flight or transferred to the first stage. It was also acknowledged that redundancy in the propulsion system of the second stage was required to enable the orbiter to be rescued in case of an engine failure in the second stage.

The conclusion (presented in 1990) was…that further research was needed. *'The STAR-H study has provided a basis from which to evaluate air-breathing launchers. This was necessary due to the thinness of the initial database which did not include any actual experience. This study, based on a robust concept, increased the confidence in the feasibility of air-breathing launchers, which could, in the future, reduce the cost and increase the flexibility of space launchers […] Large problems remain to be solved, however, on both the technical and operational aspects, before the competitiveness of this type of launchers against present-day launchers is proven.'*[145]

B2 Aerospatiale

Aerospatiale Taranis

After the various studies of future winged launchers during the early 'eighties (see Chapter 7), in late 1989 Aerospatiale came up with the RAF (Recherche Auto-Financée: Self-Financed Research) 90 Transatlantic launcher: a larger vehicle to come after Hermes. Jean-Paul Bombled, who designed the vehicle with Henri Lacaze, christened it Taranis after the Celtic (Gallic) god of sky and thunder.

Taranis was a fully reusable, two-stage-to-orbit vehicle, taking-off vertically and landing horizontally. It was designed to orbit a 6.9-ton (7-tonne) load, comparable to what was required of most of similar designs of the time. Only the second stage (orbiter) was manned. Contrary to most other designs, which were intended for separation around Mach 6 or 7, Taranis was to separate at Mach 13 at 295,275ft (90,000m), the first stage gliding down across the Atlantic (hence the 'Transatlantic' nickname) to land at Dakar (for an equatorial trajectory), Bermuda (for insertion into polar orbits) or in the Azores for any reclined orbit. The second stage was designed to be the smallest possible but could accommodate additional boosters (non-recoverable). Acceleration at take-off was given at 38.5ft/s (11.74m/s) or 1.3G. During separation the first stage was powered by only one or two of its bank of engines, so that the second stage would easily overtake the first stage and would not be damaged by its engine jet. Small rockets on the second stage move it away from the first and would later be disposed of *'so as to not kill the aerodynamic efficiency of the [second] stage during the return glide flight.'*[146] Various emergency scenarios were examined:

- One of the engines of the first stage fails. Depending on when this occurs, it was estimated that the mission could still be carried out. If early in the flight, then a steep climb should be carried out, while fuel is jettisoned and the two stages return (still docked?) to Kourou. If occurring later in the flight, 'inverted cross-feeding' (fuel is made available to all remaining engines in both stages) would enable a diversion to Recife, Cape Verde islands, Bermuda or Azores, depending on the trajectory.

BELOW Artist's impression of Taranis at take-off. Designer Henri Lacaze thought it was better and more economical to leave the atmosphere as quickly as possible, hence the vertical take-off for winged vehicles. *Aerospatiale*

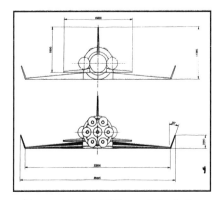
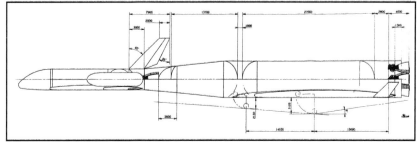
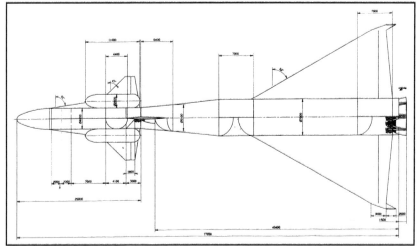

ABOVE AND RIGHT Shape 0.2 of what was still the RAF 90 Lanceur Transatlantique in September 1989. Note the extanded landing gear in the side view. *Aerospatiale via H Lacaze*

- Complete shutdown of the first stage. Fuel is jettisoned from the first stage and the two stages, still docked, return to Kourou on the sole power of the second stage.
- Second stage shutdown while still docked with the first stage. In this case, fuel from the second stage would be used by the first stage engine ('inverted cross-feeding') for a return to Kourou.
- Second stage shutdown after separation. This situation was to be investigated more precisely with the hope of finding ways to bring back the second stage to an airfield.

The following advantages and failings were identified:

Advantages:
- Better aerodynamics of the two stages
- Separation speed lower than in previous designs
- One less engine needed for the first stage
- Easier piloting
- Ability to quickly remove the second stage in an emergency
- Ground operation simplified
- Lighter, cheaper launcher

BELOW 'Ghost' view of shape 0.2. Legend: LH2= LH_2; LOX = LOX; Vulcain = Rocket engine Vulcain; Derive = Fin and rudder; Jupe ½ = Interstage skirt; Jupe L/R = Skirt between two sections of Stage 1; Bâti-moteur = Engine frame support; Jupe AR = Rear skirt. *Aerospatiale via H Lacaze*

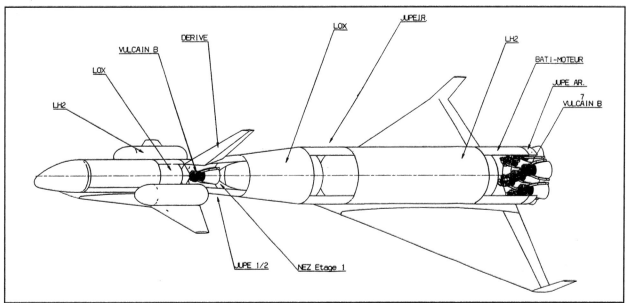

ABOVE An illustration of the RAF 90 Lanceur Transatlantique with the second stage atop the first. *Aerospatiale via H Lacaze*

BELOW Three-view of the 'shape 1.1' from January 1990. Note the crew seats in the side view and plan view. *Aerospatiale via H Lacaze*

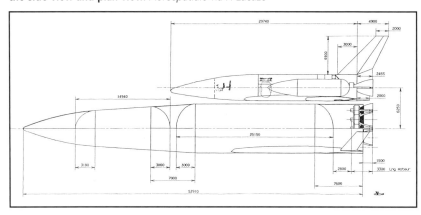

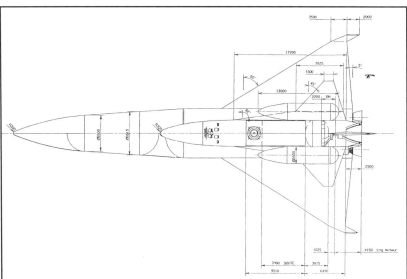

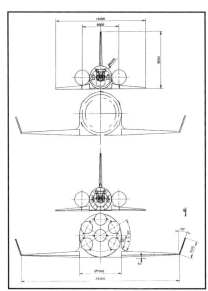

Problems:

- Second stage engine needs to be able to operate in atmosphere and then in vacuum
- Long operating time for the second stage engine
- Separation occurs with both engines running
- An additional emergency possibility: second stage engine failure during the early phase of flight
- Complexity of the 'cross feeding' fuel lines
- Crew compartment near first stage tanks
- Sideway interface efforts
- Wider cross-section leading to increased drag
- Mechanical interfaces required between the two stages

All these were studied in depth and the final conclusions were that many technologies needed to be mastered:

- New materials (alloys)
- Engine nozzles that could work in both atmosphere and vacuum
- Soft and rigid fuel lines with long travel in all dimensions
- Secondary injection in a rocket engine on a cardan mounting

More than that, a few difficulties were qualified as 'slightly hard' to resolve:

Aerospatiale Taranis 1st stage *	
Status	Project
Span	91ft 10in (28.00m)
Length	181ft 9in (55.41m)
Gross Wing area	6,360sq ft (591sq m)
Gross weight	71,364lb (32,370kg) (dry, without orbiter)
Engine	6 × rocket
Maximum speed/height	Mach 13 at 295,275ft (90,000m)
Load	–

Aerospatiale Taranis 2nd stage *	
Status	Project
Span	52ft 6in (16.00m)
Length	105ft 8in (32.20m)
Gross Wing area	2,142sq ft (199sq m)
Gross weight	37,666lb (17,085kg) (dry)
Engine	1 × rocket
Maximum speed/height	–
Load	13,228lb (6,000kg)

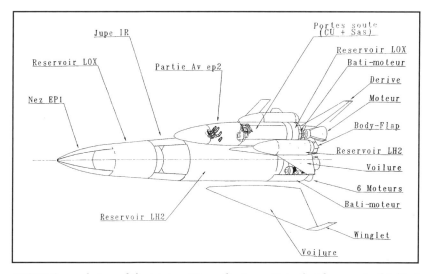

ABOVE General view of the Lanceur Transatlantique. Note that for once in this kind of study, the crew seats are illustrated. Legend: Nez EP1 = Nose Propulsion Stage 1; Reservoir LOX = LOX tank; Jupe IR = skirt; the meaning of 'IR' remains unknown; Partie Av ep2 = Front section, Propulsion Stage 2; Portes soute CU + sas = Doors of cargo bay. Payload and airlock, Reservoir LOX = LOX tank; Bâti-moteur = Engine frame; Dérive = Fin+rudder assembly; Moteur = Engine; Body-flap = Body flap (flap at the rear of the fuselage, similar to Rockwell Orbiter); Reservoir LH2 = Liquid hydrogen tank; Voilure = Wing; 6 Moteurs = 6 Engine; Bâti-moteur = Engine frame; Winglet = winglet; Voilure = Wing; Reservoir LH2 = Liquid hydrogen tank. *Aerospatiale via H Lacaze*

- Diffusion of interface efforts between the two stages.
- Ejection (separation) sequence in case of a complete shutdown of the first stage.
- The various lead-through openings required in the thermal protection of the second stage.
- Where to put the rockets required for the separation of the second stage from the first?

These questions – in particular those related to the separation between the two stages – had been latent since the sixties but had, up to that point, been mainly forgotten. '*Early on, the key point had been studying the best architecture; separation was never detailed, it was always assumed separation would be troublesome,*'[147] remembered Henri Lacaze, who had followed the question since the earlier Transporteur Aérospatial studies.

Taranis was followed by STS-2000.

Aerospatiale STS-2000

This name was an umbrella designation for studies into an air-breathing launcher, designed in parallel with Taranis. While Taranis had a vertical take-off, STS-2000 was a horizontal take-off, air-breathing vehicle. Both single-stage-to-orbit and two-stage-to-orbit were designed with the intention of putting a 6.9-ton (7-tonne) payload into orbit after taking off from a standard runway.

For the two-stage-to-orbit version, the first stage would take-off from a conventional runway (possibly in Europe), position itself at the appropriate latitude and accelerate to Mach 6 and an altitude of 18.6 miles (30km). The separation would take place at this time, the second stage climbing to over 62 miles (100km) while accelerating to Mach 25 then continue up to 250 miles

BELOW Later illustrations of the Lanceur with the two stages stacked. *Aerospatiale via H Lacaze*

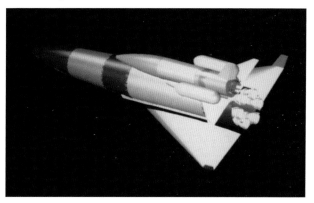

* Data come from the written section of the RAF 90 report and differ slightly from the drawings included in said report.

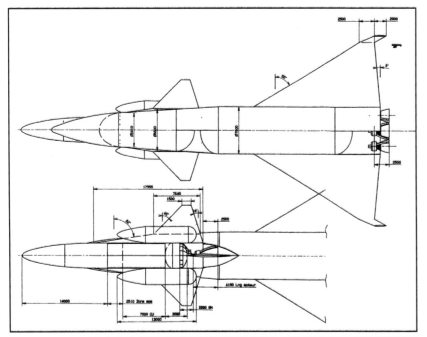

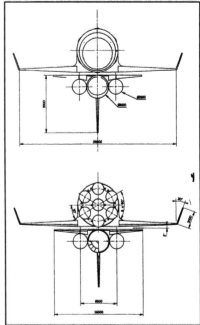

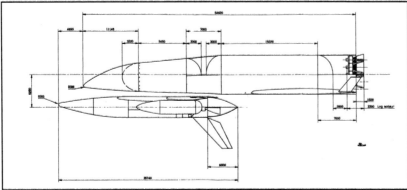

ABOVE Three-view of the shape 2.0 from February 1990. *Aerospatiale via H Lacaze*
BELOW 'Ghost' view of the shape 2.0. Legend: **Nez EP1 = Nose first stage; Reservoir LOX = LOX tank; Jupe IR = Skirt between two sections of the first stage fuselage; Reservoir LH2 = LH$_2$ tank; 7 moteurs = 7 engines; Bâti-moteur = Engine frame; Winglet = Winglet; Voilure = Wing; EP2 (1) = Second stage.** The note regarding EP2 indicates that the insides are identical to those of previous shapes. *Aerospatiale via H Lacaze.*

(400km) to be inserted into orbit. After separation the first stage would land in Kourou (where a new landing strip was to be built for Hermes and could be used for the STS-2000). After its sojourn in orbit (no duration was specified), the second stage would return to Earth by gliding to land at *'any airport.'*

Various engine types were considered, in association with engine specialists (un-named but obviously SNECMA and SEP would have been involved). For the single-stage-to-orbit, were the propulsion system had to operate from 0 to Mach 25, the turbo-ram-rocket type of engine was favoured. In all variants, the rocket engines would use LH/LO.

BELOW Size and performance comparison of Ariane 4, Ariane 5, Taranis and STS-2000. *MBDA*

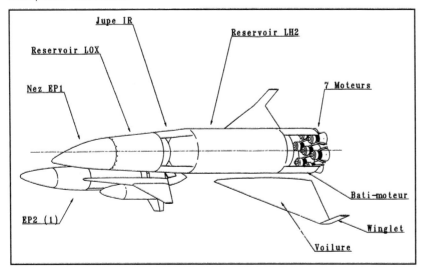

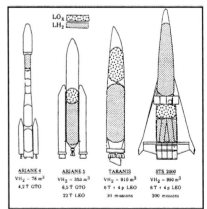

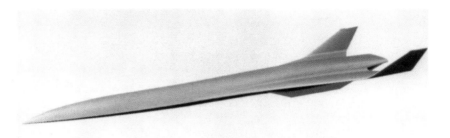

RIGHT Undated Aerospatiale study for a Mach 5 vehicle. The overall shape suggests this study was part of the STS-2000 umbrella programme. *Author's collection*

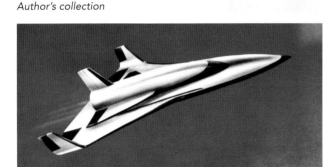

ABOVE Promotional (and somewhat stylised) artwork depicting a single-stage-to-orbit STS-2000 taking off from an airport (note the control tower in the lower left corner). *Aerospatiale*

ABOVE SSTO STS-2000 in orbit as viewed by an Aerospatiale artist. *Aerospatiale*

BELOW Interesting line drawing of an SSTO variant depicting the aircraft in cutaway. Most of the fuselage contains fuel tanks with a cockpit section at the front and a small cargo bay between the tanks and the engines. *ONERA*

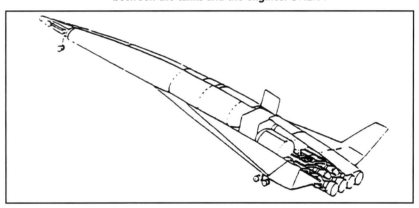

Aerospatiale STS-2000 2STO 1st stage	
Status	Project
Span	103ft 1in (31.42m)
Length	196ft 10in (60.00 m)
Gross Wing area	–
Gross weight	449,750lb (204,000kg)
Engine	–
Maximum speed/height	Mach 6 at 98,425ft (30,000m)
Load	266,750lb (121,000kg)
Aerospatiale STS-2000 2STO 2nd stage	
Status	Project
Span	52ft 6in (16.00m)
Length	124ft 8in (38.00m)
Gross Wing area	–
Gross weight	266,750lb (121,000kg)
Engine	–
Maximum speed/height	Mach 25 at 328,075ft (100,000m)
Load	15,432lb (7,000kg)

Aerospatiale STS-2000 SSTO	
Status	Project
Span	94ft 9in (28.89m)
Length	239ft 6in (73.00m)
Gross Wing area	–
Gross weight	745,150lb (338,000kg)
Engine	Turbo- ram-rocket
Maximum speed/height	Mach 25 at 328,075ft (100,000m)
Load	15,432 lb (7,000 kg)

BELOW Three-view drawing and weight table of the proposed SSTO STS-2000. *Aerospatiale*

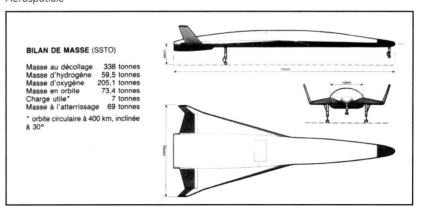

ABOVE Other promotional art illustrating the two-stage variant of STS-2000. *Aerospatiale*

BELOW Computer-generated simulations are always very colourful. Here are depicted the isobaric lines of the STS-2000, calculated at Mach 4, the separation speed. *Aerospatiale*

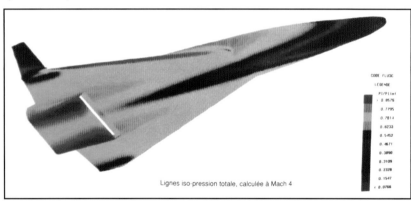

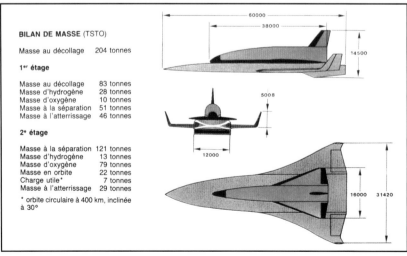

Aerospatiale Oriflamme

Oriflamme was designed by Alain Wagner and Jean-Pierre Bombled of Aerospatiale as a SSTO using air-breathing propulsion up to Mach 12. The only mission envisioned for Oriflamme was to take-off from Kourou to deliver 15,432lb (7,000kg) and four passengers to a space station on a LEO orbit. Take-off would be assisted (but it was not specified by which means).

The constraints/objectives were:

- Length less than 229ft 8in (70.00m)
- Taking into account an expansion deflection nozzle with a length of 55ft 9in (17.00m)
- No thermal flow above 2,000°C

The design team elected to use three separate engines: single flux turbojets, fixed geometry dual mode ramjets and cryogenic rocket motors. Turbojets were chosen because they are well known engines which could be fuelled by kerosene. Kerosene has a lower specific impulse than hydrogen but is denser and could be stored in the wings meaning a smaller fuselage. The fixed geometry dual mode ramjets were selected because they were light and exhibited 'from available data' good propulsion performance, but the design team acknowledged that more propulsion systems would need to be evaluated.

The report noted that a computer programme named 'TOPLA' was used to optimise the trajectory, the number and kind of engines, the propulsion mode and the switch from one mode to another. The results were rated '*disappointing*' mostly because they did

LEFT Three-view drawing and weight table of the proposed two-stage variant of STS-2000. *Aerospatiale*

BELOW Another view of the computer model created by Aerospatiale. *Aerospatiale*

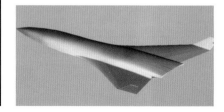

CHAPTER TWELVE
AROUND AND BEYOND HERMES

ABOVE Colour computerised portrayal (employing 1990's technology) of Oriflamme. *Aerospatiale via H Lacaze*

RIGHT Two-view and dimensional data of the Oriflamme project. *Aerospatiale via H Lacaze*

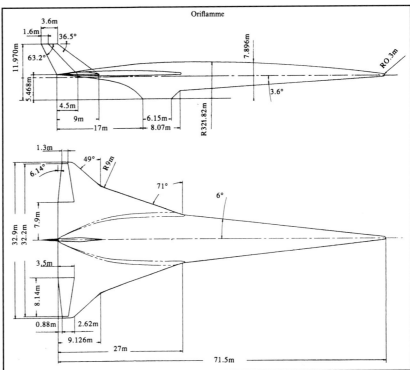

Aerospatiale Oriflamme	
Status	Project
Span	107ft 11in (32.90m)
Length	234ft 7in (71.50m)
Gross Wing area	–
Gross weight	745,150lb (338,000kg)
Engine	Turbojets + rockets + dual-mode ramjet
Maximum speed/height	Mach 12
Load	15,432lb (7,000kg) and 4 × astronauts

not reinforce the design team's opinion. *'Instead of our implicit object, to reach Mach 12 with pure air-breathing propulsion, TOPLA dictated we should have some thrust augmentation with the rockets beyond Mach 9. In fact, Oriflamme may reach Mach 12 in pure air-breathing mode but it is more efficient in terms of payload – due to a low net scramjet thrust – to call upon the rockets earlier.'*[148]

Engines were grouped in a nacelle underneath the fuselage. Three different nacelle concepts were assessed before the one used was selected, but here too, the studies could have continued. The nacelle obliged designers to put the wing in a high position to accommodate the expansion deflection nozzle of the ramjet. Re-entry was deemed easy—indeed the report satisfied itself in announcing, *'No particular problem was detected with regard to the cross-range requirements and equilibrium temperatures'.*

The internal volume of the vehicle was mostly taken up with tanks:

- Front liquid hydrogen tank: trilobe tank of 21,200cu ft (600cu m).
- Liquid oxygen tank made of three spheres: capacity 4,725cu ft (134cu m).
- Rear liquid hydrogen tank: trilobe tank of 6,750cu ft (191cu m).
- Kerosene tanks in the wings: capacity 2,025cu ft (57cu m) each.

Trilobe tanks were selected because it was possible to design internal walls so that each lobe worked as if it was part of an entire circular cross section tank, leading to a lower mass. Conformal tanks of elliptical cross section were not used because the internal pressure would tend to make them circular, and reinforcing them to carry the bending loads would lead to the use of a *'prohibitive amount of material'.*

In its conclusion the study recommended that detailed design, manufacturing of test articles and testing of the engine nacelle – and particularly of its rear nozzle – be performed. It was indicated that work on the integration of the dual-mode ramjets had been transferred to the tactical missile division

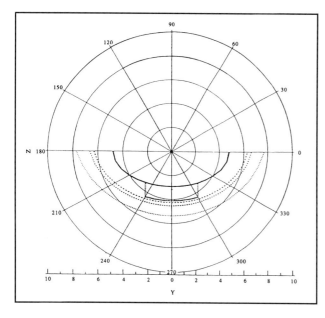

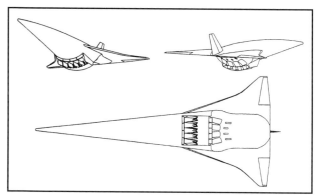

LEFT Study of the shockwaves on the lower fuselage and intake of the nacelle. The closely-dotted line is for Mach 7; wide-dots for Mach 10; and dashed line is for Mach 12.
Aerospatiale via H Lacaze

BELOW Various line drawings of the Oriflamme.
Aerospatiale via H Lacaze

of Aerospatiale. '*Feasibility of this vehicle still requires technological breakthroughs, more particularly in the fields of materials/structure and supersonic combustion*' was the ultimate conclusion of the report.

Aerospatiale Radiance

This project, led by E Boyeldieu, C Billant and A Wagner of Aerospatiale, was intended to design a fully-reusable TSTO with an air-breathing first stage. It was dubbed 'Fast TSTO' because separation occurred at Mach 12. It was viewed as complementary to Taranis, in that Taranis studied a rocket powered TSTO with a separation occurring at Mach 13.

Its main mission was to take-off from Kourou and rendezvous in low Earth orbit with a space-station (the un-named space station is obviously the future ISS) for delivery of a 13,228lb (6,000kg) payload and up to six people. However, other missions were also envisioned:

■ Delivery to a geostationary transfer orbit (GTO). Radiance was to deliver 16,535lb (7,500kg) and two people to a 124-mile (200-km)/5.23° orbit. From then an additional stage would then place the payload into a GTO orbit. It was suggested that Radiance should stay in orbit until GTO insertion was achieved and be able to bring back to Earth the operational payload as a contingency.

■ Sun-synchronous orbit (SSO): Radiance would put a payload of 11,023lb (5,000kg) into a polar orbit (124 miles [200km], 98°) and subsequently recover it. This orbit could later be raised by the satellite (in the payload) itself.

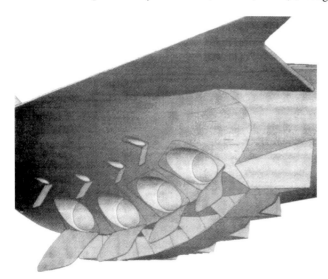

ABOVE Representation of the expansion-deflection nozzle. Ports for rockets and turbojets are shown open, they would be closed with the ramjet running.
Aerospatiale via H Lacaze

ABOVE A rather poor picture (extracted from the Oriflamme report) of a test article of the glass matrix composite corrugated sandwich intended to offer thermal protection to the underside of Oriflamme. *Aerospatiale via H Lacaze*

RIGHT Cutaway of Oriflamme showing various tanks and the payload compartment. *Aerospatiale via H Lacaze*

BELOW Overall dimensional drawing of the Radiance vehicle. *Aerospatiale via H Lacaze*

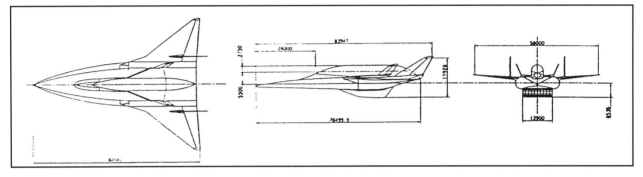

- Rendezvous with the Russian *Mir* space station and deliver or recover 11,023lb (5,000kg) of payload and up to six astronauts (217 to 248 miles [350 to 400km], 51.6°).

Radiance was composed of two stages:

i A lower, air-breathing first stage powered by ten turbojets using kerosene fuel. They were active up to Mach 3.5, at which speed they shut off and ten dual-mode ramjets cut in, accelerating it to Mach 12. After separation above the Cape Verde islands it would glide to land at Rota airfield in Spain. Dry mass of the first stage was estimated as 458,875lb (208,140kg).

ii An upper rocket-powered second stage, integrated with the back of the first stage via fairings. Propulsion was through an advanced, high-pressure rocket engine with an auxiliary propulsion system for orbital manoeuvring and attitude control using gaseous hydrogen and oxygen. Dry mass of the second stage was estimated as 84,194lb (38,190kg).

An important aspect of Radiance was the low value chosen for the first stage nose radius (2¾in [7cm]) and the wing leading edge radii (³/₈in [1cm]). This question of the nose and wing leading edge radii was important to control the thermal flows on the surface of the vehicle, as pointed out by Michel Rigault, who stated, '*The nose radius directly impacts the local thermal flows. Increasing the radius, lowers the local heat flows. Yet this radius also affects the hypersonic aerodynamic qualities of the vehicle, and therefore its cross-range performance.*'[149]

This reduction of radius was possible through the use of a product called Aerocoat which had been developed by Aerospatiale (in association with ONERA).[150] The thermal constraints supported by the nose of the first stage at the end of the ascent were described as 'most critical' with the temperature reaching up to 2,260°C. The thermal protection of the first section of the first stage was described as '*Aerocoat using fluid vaporisation to keep the wall temperature of the niobium shell lower than 1,500°C on a length of 3ft 3in (1.00m), an anti-oxidation coated carbon-carbon stiffened shell on a length of 29ft 6in (9.00 m), and an intermetallic [titanium alloy] skin stronger shell, possibly reinforced [with silicon carbide] for the third part on a length of 10ft 10in (3.30m).*'[151]

Silicon carbide shingles were to be used to reduce the thermal conduction to the slush hydrogen tank. The rest of the fuselage was '*a shell stiffened with rings and stringers in intermetallic [titanium alloy]. The shell [was] thermally protected with multiple layers of rhodium separated by silica cloth; these layers [were] protected from airflow with [silicon carbide] shingles.*'

The second stage followed the same logic, with a nose in carbon-carbon composite and the rest '*of the same conception previously detailed for the first stage.*' Mass and centre of gravity studies were computed by 'slice', each stage being artificially separated into sections to ease computing.

And the conclusion? It was acknowledged that the vehicle's critical sections were '*built from materials which belong to future, very advanced technology […] Since these materials are in a very early development stage, there is still much to do to bring them to the stage where they can reliably be planned for manufacture of very large structural parts.*' So? Months of work, ten pages of a highly scientific report to conclude this was all metallurgical science-fiction.

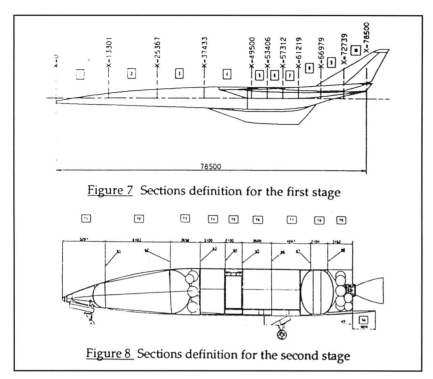

Figure 7 Sections definition for the first stage

Figure 8 Sections definition for the second stage

ABOVE For detailed studies of mass, the two stages were arbitrarily separated into sections, and for each section mass and centre-of-gravity calculations were performed. The view of the second stage gives indications of the inside of the vehicle, down to the landing gear. *Aerospatiale via H Lacaze*

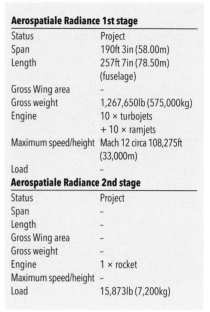

Aerospatiale Radiance 1st stage	
Status	Project
Span	190ft 3in (58.00m)
Length	257ft 7in (78.50m) (fuselage)
Gross Wing area	–
Gross weight	1,267,650lb (575,000kg)
Engine	10 × turbojets + 10 × ramjets
Maximum speed/height	Mach 12 circa 108,275ft (33,000m)
Load	–
Aerospatiale Radiance 2nd stage	
Status	Project
Span	–
Length	–
Gross Wing area	–
Gross weight	–
Engine	1 × rocket
Maximum speed/height	–
Load	15,873lb (7,200kg)

B3 Dornier

Dornier EARL

In 1986 the German Ministry of Research and Technology (BMFT), like the CNES in France, sponsored a study for a fully-recoverable launcher. Earlier exercises had suggested that by mid-2000, the Ariane 5 launcher would have an increased launch rate which would make a consumable launcher inefficient while the advanced vehicles like Sänger or HOTOL would not yet be ready—so an interim, fully reusable launcher was needed but one still employing proven technologies. That was Dornier's (the company still deeply involved in Hermes and notably participating in the 'lifeboat' version offered to NASA) EARL, the European Advanced Rocket Launcher.

Dornier based its work on the earlier Future Launcher System study in which it had been involved (see Chapter 7). By 1988, Dornier had designed a family of launchers to cater for the various load options.

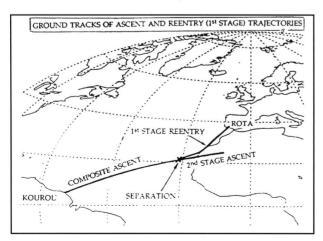

LEFT Ascent and return trajectory for the first stage. The landing in Spain is indicated here. *Aerospatiale via H Lacaze*

The main missions studied paralleled those assigned to Hermes at the time and were:[152]

- Manned flight with a 4.9-ton (5-tonne) payload to MTFF with a crew of two pilots and two payload specialists for a duration of four days. The Earth-orbit capacity was also of 4.9 tons (5 tonnes). The payload volume was 2,825cu ft (80cu m) and maximum acceleration was 3.5G. Escape solutions for the crew were to be provided throughout the whole flight envelope by the use of a rescue capsule (similar to the one then envisaged for Hermes). It was hoped to have five manned flights per year.

- Up to 17.7 tons (18 tonnes) could be put into low Earth orbit, in the form of 'bulky infrastructure sub-assemblies'—thus, space station elements. This configuration was unmanned.

- 6.9 tons (7 tonnes) in geostationary transfer orbit (GTO). It was planned to have between seven and ten unmanned flights per year.

ABOVE Stylised artwork of EARL at take-off and landing. *Dornier via H Lacaze*

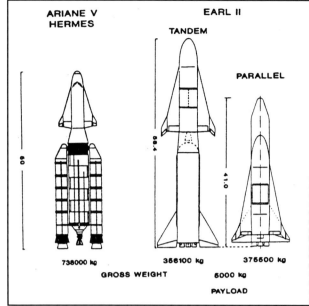

ABOVE RIGHT Comparison of the two main variants of EARL (stacked and tandem) with Ariane 5 and Hermes. *Dornier via H Lacaze*

Both stacked and parallel launch (later labelled 'EARL II') configurations were studied. With the following results:

Stacked configuration (also named 'tandem') advantages were:

- Simpler technicalities
- Lower drag
- Versatility of the upper stage

However, it was found to be very susceptible to side forces (strong winds at take-off had been a heavy constraint for Ariane 5 and Hermes).

On the other hand, parallel configuration was better on the following points:

- All engines could provide thrust at take-off
- Structural mass was 'possibly' lower
- It handled wind forces better

While having a few handicaps as well:

- Larger aerodynamic drag

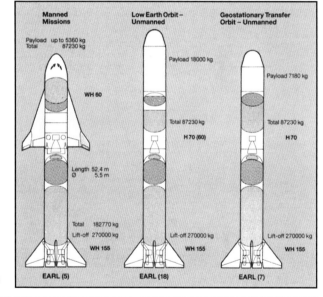

RIGHT Tandem variants of EARL with details of the rocket engines and fuel tanks. *Dornier via H Lacaze*

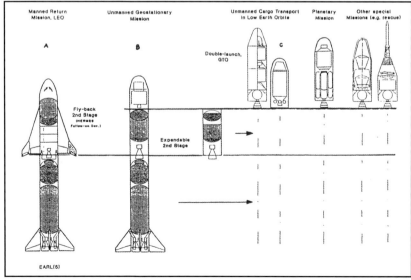

RIGHT Another illustration of tandem variants of EARL describing a large number of payloads for a variety of missions. *Dornier via H Lacaze*

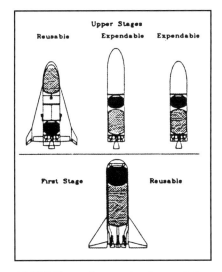

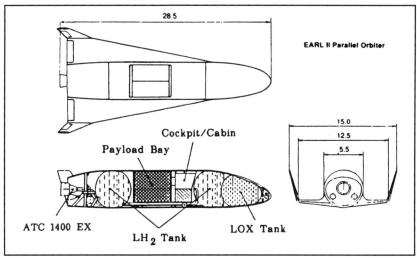

ABOVE Illustration putting forward the different payload fairings of the tandem EARL. *Dornier via H Lacaze*

ABOVE The orbiter was different for the parallel (shown) and the tandem variants. Here one can even see the doors of the cargo bay. *Dornier via H Lacaze*

BELOW Sectioned views showing inner details—notably the small turbojets used for the return flight; position of the cabin, of the cargo bay and of the landing gear. *Dornier via H Lacaze*

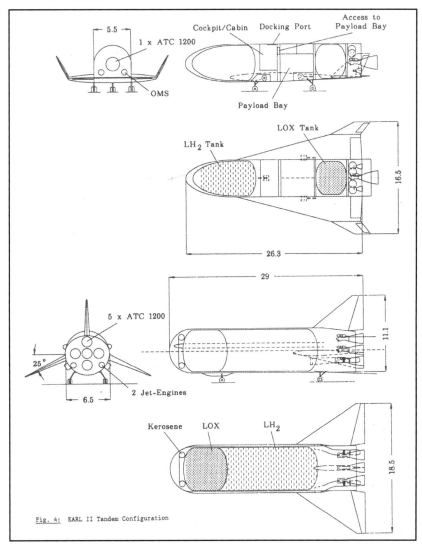

- Lateral trim losses were computed as being very high
- Upper stage engines must operate in both atmosphere and space

The original intent appears to have been to recover the first stage by gliding back home ('non-active fly-back trajectories') but this was found impossible due to the necessary high-G turn manoeuvres. For this reason, two small turbojets of unspecified type were buried in the midst of the rocket engines with side-opening air intakes.

The second stage orbiter could glide back to the launch site (obviously Kourou was in the mind of Dornier's designers) after re-entry. It was hoped it could even reach a southern European landing site.

EARL designers intended to use technologies already extant or under development:

i Integral fuel tanks (Ariane 5)
ii Rocket engines (Shuttle, Ariane)
iii Structural design of the orbiter (Hermes, HORUS)
iv Thermal protection (Hermes)
v Crew escape system (Hermes)
vi Hypersonic flight (Sänger)

RIGHT And here is shown the corresponding first stage. *Dornier via H Lacaze*

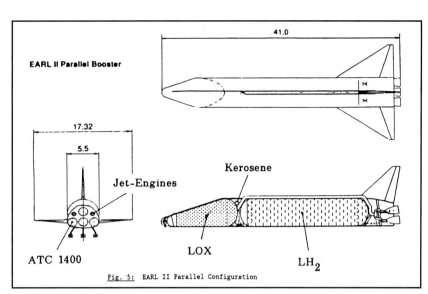

As for STAR-H, choices of solutions were prepared, notably for the propulsion, but the precise engine was not selected. For example, four rocket engines were proposed, the first three for ground-to-space operation: Shuttle SSME, Ariane HM-60 Vulcain, Dual-expander Engine; the fourth for space operation only, MBB ATC-700. The report ended by claiming that *'To provide flight readiness of EARL in due time, taking up a dedicated technology program nearly at once would be necessary. Such programme should last about five years from 1989 onwards to prepare a decision starting up development around 1994/95.'*[153] The intention was to have the vehicle operational around 2005–2008. The option was not taken up by ESA.

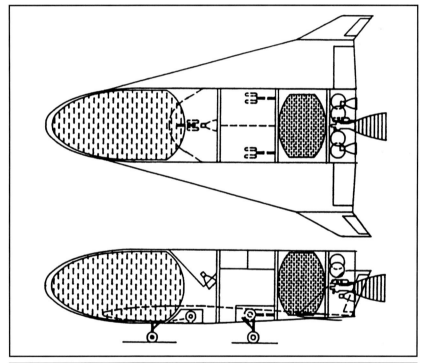

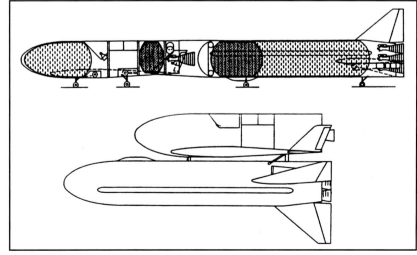

MIDDLE Here, the first stage. The long ducts for the auxiliary turbojets appear to have doors to close them during take-off. *Dornier via H Lacaze*

RIGHT The two tandem stages docked together. With the landing gear down it exhibits a very strange configuration which does not appear to have any actual use. *Dornier via H Lacaze*

B4 Ecole Polytechnique Féminine

ASUR

In 1992, the *Ecole Polytechnique Féminine*, a French engineering school for women, worked with the Ohio State University on a project for students, the goal of which was to design a Mach 6 hypersonic launcher. They based their work on Star-H (see above) and Sänger II (see Chapter 9). The result was ASUR (French for 'azure' the blue sky, and also an anagram of USRA, the Universities Space Research Association). The project team was helped by Aérospatiale and ONERA.

Known parameters when the project was initiated were the weight of the orbiter 299,075lb (136 tonnes) and the specific impulse of the engine, 2,000s. Unfortunately, neither the engine nor the orbiter was detailed in the study although the use of a 'turbo-expander-rocket' was advised. A turbo-rocket-ramjet could also be used. It was, however computed that five engines would be needed to overcome the drag rise in the lower supersonic mode. Each engine would have its own ramp and intake as in Sänger II.

Geometry of the vehicle was obtained from the fuel volume needed. Weights of the specific elements of the launcher were based on similar elements of Concorde. Carbon-composite materials were selected for those parts having to withstand the high temperature gradients of hypersonic flight. Heat fatigue in a reusable vehicle was acknowledged and likewise, the problems associated with discontinuity in temperatures leading to deformations and possible cracks were recognised. The need to protect fuel tanks and the steel legs of the undercarriage from high temperature was indicated.

An interesting approach was that the vehicle was constantly in air-breathing mode and therefore only embarked liquid hydrogen as fuel. '*Instead of LOX, ASUR uses oxygen from the air because it flies below 22 miles (35km) and we consider the atmosphere has enough density at those altitudes.*'[154] The wording suggests oxygen is extracted from the atmosphere during the flight, evoking the mysterious engine used on HOTOL.

Take-off would be from any airfield with a 1.5 mi (2.4 km)-long runway, but a coastal region was recommended to spare the local inhabitants rocket noise and sonic booms. Separation between the launcher and the undefined orbiter would occur at Mach 6 at an altitude between 95,000 and 100,000ft (28,950 and 30,475m). The separation manoeuvre would see the ASUR in a slight dive while the orbiter continued its trajectory straight ahead. Forcing the separation with mechanical devices was considered but rejected on the ground that '*jacks won't withstand very high temperatures*'—which appears an odd reason, as the jacks would most probably be recessed in the dorsal section of ASUR, one of the coldest areas of the whole vehicle.

Return could be made to the original take-off place. In the case the orbiter could not be launched, ASUR would use the orbiter's fuel before landing. The report concluded (too?) optimistically, '*ASUR belongs to a new category of reusable launchers. It opens new horizons for space conquest.*'

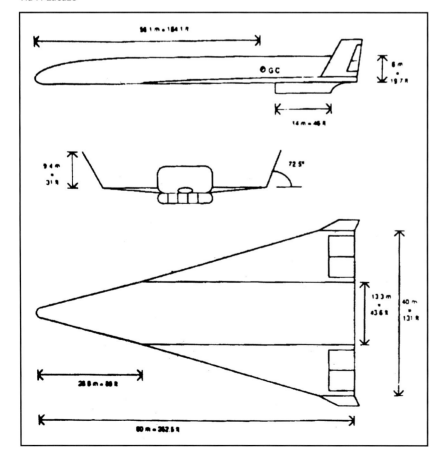

BELOW Three-view drawing of Ecole Polythechnique Féminine ASUR launcher.
Via H Lacaze

ASUR	
Status	Study
Span	131ft (40m)
Length	–
Gross Wing area	17,222sq ft (1,600sq m)
Gross weight	374,775lb (170 000kg) (without orbiter)
Engine	5 × turbo-expander-rockets
Maximum speed/height	Mach 6 at 98,425ft (30,000m)
Load	299,830lb (136,000kg)

B5 ESA Views

FESTIP

France declined to participate actively in FESTIP. According to Michel Rigault, '*France did not participate in FESTIP because CNES was not convinced by air-breathing, reusable launchers.*'[155] Obviously, the experiences of Star-H, Taranis and others were not particularly encouraging regarding the feasibility of such a concept. Yet, it may be worth mentioning here some of the projects which evolved from it.

FESTIP was an ESA program initiated in 1994 to study reusable launchers and to assess the technology required. It developed over four years, with the German Sänger, British HOTOL and Aerospatiale Taranis projects sometimes attached to it.

Some of the possible development lines were eliminated from the beginning:

- Air breathing SSTO. Technology was just not available.
- Concepts based on air-launching from an existing aircraft. Capacity of the largest possible mothership, the An-225, was only 4.9 to 6.9 tons (5 to 7 tonnes) in equatorial LEO, whereas 6.9 tons [7 tonnes]) was the FESTIP target.
- Concepts based on parachute recovery—which was found incompatible with the masses to be recovered by the various concepts.[156]

Eight concepts were finally selected (FSSC may mean 'future space system concept'):

i   FSSC-1. SSTO rocket winged body, vertical take-off/horizontal landing.

ii FSSC-3. SSTO rocket, vertical take-off, vertical landing. The absence of wings was found to be more an added complexity than an asset.

iii FSSC-4. SSTO rocket, winged body, horizontal take-off (from a sled) and landing (on undercarriage). The sled idea (as proposed for HOTOL) was found to be very innovative.

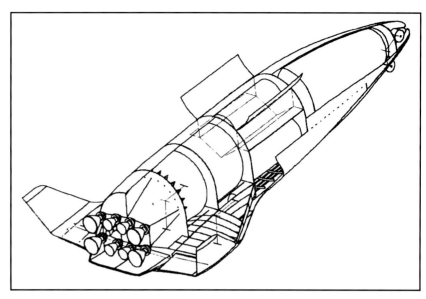

ABOVE FSSC-1, a single-stage-to-orbit (SSTO) study. An attribute common to all the space projects is the small size of the payload bay compared with the fuel tanks. *ESA via author*

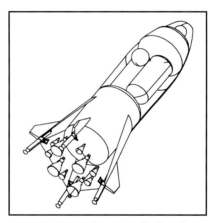

ABOVE FSSC-3, the vertical take-off SSTO, was deemed to generate more design problems than advantages. *ESA via author*

iv FSSC-5. SSTO rocket, lifting body shape, vertical take-off and horizontal landing. One variant used an aerospike engine. This design was inspired by the US Venture Star design and was found impossible to design practically (the aerospike engine just encompassed too many unknowns) and economically (staged combustion engines were not 'very promising').

RIGHT This cutaway of FSSC-3 shows the liquid hydrogen tanks in blue and the liquid oxygen tanks in green. Note the asymmetrical (side located) payload bay. *ESA via author*

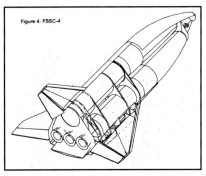

Figure 4. FSSC-4

LEFT The main design feature of FSSC-4 was its take-off via a sled... which remained un-illustrated! *ESA via author*

RIGHT This rendering illustrates the aerospike variant of FSSC-5, deemed too complex for European technology. *ESA via author*

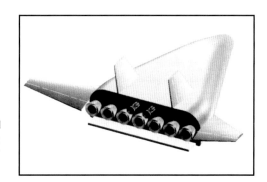

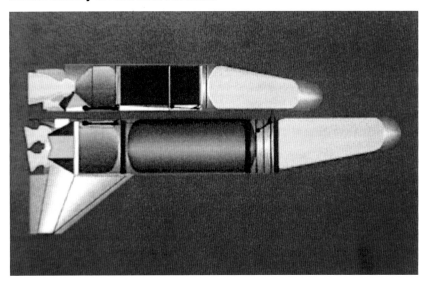

Figure 5. FSSC-5 aerospike — *Figure 6. FSSC-5 rocket*

ABOVE Both variants of FSSC-5 (aerospike or common rocket engines) were evaluated and pronounced impractical. *ESA via author*

BELOW Cutaway of FSSC-9. *ESA via author*

v FSSC-9. TSTO fully reusable rocket. This was a vertical take-off, horizontal landing vehicle deemed to be technically 'within reach' but only marginally interesting on economic grounds. It would have needed further optimisation. It evolved from the Dornier EARL II with some additional input from the Aerospatiale Taranis project.

vi FSSC-12. A TSTO, fully reusable vehicle with an air-breathing first stage. It was appreciated as being technically feasible but with high development and operating costs. Originally based on MBB HORUS/Sänger project.

vii FSSC-15. Suborbital single stage rocket. Technically feasible and with a potentially good economical balance.

viii FSSC-16. A family of TSTO vehicles going from semi-reusable to fully-reusable. The semi-reusable variant would be feasible 'in the near term' but has questionable performance gain to offer when compared with Ariane 5. The fully reusable 'Siamese' version would require improvements in technology.

BELOW FSSC-15 was not intended to reach orbit but only an altitude and a speed sufficient to launch a payload on a second-stage booster. *ESA via author*

BELOW Two variants of FSSC-12, the shape derived from MBB Sänger and Aerospatiale Taranis projects. *ESA via author*

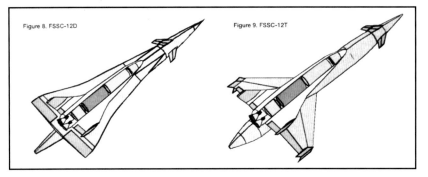

Figure 8. FSSC-12D — *Figure 9. FSSC-12T*

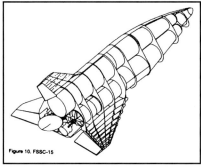

Figure 10. FSSC-15

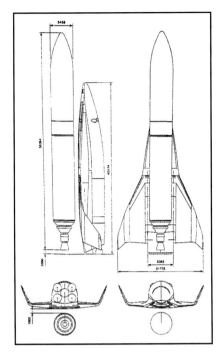

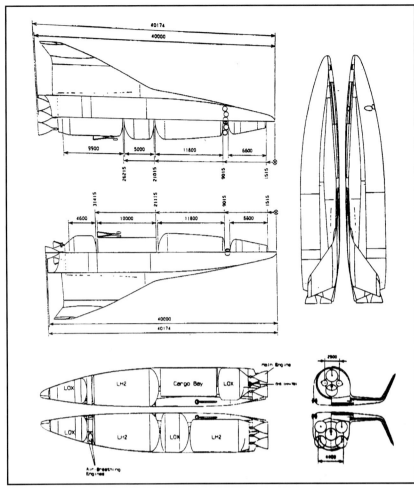

ABOVE FSSC-16 semi-recoverable variant. *ESA via author*

RIGHT And here is the fully-recoverable variant of FSSC-16. *ESA via author*

BELOW Close-up of the unusual configuration of the turbojets used to power FSSC-16 when returning to base. The upper scoop and the lower nozzle were retracted during the whole space part of the mission. *ESA via author*

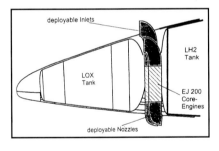

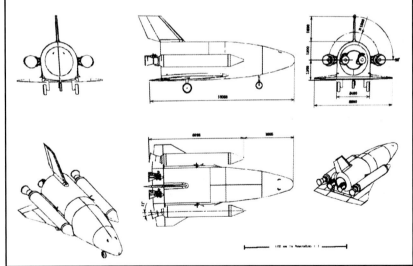

RIGHT Multiple drawings of the proposed EXTV demonstrator. It is shown here in its advanced form with solid boosters to give it an orbital capability. *ESA via author*

FSSC-15 and -16 were the preferred designs to emerge from the four years of studies.

SSTOs were found to be very attractive commercially, but requiring a technology which was not available in Europe at the time. A FESTIP EXTV was proposed as a demonstrator to acquire experience in the recurrent use of high-speed, reusable rocket vehicles.

A further development could add solid fuel boosters and thermal protection to actively test re-entry.

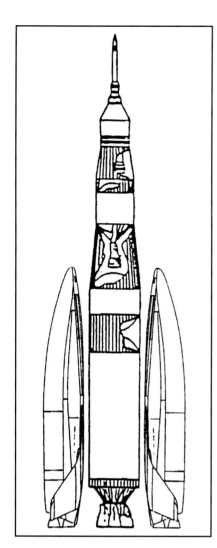

ABOVE A CNES Ariane variant proposed for lunar missions. The core stage would have used five Vulcain engines (only one is used on the Ariane 5) and the usual four solid boosters would be replaced by two reusable winged boosters. *ESA via author*

C – Meanwhile on the Hypersonic Ramjet Front…

C1 – PREPHA

The Programme de la Recherche pour la Propulsion Hypersonique Avancée (PREPHA = Research Programme on Advanced Hypersonic Propulsion), was initiated soon after the ending of Hermes and associated Aerospatiale MATRA, Dassault, SNECMA, SEP and ONERA projects. It ended in 1999 and has since been succeeded by other programmes, all oriented toward the design and testing of advanced hypersonic ramjets suitable for a European SSTO vehicle. From 1997 it was joined by a Franco-German programme associating ONERA and DLR, JAPHAR. JAPHAR's goal was to *'define'* an experimental vehicle able to fly up to Mach 8. PREPHA's main outcome was the Chamois (mountain goat) combustion chamber (*combustor*). Chamois was tested repeatedly up to Mach 6 at durations up to 10 seconds.

Following PREPHA and JAPHAR, other programmes have been pursued by ONERA and MATRA (which became part of MBDA): WRR (variable geometry, in collaboration with Russia), PROMETHEE (tested in 2002), A3CP, PTAH-SOCAR (fuel-cooled structures) and LEA to design conceptual hypersonic vehicles. LEA was to use Russian technology: a booster adapted from the Raduga missile and a Tupolev Tu-22M as mothership. Numerous tests were carried out as part of this programme including ground running of the scramjet, computer analysis of the aerodynamics of the integrated vehicle (fuselage + engine) and wind tunnel studies at TsAGI.

ABOVE A conceptual vehicle studied as part of PREPHA. *MBDA*

BELOW Two ramjet-powered aircraft studied as part of PREPHA and following programmes. *MDBA*

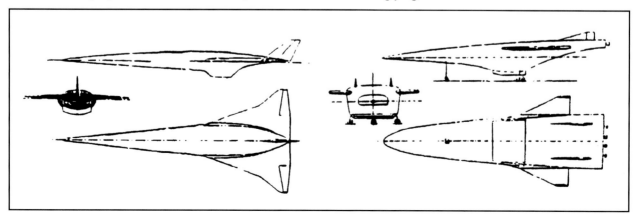

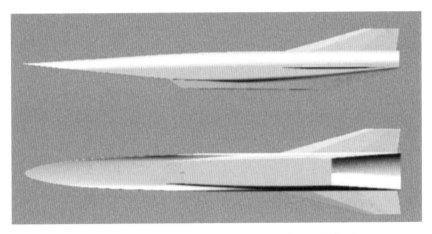

ABOVE PREPHA was followed by JAPHAR. This image shows the basic concept vehicle studied during the programme. *MBDA*

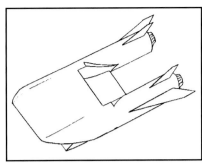

ABOVE After JAPHAR came PROMETHEE (Prometheus) and the PIAF experiment ('piaf' is a vernacular word for sparrow). As part of this programme advanced designs were studied like this one in which the 'twin fuel tanks' square off the scramjet. *MBDA*

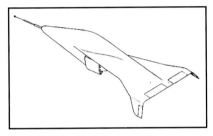

ABOVE Another design emerging from this programme was this one based on a double-cone fuselage with a semi-annular scramjet beneath. *MBDA*

RIGHT One of the most appealing designs was the Axi-symmetric vehicle with the scramjet completely integrated around the double-cone fuselage. *MBDA*

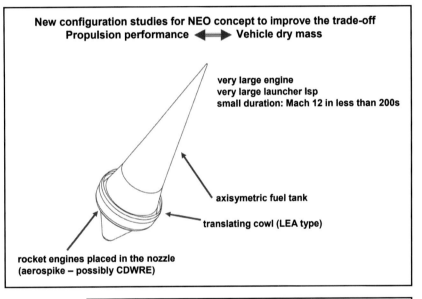

ABOVE In the early 21st Century these programmes culminated in the LEA concept. *MBDA*

RIGHT Here is the cutaway drawing of the LEA vehicle. This would have been a small machine, about the length of an air-to-air missile. *MBDA*

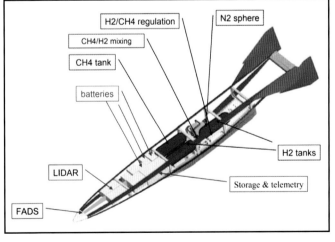

This research appears to have been intended mostly for military application. Not only has French Defence Minister Florence Parly announced (in 2019) the beginning of the flight test programme of V-max a hypersonic glider, but the 2019 *Feuille de Route* (Roadmap) for ONERA included an (unpublished) sheet about a *aéronef de combat hypersonique* (hypersonic combat aircraft).

D – Beyond Earth Orbit

La Base Lunaire

In 1997, Aérospatiale's Henri Lacaze, former Hermes project manager, wrote about a possible Moon base. This was a review and criticism of previous programmes but also a concise attempt to study the feasibility of the project. Lacaze assessed that the project should begin with a simple 'outpost' with a crew of between three and eight, each staying six months …but even such limited and humble beginnings were, he said, very different to a succession of short, limited stays.

For the main motive of establishing a lunar base *per se*, Lacaze saw the observation of the Universe from the Moon away from atmosphere and radioelectric pollution. While he acknowledged that for these reasons, scientists would prefer to establish the base on the dark side of the Moon, to make access and communication to the base easier (hence for basic safety reasons), he advocated locating the base on the lunar equator and on the near side.

The technical objectives of this base should be, at the minimum:

i Equip with appropriate delivery links from Earth and back
ii Shelter the [astronauts] and provide them with life support and energy
iii Enable configuring and reconfiguring, partly manually, partly via robots
iv Provide communication with Earth and lunar mobiles
v Permit in situ *scientific experiments*
vi Support movement of people by foot (EVA ancillaries) or vehicles and transport equipment[157]

Basing his theories on the launch facilities available at the time, he established that with a maximum low Earth orbit (LEO) mass of 19.7 tons (20 tonnes), the mass of each elementary module of the base should not be over 9.8 tons (10 tonnes). To save weight while giving a sufficient protection from cosmic rays, he suggested covering the station's living modules with regolith, possibly adding a refuge dug deeper, in case of solar flare eruption. For powering the station, he considered solar panels inefficient in the long term and advised a small nuclear reactor. The reactor would be partially dug-in and about 650ft (200m) from the base.

For the vehicle to transport equipment to the base he distinguished between the vehicles bringing heavy equipment and the crew ferry. The first one would be one-way only and would stay on the Moon, being deactivated (depressurisation, emptying the fuel tanks) then pushed away from the normal landing area. Lacaze did not suggest stripping it for spare parts— but that is most probably what the astronauts would do. The crew ferry could also be of two types:

i Either a vehicle which would not linger on the base, bringing in new astronauts and returning to Earth a few days later with the crew at the end of its turn of duty (for the simple reason that there would not be accommodation on the base for two teams at the same time), in which case the use of cryogenic fuel is possible.

ABOVE For his paper, Henri Lacaze retained a small 10-tonne module as the core element of the Moon Base. This design actually came from a paper signed Parkinson and Perino. *Via H Lacaze*

BELOW Drawing of a two-module outpost. Legend: 1 Habitat, 2 Laboratory, 3 Soft material tunnel, 4 Refuge against solar flares, 5 Support equipment, 6 Airlocks (two), 7 Handling arm (two), 8 Radiator (another one on the opposite side), 9 Solar panels (multiple, two illustrated), 10 Vehicle hangar and 11 Rover vehicle. *H Lacaze*

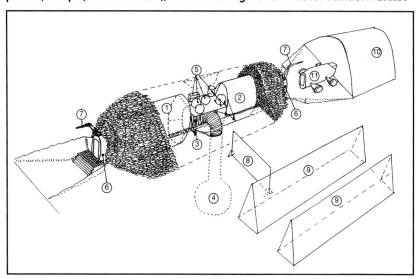

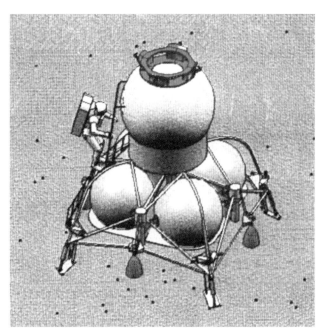

ABOVE Henri Lacaze's depiction of a vehicle shuttling between low lunar Orbit and the Moon's surface, based on the system featuring exchange of fuel tanks in Lunar orbit. *H. Lacaze*

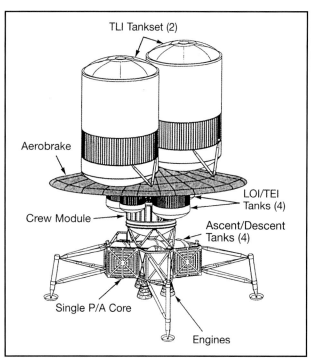

ABOVE 'Advanced' Earth-Moon vehicle with a 'shuttle' featuring drop-tanks and aerobraking and lander combined together. From a paper by Iwata *et al.* Via H Lacaze

ii Another case would be a vehicle remaining permanently on the base launch pad, to be used in an emergency evacuation. Because of the weight of fuel stock, it would be advisable to generate fuel *in situ*. The production of hydrogen appeared easy conceptually, but liable to encounter many teething troubles. Lacaze suggested the use of aluminium as fuel, while acknowledging this was 'futurist.'

He pointed out that it would be interesting to develop an Earth-Moon shuttle which could either carry a heavy load for one-way mission or a crew cabin for a true shuttle operation with return to LEO.

Lacaze estimated the costs at $50,000 million; and a yearly cost of about $4,000 million. He concluded, '*Such a programme can be envisaged only within the framework of a worldwide organisation. In addition, it is clear that the level of effort necessary to carry out the prerequisite technology developments is high. This should be split between the various space agencies, each of them*

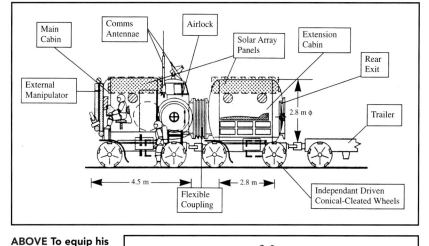

ABOVE To equip his Moon outpost with an exploration vehicle, Lacaze again used the work of Parkinson and Perino. *Via H Lacaze*

RIGHT This Moon hopper vehicle (*véhicule de franchissement*: vehicle to cross over obstacles] was designed by Roger Bonnet. Lacaze included it in his Moon Outpost study. *Via H Lacaze*

specialising, for this programme in well-defined areas. Whether a reasonably streamlined worldwide organisation, or at least co-ordination can be established for a Lunar base programme remains a matter of conjecture.'

Today the thinking continues with consideration being given by the Europeans to a 'Moon Village' which could be home for scientists of various specialities.

---

[140] *Comparison of Combined Cycle Propulsion Systems for TSTO Airbreathing Launchers,* by M Rigault, J Baldeck, S Pothier of Dassault Aviation and C Pauron, R Zendron of Hyperspace. Presented at the 42nd Congress of IAF-91-269, 1991.

[141] The same document used for the present book *Sur Une Conception Possible du Transporteur Aérospatial,* presented by H Desplante and P Perrier at the Eurospace Technical Days, January 1964 was quoted.

[142] Mail exchange with the author, October 2020.

[143] Ibid.

[144] Comparison of combined cycle…IAF-91-269.

[145] *L'Aéronautique et l'Astronautique,* No 142, 1990-3 (March 1990).

[146] *RAF 90 Lanceur Transatlantique.*

[147] Interview with the author 8th February 2019.

[148] IAF 91-207 *Oriflamme: a reference airbreathing launcher to highlight areas that are technological drivers for hypersonic vehicles,* by Alain Wagner and Jean-Pierre Bombled, Aerospatiale, 1991.

[149] Mail exchanges with the author, October 2020.

[150] Funnily, the 'Aerocoat TH' brand name, registered by Aerospatiale in 1981, had expired in 1991.

[151] IAF 93-616 *Radiance: A High Staging Speed Airbreathing First Stage TSTO Launcher,* by E Boyeldieu, C Billant and A Wagner, Aerospatiale, October 1993.

[152] *Proceedings of the 2nd European Aerospace Conference on Progress in Space Transportation,* May 1989.

[153] *EARL: Proposed European Launcher Concept for the Post-Ariane 5/Hermes Era,* W Westphal, Dornier System, October 1988.

[154] *ASUR.* Proceedings of the NASA/USRA Advanced Design Program 7th Summer Conference.

[155] Mail exchange with the author, October 2020.

[156] *Possible Future European Launchers, A Process Of Convergence,* C Dujarric, ESA, 1999.

[157] *Nouvelle Revue d'Aéronautique et d'Astronautique,* No 6, 1997.

Sources

Primary Sources

Recent Results In Rocket Flight Technique by Eugen Sänger (translated from Flug Sonderheft, 1 December 1934), NACA, April 1942

A Rocket Drive For Long-Range Bomber (Uber einen Raketenantrieb für Fernbomber) by Eugen Sänger and Irene Bredt, Deutsche Luftfahrt Forschung August 1944. Translated by M Hamermesh, Radio Research Laboratory 1952

Activity During 1963; unsigned CNES report, October 1962

Design Problems For a One-Stage Transporter; J Lambrecht, NASA, November 1963

Transporteur Aérospatial; brochure GAMD, undated

Sur Une Conception Possible du Transporteur Spatial by H Deplante & P Perrier GAMD, Conference at the Journées Techniques Eurospace, Brussels, 23-24 January 1964

Turbo-fusées pour Véhicules Hypersoniques by A A Lombard and J C Keenan, Rolls-Royce by SNECMA 1964

Transporteur Aérospatial/Raumtransporter/Aerospace Transporter by pamphlet by Nord-Aviation/ERNO/ SNECMA, distributed at 26th Paris Air Show, 1965

Review of European Aerospace Transporter Studies by H Tolle, Entwickslungring Nord ERNO, 1967

A West German Approach to Reusable Launch Vehicles by Jürgen Lambrecht and Emil Schäfer, 1967

L'avion Lanceur de Satellites by Ingénieur en chef de l'armement Thierry Moulin, Icare No 50

Systèmes de Propulsion Pour Avions a Grandes Vitesse by J Tron, Sud-Aviation, 6 May 1968

Optimisation des Paramètres de Conception d'un Véhicule Atmosphérique; Point des travaux effectués dans le cadre du contrat 902/67/CPE. Ministère des Armées, Centre de Prospective et d'Evaluations, 20 May 1968

Etude Préliminaire sur les Lanceurs Spatiaux a Propulseur Atmosphérique; marché 64-34 340-00-480-75-01, Nord-Aviation SNECMA, September 1966

Etude Préliminaire sur les Lanceurs Spatiaux à Propulseur Atmosphérique; Ministère des Armées, Centre de Prospective et d'Evaluations, 13 September 1968

Procès Verbaux d'Essais Avion Hypersonique; ONERA (various), 1968-1969

Aérodynamique d'un Avion Propulsé à Mach 7 by René Ceresuela, Ve Colloque d'Aérodynamique Appliquée de l'AFITAE, Poitiers, France, 1968

Propulseur Atmosphérique pour Premier Etage de Transporteur Spatial by Jacques Dupin (Technical Director, Nord-Aviation), undated

Recherche d'une Solution Optimale à Statoréacteur à Géométrie Fixe de Mach 3 à Mach 7 Avec Combustion Subsonique Puis Supersonique by R Marquet and C Huet, ONERA, undated

Utilisation d'un Alliage de Niobium dans la Réalisation d'un Véhicule Hypersonique by Perez, Syre, Billon, Pichoir and Guyot, 1970

Beta a Single-Stage Reusable Ballistic Space Shuttle Concept by Dietrich E Koelle, MBB, May 1970

Thermal Ground-Testing of Concorde and VERAS or Improvement in French Test Methods and Facilities by G L Leroy, N'Guyen (SNIAS), M Perrais (CEAT), H Loiseau (ONERA), 8th Space Congress Proceedings, 1971

Future Launch Systems, final presentation by Aerospatiale, 15 November 1984

Aerospace Transporter and Lifting Body Activities in Europe and Potential Participation in the Development of the Space Shuttle Orbiter by M Fuchs, J Haseloff, G Peters, ERNO Raumfahrttechnik GmbH, Bremen, 1971

Post Apollo Programme (PAP); ELDO Information undated (1972?)

Lanceurs Futurs by C. Fazi, Aerospatiale, 18 March 1986

Sänger progress report; 1986

Space Transportation, the Key to the Utilisation of Extraterrestrial Resources by H H Koelle, Technical University of Berlin, 1986

Sänger: an Advanced European Two-stage Space Transport System; MBB, 1987

MTFF Operational Design Features by H Friedrich of MBB/ERNO and A J Thirkettle ESA/ESTEC, April 1987, 24th Space Congress proceedings

Europe Spatiale (L') Cap sur le 21e Siècle, ESA, 1987

Sänger progress report; 1988

A Resupply Scenario for the Columbus *Man-Tended Free-Flyer* by H J Koopman, MBB-ERNO, April 1988, 25th Space Congress proceedings

Evolution of Air-Breathing Propulsion Concepts Related to the Sänger Spaceplane by M Albers and S Proske, MTU and P A Kramer, N H Voss and H Krebs, MBB, 1988

EARL: Proposed European Launcher Concept for Post-Ariane 5/Hermes-ERA by W Westphal, Dornier System GmbH, 39th Congress of IAF October 8-15 1988

Hypersonic Experimental Aircraft Technology Demonstrator HYTEX; MBB, undated

Sänger progress report; 1989

French Study on Air-breathing Launchers: STAR-H (a) by M Rigault, Dassault, 40th Congress of the International Astronautical Federation, Malaga, Spain 1989

Sänger progress report; 1990

Your Spaceflight Manual by David Ashford and Patrick Collins, Headline Book Publishing PLC 1990

STS-2000; Aérospatiale 7/1990

Star H: Etude de Système de Transport Spatial à Propulsion Aérobie by Michel Rigault, L'Aéronautique et l'Astronautique No 142, 1990-3

Lanceur Aerobie Bietage Star-H; Dassault Espace, undated (February 1990?)

Space Rescue System General Overview, Final Review 18-19 October 1990; Aerospatiale-Dornier-Dassault

Programme RAF 90 Lanceur Transatlantique; unsigned, Aerospatiale, 1990

Comparison of Combined Cycle Propulsion Systems for TSTO Air-breathing Launchers by M Rigault, J Baldeck, S Pothier (Dassault Aviation) and C Pauron, R Zendron (Hyperspace), 442nd Congress of the International Astronautical Federation, Montreal, Canada, 1991

IAF 91-207 Oriflamme: a reference air-breathing launcher to highlight areas that are technological drivers for hypersonic vehicles by Alain Wagner and Jean-Pierre Bombled, Aerospatiale,1991

La Propulsion des Lanceurs Spatiaux à l'Horizon 2010; ESA SP-327, 1991 December

Billiger in den Orbit?; TAG Bericht AB001, Germany, April 1991

Project Christofferus; Institut für Leichtbau und Seilbahntechnik, Switzerland, 1992

German Hypersonics Technology Programme; status report, 1992

Technikfolgen-Abschätzung zum Raumtransportsystem 'Sänger'; TAG Bericht AB014, Germany, October 1992

PLATO European Platform Orbiter Concept by Stephan Ransom and Dipl-Ing Rafner Hoffmann, MBB, undated

Sänger the Reference Concept of the German Hypersonics Technology Program by S Weintgarner, Deutsche Aerospace AG, 1993

The German Hypersonics Technology Programme Status 1993 and Perspectives by H Kuczera, H Hauck and P Sacher of Deutsche Aerospace, and P Krammer of MTU, 1993

IAF 93-616 Radiance: a High Staging Speed Air-breathing First Stage TSTO Launcher by E Boyeldieu, C Billant and A Wagner, Aerospatiale, October 1993

ASUR; Ecole Polytechnique Feminine, unsigned, undated (possibly 1993)

Cost Optimized Stage Separation Velocity of Winged TSTO Launch Vehicles by Dietrich E Koelle, Trans Cost Systems, 1994

Performance and Design Analysis of Ballistic Reusable SSTO Launch Vehicles by Dietrich E Koelle, undated, early 1990s

Engineering Design of Engine/Airframe Integration for Sänger Fully Reusable Space Transportation System by Peter Sacher, Aerospace Consulting, undated, probably 1995

Décrocher la Lune by Jean H Lacaze, Nouvelle Revue d'Aéronautique et d'Astronautique No 6, 1997

Space Transport Systems for the 21st Century by F Engström, ESA Bulletin 95, August 1998

Description of the FESTIP VTHL-TSTO System Concept Studies by Martin Bayer, undated

French Flight Test Program LEA Status by François Falempin, MBDA France, and Laurent Serre, ONERA, RTO-EN-AVT-185 undated

Current MBDA R&T Effort on Ram/Scramjet and Detonation Wave Engine by François Falempin, MBDA France, RTO-EN-AVT-185 undated

Annual reports, Dassault: various years

Secondary Sources

Aerospace Project Review; various issues

Ariane by Alain Souchier and Patrick Baudry, Flammarion collection 'l'Odyssée' 1986

Assemblée Nationale (French), *Comptes-rendus de Séances*, 1986-1993

Au-delà du Ciel; magazine, various issues 1958-1962

Air & Cosmos; magazine, various issues 1963-2000

British Secret Projects 5: Britain's Space Shuttle by Dan Sharp, Crécy Publishing, 2016

Development of Winged Re-entry Vehicles 1952-1963 (The) by John V Becker, 23 May 1983

Dream Machines (The) by Ron Miller, Krieger Publishing, 1993

Etude du Statoreacteur Supersonique et Hypersonique en France de 1950 à 1974 by Roger Marguet, Pierre Berton, Francis Hersinger (ONERA), undated

French National Space Programme 1950-1975 (The) by Bruno Gire and Jacques Schibler, Journal of the British Interplanetary Society, Vol 40, 1987

French Secret Projects Vol 1 by J C Carbonel, Crécy Publishing, 2016

French Secret Projects Vol 2 by J C Carbonel, Crécy Publishing, 2017

Frontiers of Space by Philip Buono & Kenneth Garland, Blandford, 1969

Gerfaut et Griffon by Serge Kaplan & Philippe Ricco, Avia Editions, 2006

Hermes Europe's Dream of Independent Manned Spaceflight by Luc van den Abeelen, Springer, 2017

International Missile & Spacecraft Guide by Frederick I Ordway, III & Ronald C Wakeford, McGraw-Hill Book Company Inc, 1960

Mémoires d'Usine; Comité d'Etablissement de l'Aérospatiale Chatillon s/s Bagneux. 1985

Nouvelle Revue d'Aéronautique et d'Astronautique; various issues

Notes et Souvenirs sur la 'Fondation' de l'ONERA by Jean Dubois, undated

Projet Sänger: Transporteur Spatial pour l'An 2000 (Le) by Ernst Högenhauer, Fusion No 20, 1987

Challenge to Apollo: the Soviet Union and the Space Race 1945-1975 by Asif A Siddiqi, NASA report 2000-4408

Projet VERAS (Le) by Pierre Kalbari, Bulletin de l'Association des Anciens de l'ONERA, April 2009

Rêve d'Hermes by Philippe Couillard, 1993

Silver Bird Story (The): A memoir by Irene Sänger-Bredt; 4th history symposium of the International Academy of Astronautics, Constance, Germany, 1970

Spaceflight; magazine of the British Interplanetary Society, various issues, 1984-1995

Sénat Comptes Rendus de Séances 1990-1993

Thinking Ahead With… Philippe Poisson-Quinton; La Lettre de l'Association Aéronautique et Astronautique de France, No 9 October 2004

New Commercial Opportunities in Space by D M Ashford, Aeronautical Journal, February 2007

Trident, la Quête du Haut Supersonique chez Sud-Aviation (Le) by Jean-Christophe Carbonel, Artipresse, 2017

Vaisseaux Spatiaux by Ron Miller, hors collection 2016

On the Internet:

http://astronautix.com
http://bristolspaceplanes.com
http://www.capcomespace.net
http://xplanes.free.fr
http://www.russianspaceweb.com/index.html

Index

Index of Vehicle Types

Aerospace transporter
(see also Raum Transporter,
Transporteur Aerospatial).........6, 9, 11,
23, 30, 43, 44, 81, 105, 125, 165,
166, 167, 201

Aerospatiale
   Ariane launcher 31, 88, 104, 105,
106, 107, **108**, 109, **110**, **111**, 113,
114, 118, 119, **123**, **124**, **125**, 126,
128, 129, 131, 132, 134, 135, 136,
137, 139, 140, 141, 142, 143, **145**,
147, 148, 149, 151, 152, 153, 157,
160, 161, 162, 166, 167, 169, 170,
171, 173, 175, 177, 178, 182, 183,
184, 188, 189, 192, 193, 194, 195,
198, 199, 202, 204, **208**, 214, **215**,
216, 217, 220, **222**, 226
   Ariane (Lunar) **222**
   ATSF (Avion de Transport
Supersonique Futur).....................120
   Future Launch System (FLS)...**110**, **114**,
116, **118**
   Hermes **109**, 120, 121, 122, 123, **125**,
146, **148**, **157**
   Hermes SRS 145, **146**, **147**
   Lanceur Transatlantique **207**
   Oriflamme 196, 210, **211**, **212**,
213, 226
   Radiance............196, 212, **213**, 214, 226
   RAF 90 ....**205**, **206**, 226
   STS-2000 ....122, 180, 196, 207, **208**,
209, 210
   Taranis...............122, 196, **204**, 207, 212,
219, 220
   Viro.....................127, 128, **129**, **130**, 131

Airbus
   A300 ....**157** (with Hermes)
   A310 ....**129** (with Viro)
   A320 ....**156** (with Hermes),
157 (with Hermes)

Antonov
   An-225 ....**164**, **165**

Armstrong-Withworth
   Pyramid spacecraft........................**40**, **41**

Arsenal de l'Aéronautique
   Ars 1910 ....**19**, 54
   Ars 2301 ....**19**
   RO-10 missile.......................................64
   SS-10 missile..64

BAC (British Aircraft Corporation)
   Mustard **42**, 43, 64

BAe (British Aerospace)
   HOTOL 9, 32, 44, 110, 116, 130,
139, 141, 159, **160**, **161**, **162**,
163, **164**, 168, 172, 178,
202, 214, 218, 219
   HOTOL 2 ....**165**
   Interim HOTOL **165**, **166**

BDL (Bharat Dynamics Ltd)
   Avatar...180
   Hyperplane **179**, **180**, 188

Bell Aircraft
   BoMi (Bomber-Missile)
   MX-6 ....23, **25**, **26**
   Brass Bell **23**, **24**
   Narval CT-41 61
   RoBo (Rocket Bomber).................23, **25**
   Suborbital Transport Aircraft............**26**
   X-1 22

Bristol Siddeley
   Eurospace Space Launcher.................**42**

Bristol Spaceplanes
   Spacebus...**167**
   Spacecab **165**, **166**, 167

Boeing
   747 'Jumbo Jet' 174, 203
   B-52 ....**35**, 36
   X-20 Dyna-Soar..........**10**, 22, 23, 25, 26,
27, **28**, 30, 64, 108,
110, 158, 167

Bölkow
   Eurospace RaumTransporter
proposal **33**

CNES
   Ariane IV/Hermes **108**
   Ariane V/Hermes..............**110**, **111**, 126
   Ariane
   (recoverable variants)....**118**, **119**, 120
   Hermes (early) see also Aerospatiale
and Dassault................**106**, **107**, **109**,
112, 113
   Hermes (post contract allocation)..**132**,
133, **134**, **135**, **136**, **137**, **138**, **139**,
149, 150, 151, 152, 153, **154**, **155**,
156, 159, 160, 161, 166, 167, 169,
170, 171, 172, **174**, **175**, 176, 178,
179, **181**, 183, 184, 185, 188, **189**,
190, **191**, **192**, 193, 194, 195, 196,
197, **198**, **199**, **200**, **201**, 202,
203, 204, 208, 214, **215**,
216, 222, 226
   Hermes 2000 / X-2000......**192**, **193**, 194
   Minos **105**, 106, 109
   Solaris 105, 106, 109
   Spot 135, 152, 153

Chelomey
   Raketoplan **21**, 22

Convair
   Atlas SM-65 ....41, 43, 70

Dassault (Avions Marcel Dassault/
Dassault Aviation)
   Hermes............123, 124, 125, **126**, **127**,
128, 133, **134**, 139, **140**,
154, **156**, 158
   Maia 150, 152, **153**, 193
   MD-620 missile 83, **86**, 123

Mercure 133
Mirage IV 53, 86
Mirage 2000 ....126, 154
Mirage 4000 126
Mirage VTOL 133
Project 'Launac' 193
Spacesuit for EVA **150**, **151**
Star-H 79, 110, **196**, **201**, **202**, **203**,
204, 217, 218, 219
Super-Caravelle 133
Transporteur Aero Spatial (TAS) **63**,
81, **82**, **83**, **84**, **85**, 86, **87**, **88**, 167

de Havilland
   Blue Streak 40, 41, 42, 43

Dornier
   EARL................196, 214, **215**, **216**, **217**,
220, 226

Ecole Polytechnique Féminine
   ASUR...**218**, 226

ELDO
   Europa rocket launcher 29, 44, 88, 104
   Space tug..34

ERNO
   Aerospace Transporter study
see Nord-Aviation
Transporteur Aérospatial
   Bumerang/LB series **32**, **33**
   MTTF (Man-tended, free flyer).......**197**,
198, **199**
   Shuttle...**31**

ESA
   Columbus...........128, 129, 139, 141, 148,
150, 166, 176, 189, 190, 194,
195, 196, 197, 198
   EXTV...**221**
   FESTIP (Future European Space
Transportation Investigation
Programme ESA Programme)...110,
141, 176, 188, 193, 196, **219**, 221
   IXV 121, 131, 135, 152, 193, **194**
   Plato 177, 178
   Polar Platform....................139, 197, **198**

Euromissile
   Milan..179

EVA suits 140, 144

Hawker-Siddeley Dynamics
   Two-stage to orbit design....................**43**

JPL (Jet Propulsion Laboratory)
   Explorer 1 satellite..............................29

Junkers
   G-23 11
   Raum Transporter............**32**, **34**, 35, 168
   RT-8 ....**36**

Kawasaki
   HOPE..**184**, **185**

HOPE-X 185, **188**
Latécoère
    MASALCA 54
Leduc
    022 47
Lockheed
    C-5 Galaxy 97
    SR-71 204
LRBA
    PARCA (Projectile Autopropulsé
        Radioguidé Contre Avions) 47, 48
    Véronique sounding rocket 8
Martin (Glenn) Aircraft
    PRIME (X-23) 153, 158
    Titan rocket launcher 25, **28**
Matra
    R-431 54
    R-511 .... 51, 92
MBB (Messerschmitt, Bölkow, Blohm)
    ATC-500/ATC-50 (engine) 169, 171
    ATC-700 (engine) 217
    ATC-1200 (engine) 173
    Beta II 110, 116, **124**, 131
    Cargus 169, **170**, 171, 172, **173**, 174
    European Hypersonic Transport
        Vehicle (Sänger II as SST) 171,
        172, **174**
    Hytex 174, **176**, **177**
    Istra 110, 116, **124**
    Itustra 110, 116
    Horus 152, 168, 169, **170**, 171, 172,
        173, 174, 216, 220
    MTFF (Man-tended, free flyer) 141,
        143, 144, **197**, **198**, **199**
    Sänger II spaceplane ..36, 110, 139, 141,
        161, **168**, **169**, **170**, 171, 172, **173**,
        174, **175**, 176 **178**, 193, 194,
        218, 219, 220
    TRA-400 engine 169, **170**
McDonnell Aircraft
    ASSET (Aerothermoelastic
        Structures Systems
        Environmental Tests) 28, 102,
        153, 158
    Gemini Spacecraft 23
    Mercury Spacecraft 23, 25, 28, 106, 137
McDonnell-Douglas
    Shuttlecraft project 32, 33
Messerschmitt
    Me 262 .... 13, 17, **18**
Meyer-Piening
    Christofferus 188
Mitsubishi
    H-II **184**
    HOPE **186**, **187**
MSFC
    Saturn V 64
Myasishchev,
    Article 48 .... **20**

NASDA
    ALEX 184, **187**, 188
    OREX 184, 185, **187**

Nord-Aviation
    625 60
    Coralie (Europa rocket stage) 88
    CT-41 .... 9, 59, **61**, **62**
    Griffon 54, **55**, **56**
    Mach 4 aircraft **57**, 59
    Mistral 38, 63, **65**, **66**, **67**, **68**, **69**, 70,
        71, 72, 73, 88, 169
    Mistral SST 72
    Sirius (ramjet) **60**
    SS-10 missile 64
    SS-11 missile 179
    STS-450 60
    Transporteur Aérospatial **29**, 30, **37**,
        38, **39**, 70
    Vega (ramjet) **60**
    Veras 9, 30, 33, 50, **98**, **99**, **100**,
        101, 102, 103, 106, 122, 158
North American
    Apollo capsule 23, 28, 32, 84, 105,
        119, 141, 144, 183
    B-70 204
    X-15 .... 23, 153
NPO Molniya
    MAKS 181, **182**, **183**

ONERA
    2140 47
    2140 twin ramjet **48**
    2141 .... **48**
    Ardaltex 47
    Avion L1 .... **78**, **79**
    Avion L3 .... **80**
    Bérénice **50**
    Esope 51
    JAPHAR 222, **223**
    LEA **223**
    NA250 .... **48**
    OPD-320 51
    PIAF **223**
    PREPHA 111, 194, 196, 201, **222**, 223
    Prométhée 222, **223**
    Reentry shape **90**
    Scorpion 49, **50**
    Statex **48**
    Sataltex **49**

Peenemünde
    Aggregat A4 22

Raduga MKB
    Raduga missile 174, 222
Raytheon
    Hawk missile 48, 54
Rockwell
    Shuttle/orbiter 41, 106, 107,
        110, 112, 135, 137 (Challenger),
        139 (Challenger), 140 (Challenger),
        148 (Challenger), 172, 178
        (Challenger), 197, 202, 207
Rolls-Royce
    RB 545 Swallow engine 160, 161, 163
    Space Launcher System 45
    Turbo-rocket **64**

Salyut OKB
    Space Station Mir 138, 148, 189,
        194, 213
Sänger
    Antipodal Bomber aka Silverbird,
        aka Silbervogel, aka Raketen
        Bomber 6, 10, 11, **12**, **13**, **14**, **15**,
            16, 17, 18, 19, 20, 22, 25,
            28, 19, 46, 64
Scaled Composites
    X-38 .... **146**
SEP (Société Européenne de Propulsion)
    Diogène 2 .... **113**
    Hyper-Diamant **113**
    Super-Diamant **113**
    Vulcain (launcher) **113**
    Vulcain HM-60 engine 118, 119,
        137, 173, 202, 205, 217, 222
SEREB
    Diamant rocket launcher 31, 73
    Emeraude sounding rocket 101
SNCASE
    SE 4200/4263 .... 51, **52**
    SE 4400 .... **50**, 52
    SE 4500 .... **53**
SNCASO
    SO 9050 Trident 91, **92**
    SSTO (concept) 9, 34, 70, 110,
        116, **124**, 159, **180**, 196, 202, **209**,
        210, **219**, 222
Sud-Aviation
    Concorde 56, **65**, 71, 72, 76, 91,
        96, 99, 120, 133, 140, 166,
        167, 175, 218
    EM 3 .... **93**
    IM 3 91
    LSB **93**
    LSM 1 .... **94**
    LSM 2 .... **89**, **94**
    LSM 3 .... **95**
    PLT180 .... **72**, 74
    PLT 186 .... **73**, 75
    PLT 192 .... 93, **94**, 96
    PLT 204 .... 93, **94**
    PLT 219 .... **73**, 75, 76
    PLT 220 .... **74**, 76, 77
    PLT 221 .... **77**, 78
    PLT 222 .... 93, **95**, 96, **97**
    X-417 54
    X-422 54

TAS (concept) **63**, 79, 83, 105, 111,
        112, 123, 124, 193, 202, 203
Transport Alliance:
    C-160 Transall 32, 33
Tula KPB
    9M113 Konkurs missile 179
Tupolev
    Tu-22 222
Tsybin
    PKA **20**, 21

USAF:
    Hyward (hypersonic weapons
        and R&D) 23

USSR:
   BOR (Soviet spaceplane programme) **181**, 191
   Buran (soviet shuttle programme) ...145, 147, 148, 157, 165, 181
   Soyuz (Soviet/Russian spacecraft programme).........106, 146, 148, 181, 183, 189, 194
   Zemiorka R-7 (soviet missile programme).....20, 21

Index of People and Organisations

Abdul Kalam, A P J................................179
Aeritalia 115, **121**, 140, 197
Aérospatiale..........9, 90, 100, 105, 106, 108, 110, 111, 114, 116, 118, 119, 120, 121, 122, 123, 124, 125, 126, 127, 128, 129, 130, 131, 132, 133, 136, 137, 140, 142, 145, 146, 147, 148, 149, 150, 151, 154, 156, 157, 179, 180, 181, 182, 183, 191, 193, 196, 201, 204, 207, 208, 209, 210, 211, 212, 213, 214, 218, 219, 220, 222, 226
Air & Cosmos (magazine) 30, 31, 102, 107, 125, 131, 152, 158, 163, 178, 195
Alaplantive, M. 59
Allègre, Claude...195
Allest (d') Frédéric..105, 106, 129, 130, 131
Ananoff, Alexandre 6
Arianespace............................129, 151, 156
Armstrong-Withworth 40, 41
Arunachalam, V. S. 179
Arsenal de l'Aéronautique 18, 19, 54, 64
Ashford, David................165, 166, 167, 178
Astronaut(s) (includes Cosmonaut-s and Spationaut-s)............6, 9, 20, 21, 22, 36, 40, 41, 43, 104, 106, 107, 109, 126, 131, 134, 135, 136, 137, 138, 139, 140, 142, 143, 144, 145, 147, **148, 149, 150**, 154, 155, 156, 159, 166, 173, 188, 189, 197, 199, 200, 211, 213, 224
Aviation Magazine 30, 40, 87, 102, 131, 189, 195

BAe (British Aerospace) 9, 110, 159, 160, 162, 165, 166, 168, 169, 171, 172
Barré, Jean-Jacques 6
Baudry, Patrick...............109, 131, 134, 137, **148, 149**, 195
Beaussard, M. 18
Becker, John V. 22, 28
BFMT Bundes Ministerium für Förschung und Technologie 168, 171, 172, 174, 176, 214
Bell Aircraft 22, 23, 24, 25, 26, 61
Bellot, naval engineer...............................101
Bertrand, Joseph..6
Bidard-Reydet, Danielle.........................194
Bignier, Michel 131
Billant, C. 212, 226
BIS (British Interplanetary Society) 31
Blamont, Jacques......................................195
BNSC (British National Space Centre).....161
Boeing........25, 26, 35, 36, 64, 167, 174, 203

Boex, Joseph-Henri-Honoré.....................6
Bohn, Pierre...122
Bolkhovitinov..18
Bombled, Jean-Paul................204, 210, 226
Bondaryuk, Mark M..........................18, 20
Bonnet, René 225
Boyeldieu, E....................................212, 226
Brard, Jean-Pierre 194
Bredt, Irene 10, 11, 13, 18, 19, 64
Bristol Aerospace 165
Bristol-Siddeley.......................6, 42, 43, 44
Bristol Spaceplanes 165, 166, 167

Caspart, Jean-Jacques.....................142, 149
Carbonne-Lorraine....................................98
CASA................................115, **119**, 191, 194
CEAT Centre d'Essai Aérospatial de Toulouse........................98, 101, 103
Centre d'Essais des Landes....................101
Chelomey Vladimir Nikolayevich 22
Chrétien Jean-Loup 107, 143, 154, 181
Clervoy, Jean-François 134
CNAM Conservatoire National des Arts et Métiers 195
CNES Centre National d'Etudes Spatiales 9, 104, 105, 106, 107, 108, 109, 110, 111, 112, 113, 118, 119, 120, 125, 126, 128, 129, 130, 131, 132, 133, 134, 135, 136, 137, 138, 139, 140, 141, 142, 143, 145, 149, 150, 152, 153, 171, 181, 183, 185, 191, 193, 194, 195, 201, 202, 214, 219, 222
Conchie, Peter 160
Contraves 191
Couillard, Philippe 119, 130, 131, 132, 135, 136, 140, 141, 142, 143, 146, 147, 148, 149, 150, 151, 153, 161, 171, 189, 192, 194, 195
Courtois, Michel 149
Cousin, Alain 194
CPE Centre de Prospective et d'Evaluation 9, 39, 63, 64, 70, 71, 72, 73, 74, 75, 76, 77, 84, 87, 88, 89, 90, 92, 93, 94, 97
CRI..191
Curien, Hubert 107, 136, 193, 194

Damblanc, René 6
DARA (Deutsche Agentur für Raumfahrtangelegenheiten).....150, 189
Dassault (Avions Marcel Dassault/ Dassault Aviation) 9, 30, 40, 63, 70, 79, 80, 81, 82, 83, 84, 85, 86, 87, 88, 90, 95, 105, 110, 120, 121, 122, 123, 124, 125, 126, 127, 128, 130, 131, 132, 133, 134, 135, 136, 137, 138, 139, 140, 142, 149, 150, 151, 152, 153, 154, 155, 156, 157, 158, 167, 180, 191, 193, 194, 196, 201, 202, 203, 222, 226
Département de Guyane/ Programme PHEDRE.......................194
Dermichev, Gennady.................................21
DEFA (direction des études et fabrications d'armement) 48, 62
De Gaulle, Charles, Général 60
De Gaulle, Jean...194
Deplante, Henri...79

Devaquet, Alain 138
DFVLR (German Research & Development Institute for Air & Space Travel)) 32, 136, 150, 178
DGA (Direction Générale de l'Armement)..................................59, 153
Dorand (Gyravions) 30, 64, 102, **103**
Dornberger, Walter............................22, 23
Dornier System 32, 114, 131, 140, 145, 150, 196, 214, 215, 220, 226
DRME (Direction des Recherches et des Moyens d'Essais) 39
Ducout, Pierre..195
Dushkin, Leonid S 18, 20

ELDO..........29, 30, 34, 44, 88, 99, 104, 120
EMSI (European Manned Space Infrastructure ESA Programme)......141
ERNO.................6, 31, 32, 33, 34, 37, 39, 44, 114, 118, 140, 197, 198, 199
ESA 9, 31, 33, 106, 107, 110, 111, 114, 116, 128, 129, 130, 131, 135, 137, 138, 139, 140, 141, 142, 143, 144, 145, 146, 147, 148, 149, 150, 151, 152, 153, 159, 160, 161, 162, 168, 170, 171, 176, 177, 178, 184, 185, 189, 191, 192, 193, 194, 196, 197, 198, 202, 217, 219, 220, 221, 226
ESRO 104, 142
Esnault-Pelterie, Robert 6
ETCA...191
Eureca programme........................139, 178
Eurohermespace............150, 151, 181, 183, 190, 194
Eurospace 4, 6, 9, 29, 30, 31, 32, 33, 39, 40, 42, 43, 44, 45, 70, 79, 166, 226

Fabius, Laurent 130, 132
Falempin, François 111
Félicette (cat) 6, **9**
Feustel-Büechl, Jorg................................148
FIAT 118
Flammarion, Camille 6
Flight (Magazine)....................102, 148, 178
Fokker 115, **119**, **120**, **121**, 140, 191, 194, 197
Fottinger, Hermann 11
Francis, Robert, Hugh 44

Galley, Robert..194
Gauge, Paul 9, 91
GEERS, Groupe d'Etudes Européen pour les Recherches Spatiales 29
Georgii, Walter 13
Giscard d'Estaing, Valery......................107
Gopalaswami Raghavan 179, 180
Gorbachev, Mikhail 145
Govaerts, Pierre 140
Gozlan, Albert..18
Graffigny (de), Henri..................................6
Grumman Aerospace Corporation 125, 135

Hamermesh, M. 18
Hassan, M. Q. 165
Hawker Siddeley 6, 40, **43**, 44, 45, 61, 165

Hector (rat) 6, **9**
Hedfeld, H. 11
Hergé 6
Hilton, W. F. 40, 43
Hoffmann, Rainer 177
Huet, C. 49
Husson, Jean-Claude 131

INSA de Lyon 98
Isayev, Aleksei 18
ISS/Space Station Freedom NASA
  Space Station programme 106, 128,
    140, 141,143, 145, 146, 148, 149, 166,
    174, 176, 189, 195, 196,197,
    199, 200, 212

Junkers (company) 6, 29, 32, 34, 36,
    37, 45, 168
JASDA/JAXA 184, 185, 187, 188, 195
Joxe, Pierre 190

Kaiser Wilhelm Institute 11
Keenan, John Gregory **44**, 45
Keldysh, Mstislav Vsevolodovich 18, 19, **20**
Khrushchev, Nikita 21, 22
Kichkin , S. T. 20
Koelle, Dietrich 110, 116, 124, 131, 172
Kohl, Helmut 129, 138
Korolev, Serguei P. 20, 21, 22
Kulaga, Evgeny 21

Lacaze, Henri 9, 70, 97, 101, 105, **108**,
    110, 114, 120, 125, 131, 140, 154, 158,
    161, 176, 179, 185, 196, 204,
    207, **224, 225**
Lambrecht, Jurgen 34, **37**, 45
Lasswitz, Kurt 10
Latécoère 54
Leduc, René 47
Leffe (de) Alain 9, 116, 122, 129, 137,
    138, 139, 143, 148, 153,
    154, 155, 156
Lemainque, François 125
Lewis, Gordon 161
Love, E.S. 23
Lozino-Lozinsky Gleb E 181
Lygrisse, Paul 47

Malaval, M. 62
Marguet, René 48, 49
Martre, Henri 120, 179
Matra 51, 54, 92, 106, 118, 140, 222
Marquis, Raoul 6
Martin Aircraft 25
MBB (Messerchmitt, Bölkow, Blohm) 9,
    36, 37, 110, 114, 116, 118, 124, 131,
    140, 142, 150, 152, 160, 168, 169,
    170, 171, 172, 173, 174, 175,
    176, 177, 179, 197, 198,
    199, 217, 220
McDonnell-Douglas 31, 32
Messerschmid, Ernst 140
Meyer-Piening, Hans-Rheinhard 188
Mikoyan OKB 22
Minerve (programme) 53, 54

Minister of Education, Research
  and Technology (France) 195
Ministère de la Défense
  (French Ministry of Defence) 9, 54,
    98, 100, 101, 105, 180
Ministry of Aviation (USSR) 165
Ministry of Defence (India) 179
Ministry of Science and
  Technology (France) 173
Ministry of Transport and Space
  (France) 189
Mitterand, François 107, 129, 130, 132, 138
Moulin, Thierry 64
Myasishchev, Vladimir Mikhailovich 21

NASA (National Air & Space
  Administration) 20, 21, 22, 28, 31,
    38, 47, 89, 106, 125, 130, 135, 137,
    141, 143, 145, 146, 147, 150, 178,
    181, 184, 191, 214, 226
NII-1
  (Research Institute n°1) 18, 19, 20, **21**
Nord-Aviation 6, 9, 18, 25, 28, 29, 30,
    31, 32, 33, 37, 38, 39, 47, 54, 55, 57, 59,
    61, 62, 63, 64, 65, 66, 67, 68, 69, 70, 71,
    72, 73, 75, 77, 79, 88, 89, 98, 99, 100,
    103, 122, 226
Nonweiler Terence R F 40
NPO Energia 154
NPO Molniya 181, 182, 183

Oberth, Hermann 10
OKB-1 20
OKB-23 ....21, 22
OKB-52 22
OKB-256 ....20, 21
ONERA 9, 30, 47, 48, 49, 50, 51, 54, 57,
    64, 71, 74, 76, 78, 79, 80, 89, 90, 91, 93,
    98, 99, 100, 101, 103, **155**, 156, 180,
    201, 203, 213, 218, 222, 223
ORS 191
Parkinson, Robert 159, 164, 224, 225
Parly, Florence 223
Pattie, Geoffrey 162
Péchiney 98, 99
Perrier, Pierre 79, 122, 193, 226
Poisson-Quinton, Philippe 98
Poncelet, Christian 195

Ransom, Stephen 177
Ravel, M. 62
Reagan, Ronald 160
Reisenhuber, Heinz 138
Rigault, Michel 9, 79, 83, 105, 121,
    123, 124, 125, 135, 139, 152, 154,
    171, 193, 201, 202, 203, 204,
    213, 219, 226
Rocketdyne 161
Rolls-Royce 44, 45, 64, 75, 159,
    160, 161, 162
Roubertie, Jean-Maurice 122
Rosny, J-H Aîné 6
Roy, Maurice 30, 47
Royal Aircraft Establishment 160

Saab-Ericsson 191
Salomon, M. 195
SALOON programme 144
Sänger, Eugen 6, 9, 10, **11**, 12, 13, 14,
    15, 17, 18, 19, 20, 22, 23, 25, 28,
    29, 30, 31, 34, 35, 40, 63, 64
SEPR/SEP (Société Européenne
  de Propulsion) 30, 51, 53, 64, 84, 91,
    92, **113**, 118
Serov, General 19
Servanty, Lucien 91, 92
Sharman, Helen 189
Siddiqi, Asif 20, 21
Skoda-Kauba company 13
SNCASE 9, 47, 50, 51, 52, 53, 77, **134**,
    135, 140, 156, 180, 191,
    201, 208, 222
SNCASO 91
SNECMA 30, 31, 32, 37, 39, 44, 45, 54,
    63, 66, 70, 71, 72, 88, 180,
    201, 203, 222
Souchier, Alain 109, 131
Stalin, Josef 18, 19
Sud-Aviation 9, 30, 47, 53, 54, 63, 64,
    65, 72, 73, 74, 75, 84, 89, 91, 92,
    93, 94, 95, 96, 97, 98, 133
Symphonie (programme) 104, 109

Thatcher, Margaret 160
Tokayev, Colonel 19
Tolle, H. 31
TsAGI 19, 21, 22, 145, 222
Tsybin, Pavel 20
Turcat, André **56**, 88, 140
Turner, Rex 31

University (State of Ohio) 218
University of Poitiers 98
University of Stuttgart 110, 116
USAF (US Air Force) 23, 25, 28, 158
US Navy 59, 61

Valier, Max 11
Varnoteaux, Philippe 48
Verne, Jules **6**
VFW-Fokker 197
VIAM (the All-Russian Scientific
  Research Institute of Aviation
  Material) 20
Villain, Jacques 64
Vogelpohl, G. 11
Volvo 118, 174
Von Braun, Wernher 10
Von Hoefft, Frantz 11
Von Zboroswki, Helmut 11, 45

Wade, Mark 21
Wagner, Alain 210, 212, 226
Watson, H. R. 40, **41**, 43, 45
Waverider concept 40, 43
Wehrmacht 22
Wild, Wolfgang 150

Ziebland, H. 11
Zvezda 148, 154